Climate Risk Management in Agriculture

U. C. Mohanty · Palash Sinha ·
M. M. Nageswara Rao · Dillip Kumar Swain ·
K. K. Singh

Climate Risk Management in Agriculture

Monthly and Seasonal Forecast Application

Springer
CP

U. C. Mohanty
Centre for Climate Smart Agriculture (CCSA)
Siksha 'O' Anusandhan (Deemed to be University)
Bhubaneswar, Odisha, India

M. M. Nageswara Rao
University Corporation for Atmospheric Research (UCAR) at NOAA/NWS/NCEP/EMC
College Park, MD, USA

K. K. Singh
Department of Agriculture Meteorology
Anand Agricultural University
Anand, Gujarat, India

Palash Sinha
Center for Development of Advance Computing (C-DAC)
C-DAC Innovation Park, Panchavati, Pashan
Pune, Maharashtra, India

Dillip Kumar Swain
Department of Agricultural and Food Engineering
Indian Institute of Technology Kharagpur
Kharagpur, West Bengal, India

ISBN 978-3-031-51861-4 ISBN 978-3-031-51862-1 (eBook)
https://doi.org/10.1007/978-3-031-51862-1

Jointly published with Capital Publishing Company
The print edition is not for sale in Afghanistan, Bangladesh, Bhutan, India, Sri Lanka, the Maldives, Nepal and Pakistan. Customers from Afghanistan, Bangladesh, Bhutan, India, Sri Lanka, the Maldives, Nepal and Pakistan please order the print book from: Capital Publishing Company.

This Springer imprint is published by the registered company Springer Nature Switzerland AG
The registered company address is: Gewerbestrasse 11, 6330 Cham, Switzerland

Paper in this product is recyclable.

Preface

In India, more than 50% of the population is engaged in agriculture and allied activities, contributing around 18% to the country's Gross Value Added (census report 2021). Further, more than 50% of the total agricultural production comes from a rain-fed area covering $\approx$60% of total agricultural land and is highly vulnerable to weather hazards. Seasonal weather variability in terms of droughts, flash floods, heavy rainfall, thunderstorms, cyclones, heat waves, etc. severely affect the agricultural production, particularly the rainfed agriculture and rural economy. To avoid such weather-related risks, weather-based crop planning is essential, thus, seamless accurate weather forecasting is one of the important inputs for agricultural production management.

Although several improvements have been made in understanding the science of the atmosphere and the development in a physical and statistical model prediction for weather and climate, yet, the monthly to seasonal scale prediction remains the most difficult, and challenging task for the research community. This is mainly because of the tropical region, where the interaction between small-scale physical processes with the large-scale flow is very complex, and land surface characteristics are highly heterogeneous. For example, the 2002 Indian monsoon season (June-September) was a drought year (-19% of the long period average) due to a large rainfall departure in the month of July (-49% of normal), which turned out to be the severe in recorded history. Since July month is the most vital for sowing the Kharif crop, the food grain production dropped to 183 million tons as compared to 212 million tons in the previous year because of the unprecedented intra-seasonal variation of rainfall during that year. In addition, climate change is inevitable, and intensity, pattern, duration, and frequency of climate variability in association with climate extremes are being changed which greatly impacts agriculture and allied sectors.

In this context, a major number of earning people depend on agricultural production and there is an urgent requirement to develop a better sustainable technique for advanced intimation of weather and climate information in extended range time scales (monthly to seasonal scale) for weather-based agricultural planning and management. Such kind of forecast will be helpful to mend a decision support system for sustainable agricultural production (and minimizing the losses) and the livelihood of a farming community. The mandated need stated above motivated to direct concise pathways for the development of monthly and seasonal scale weather/climate forecasts.

In this book, several aspects of understanding the climate/weather and yields for generation of climate risk management (CRM) have been discussed, including a handson practice through total ten chapters. The Chap. 1 is introductory, where demanding and urgent issues on weather/climate from the agricultural point of view are presented. The research challenges on weather-crop prediction systems are summarized, and the scientific way is systematically described to overcome the issues for betterment in agricultural risk management.

In Chap. 2, the impact of climate change and its variability on different crops has been elaborately presented. The changing pattern of climate/weather systems, particularly changes in pattern, intensity, and duration of climate/weather extremes, has an adverse effect on the crop yield production which has been explained in this chapter. In addition, the response of yields in the changing climate, including a possible way to minimize the loss of agricultural output, has been reflected for different major crops.

The life cycle of a crop varies on the type of the crop, and the required weather/ climate conditions vary at different stages of particular crop phenology as well as crop to crop. Thus, knowing the state of the atmosphere at different time scales is very much important for enhancing agricultural production. In terms of agrarian need, seamless weather/climate predictions starting from the now-casting to seasonal scales have been depicted in Chap. 3 with a brief description of the impact (usefulness or affect) of individual scale forecast on agriculture.

It is noteworthy to understand the hazardous weather events that greatly impact on the agricultural and allied sectors. The deviation from the normal climatic state is also important to understand for better management system in agricultural production. Chapter 4 describes different hazardous weather and climate events, including the detailed representation of the fundamentals of weather/climate prediction systems. All the major components for developing a numerical model for the predictions of weather/climate state, as well as different modelling systems (different dimension models for various space and time scales) have been presented in this chapter. The advantages and limitations of the prediction systems have also been provided.

The numerical weather prediction (NWP) models are the best tools for advance intimation of the weather and climate state owing to their capability to deal with different scales of physical processes; however, the skill of an NWP model depends on several issues, like initialization, the way of parameterization to represent the small scale physical processes, complex interactions processes among land-ocean-air, etc. Several eminent organizations/institutions have developed NWP at global/regional scales and disseminate real-time weather and climate predictions. However, the weather and climate predictions may vary from one to another model. The details of the different model products and their accessibility have been reported in Chap. 5. In addition, several approaches along with standard statistical methods have been explored in this chapter which is helpful to evaluate a model's performance in predicting weather/climate conditions. Chapter 6 deals with the various bias correction methods to detailed representation with examples to remove the model's systematic bias. Bias correction is one of the widely used approaches to improve the model forecast skill. Moreover, since different organizations/institutions are providing different forecasts uing various NWP models, and the skill of the models varies with time and space, thus it is important to have a single forecast using multimodel and multi-statistical techniques. This chapter has reported a reliable approach for generating a single forecast at monthly and seasonal scale (deterministic and probabilistic) from a grand ensemble (produced from multi NWP model and multi-statistical approaches such as bias correction, down-scaling, etc.).

The crop yield prediction model needs daily weather inputs. Thus, the monthly/seasonal weather forecast products are needed to be disaggregated in daily sequence. Chapter 7 has described different disaggregation methods for producing daily weather from seasonal products. Several weather generator, including the process of daily weather sequence, has been explained with examples for the benefit and understanding of the readers.

Crop models have different components representing the interactions between soil, atmosphere, crop phenology processes, and field level input management. The crop yield prediction skill depends on several issues needed for crop growth which takes care of crop cycle and production. The different components which are needed to develop a crop model and simulation approaches have been discussed in Chap. 8. This chapter also reported various types of crop models.

Chapter 9 encompasses the process of generating a crop weather calendar (CWC) that is the optimize weather conditions to receive maximum gain in agricultural productions. This CWC is helpful in understanding the risk of a particular crop for specific weather conditions. The details to generate climate risk management (CRM) based on agro-met advisory have been described with examples for different atmospheric state/phenomena. Chapter 10 contains farmers' feedback from various corners of the world for different crops and the details of the prospects of using the CRM. Chapter 11 contains a detailed methodology on each step for hands-on-practice which involves, downloading, processing of model data and observations, model evaluation and bias corrections, generation of a single forecast using multimodel multi statistical techniques, disaggregation, and preparation of CRM based on crop model products. In the end, a details procedure of hands-on-practices for observed data analysis, evaluation of model forecast, removal of systematic bias of the weather/climate forecasted model products, consensus weather forecast, and generation of CRM have been provided. All these contents are presented in such a way that students, researchers, and operational agencies can be benefitted.

This book has been prepared with a focus so that students from undergraduate, post-graduate, research scholars as well as experts, operational agencies, policymakers, crop insurance agencies, decision support systems (DSS) involved in the field of atmospheric science, agricultural science, agricultural engineering, environmental science and allied sectors can easily understand the science and associated mathematics and apply the same with the use of modelling framework for prediction and contribute positively in DSS.

The contribution of work by Prof. U. C. Mohanty, Dr. P. Sinha and Dr. M. M. N. Rao were accomplished during their association with IIT Delhi and IIT Bhubaneswar. Similarly, the contribution of Dr. K. K. Singh was accomplished during his association with IMD Delhi.

Bhubaneswar, India
August 2023

U. C. Mohanty
Palash Sinha
Dillip Kumar Swain
M. M. Nageswara Rao
K. K. Singh

Acknowledgements

We would like to express our gratitude to the Department of Agriculture Cooperation & Farmers Welfare (DAC&FW), Ministry of Agriculture and Farmers Welfare, Government of India, and for their support through the project "Development and Application of Extended Range Forecast System (ERFS) for Climate Risk Management in Agriculture" (ERFS Phase-I & II). We duly acknowledge to the India Meteorological Department (IMD), Ministry of Earth Science (MoES), Government of India (GoI), for providing observational data sets. We are deeply grateful to the authorities of Indian Institute of Technology (IIT) Delhi and the IIT Bhubaneswar for their support in facilitating the multi-institutional research program. We are grateful to the National Center for Medium-Range Weather Forecast (NCMRWF), MoES, Space Application Center (SAC), Indian Space Research Organization; Indian Council of Agricultural Research (ICAR) and several Indian Agriculture Universities/Institutes for their collaboration and valuable suggestions during the project implementation. The authors are thankful to Dr. L. S. Rathore, AVM (Dr.) A. Tyagi, Dr. K. J. Ramesh, Dr. M. Mahopatra, Dr. H. R. Hatwar, Dr. D. R. Pattnaik, Dr. D. S. Pai, Dr. N. Chattopadhaya, Dr. Kripan Gosh, Dr. S. C. Sahu and many others from India Meteorological Department (IMD), Dr. A. K. Sahai and Dr. Suryachandra A. Rao from Indian Institute of Tropical Meteorology, Pune, as well as Dr. A. K. Mitra, (Late) Dr. S. C. Kar, and Dr. V. S. Prasad from NCMRWF for their cooperation and S&T support at various stages during the implementation of ERFS. We sincerely thank the International Research Institute for Climate and Society (IRI), Columbia University, and National Oceanic and Atmospheric Administration (NOAA), National Center for Environmental Prediction (NCEP) for providing GCMs output in hindcast and real-time mode. We want to express our gratitude to Dr. A. W. Robertson, Dr. Michael K. Tippett, Dr. James Hansen, Dr. A. Ines, Dr. Simon J Mason, Dr. Shiv Someshwar, and many others from IRI, USA, Dr. A. R. Subbaiya and Mr. Itesh Dash from Regional Integrated Multi-Hazard Early Warning System (RIMES), Thailand for their helpful contributions to the project. In the several stages of this ERFS program, a significant contribution from several scientists and staffs like Prof. S. K. Dash, Dr. K. Prasad, Dr. M. Kulkarni,

Dr. N. Acharya, Dr. Pushpa Raj Tiwari, Dr. Archana Nair, Dr. Ankita Singh, Dr. R. K. S. Maurya, Mr. R. K. Rai, Mr. A. Kumar, Mr. N. K. Sharma, Dr. Bhagyashree Dhekale, Dr. Manas Mohanty have been involved and contributed to the progress of the project. We appreciate several researchers for their assistance to the authors, particularly Dr. Poonam Biswal, Dr. Sathya Kandasamy, Mr. Bishwajit from IIT Kharagpur, Dr. M. Swain, Dr. S. Mohanty, Dr. V. Barde from IIT Bhubaneswar. The authors are also grateful to all who were involved in the project ERFS in its Phase-I and Phase-II. Prof. U. C. Mohanty would like to express gratitude to the Siksha 'O' Anusandhan (Deemed to be University), for providing support and facilities in accomplishment of the final phase of the book. Dr. Palash Sinha would like to express his gratitude to the Centre for Development of Advanced Computing (C-DAC), Pune, for their support in accomplishing the work with the publishing house during the publication stages of this book.

About This Book

Sustainable agricultural production is the key issue for food security, and agricultural productivity and greatly depends on the weather and climate conditions. This book discusses several aspects in detail, focusing on understanding and modelling of weather/climate systems and crop yield productions, including integrated weather-crop prediction systems for climate risk management in agriculture. It extensively discussed urgent issues on weather/climate from the agricultural point of view and the impact of climate change and its variability on different crops and possible ways to minimize the loss of the farmers. This book describes different weather and climate hazards, including the fundamentals of weather/climate prediction systems with details of the numerical weather prediction (NWP) models and presented the need of seamless weather/climate predictions and their impact on agriculture. The status and availability of different monthly and seasonal scale forecasts worldwide is explored and how the forecasting models or products can be evaluated using statistical methods is clearly explained. The book concisely elucidates systematic model bias removal techniques and a reliable approach based on multi-statistics in producing a single forecast from the multi-model grand ensemble. Since crop models need daily weather sequence, several standard disaggregation methods for generating daily weather sequences from monthly/seasonal products are presented. This book explicitly describes several aspects that are needed for agricultural practices and crop modelling, encapsulates different components of crop models and their application, preparation methods of Crop Weather Calendar, application of disaggregated weather sequence in crop models, and generation of Climate Risk Matrices (CRM) for the benefits of farmers/researchers/scientists/operational agencies. A detailed methodology for hands-on practice, including downloading and processing data, model evaluation and bias corrections, generating a single forecast, disaggregation, and preparing CRM based on crop model products is provided. This book contains a total 11 chapters in describing all the above for students, researchers, scientists, and operational agencies who can be benefitted both theoretically and a reader can earn practical experiences in weather and crop modelling for skillfull prediction.

Contents

About the Authors

Prof. U. C. Mohanty worked more than 33 years at Centre for Atmospheric Sciences, IIT Delhi in various capacities, from Lecturer to Head and Professor, served in IIT Bhubaneswar for 9 years as a Visiting Professor, and currently serving as Distinguished Professor in the esteemed Siksha 'O' Anusandhan (SoA) Deemed to be University, Bhubaneswar. He also served the NCMRWF, Noida for more than two years as Head of the research. Under his guidance, the medium range weather forecasting system was for the first time operational in the country in June 1994. He was President from 2013 to 2015 in the Odisha Bigyan Academy, Vice President of Indian Meteorological Society (IMS), and Senior Associate of ICTP, Italy. The principal facets of his work and research including development of prediction systems are high impact extreme weather events (e.g., severe thunderstorms, heavy rainfall events, tropical cyclones), storm surges, summer monsoon process-circulation-disturbances, and location specific extreme weather events over India. His research work also involves monthly and seasonal scale prediction of precipitation and air temperature at the surface and application of the same for climate risk management in the sector of agriculture. He has published more than 325 research articles in reputed journals and 70 technical reports and book chapters. Under his guidance, 44 have already been awarded and 3 are pursuing Ph.D. programs.

Professor Mohanty has been conferred with many awards and honours: Shanti Swarup Bhatnagar Prize, National Award in Atmospheric Sciences and Technology, Biju Patnaik Award for Scientific Excellence, Sir Gilbert Walker Gold Medal, Samanta Chandra Sekhar Award, AR & DB Silver Jubilee Award, 12th Mausam Award, and Prof. M. G. Deshpande Award. He has also been honoured as Fellow of all four national science and engineering academics in India namely, Indian Academy of Sciences (IASc), National Academy of Sciences, India (NASI), Indian National Science Academy (INSA) and Indian National Academy of Engineering (INAE). He is also a fellow of Indian Meteorological Society (IMS) and Indian Geophysical Union (IGU). He has been conferred with Odisha Citizens Award-2018 in Education.

Dr. Palash Sinha is currently associated with the Centre for Development of Advanced Computing (C-DAC), Pune and contributing in developing of urban scale models. He has completed his master in science in physics, master of technology in atmospheric physics, and received Ph.D. award from Indian Institute of Technology (IIT) Delhi. During his Ph.D. program, he has been selected and sponsored by GoI through IIT Delhi for six-months visit to International Research Institute (IRI) for Climate and Society, Columbia University, New York as a Scientific Officer for developing monthly to seasonal scale forecasts in collaboration with IRI scientists. He developed dynamical as well as statistical downscaling approaches for extended range forecast that has added value for weather/climate forecasts and its application in the agricultural sectors. His post-doctoral from Department of Meteorology, Pennsylvania State University, USA was on the developed sophisticated methodologies for climate projections and estimation uncertainties in the climate drift. He also worked as visiting scholar at Purdue University, USA, Scientists at IIT Delhi and IIT Bhubaneswar. He has published more than 45 research articles in reputed international/national journals, contributed in seven books/book chapters/ technical notes, and presented 20 articles in national/international conferences. He closely guided the researchers in different aspects, was judge in research competition, is reviewer of more than 20 international and national journals, played role as an expert to teach and train students/researchers/scientists on the development and application of weather forecasts in agriculture across India.

Dr. M. M. Nageswara Rao is a UCAR Associate Scientist-II at NOAA NCEP EMC, USA. He holds a Ph.D in Meteorology and Oceanography from Andhra University, India. He is an expert in developing forecast guidance tools for weather and climate extremes (cold and heatwaves, floods and droughts, atmospheric rivers, etc.) using Statistical Post-processing Methods on GCMs outputs. He is an expert on integration of weather forecast products to crop models for developing crop yield prediction systems. He held various research positions in multi-institutional collaborative projects conducted at IIT Delhi, IIT Bhubaneswar, Andhra university and IITM Pune, India, University of Texas, Austin, NOAA NCEP CPC International Desks and EMC, USA. Dr. Rao's work has been widely recognized in the field of Atmospheric, Ocean, and Agriculture sciences, having received awards such as the UCAR Special Recognition Award-2022, IMS-Biennial Monsoon Award-2020, and UCAR Visiting Scientist Fellowship-2020. He has published more than 40 research articles in reputed journals and 5 technical reports and book chapters and presented more than 30 research papers at international conferences/workshops. Dr. Rao actively participated in conducting numerous training-workshops to develop capacity building on integration of forecast products with crop models and he served as reviewer for more than 30 internationally reputed journals.

Prof. Dillip Kumar Swain is actively involved in teaching and research on Climate Change Adaptations & Mitigations in Agriculture, Organic Farming, Sustainable Agricultural Production, and Crop Modeling & Simulation at Agricultural and Food Engineering Department, Indian Institute of Technology (IIT) Kharagpur. Prof. Swain has developed climate change adaptation options for crop production, organic farming and vermicomposting technology for sustainable agricultural production and quality improvement, Resource management in rice-based cropping system, Organic hydroponics, etc. He is also putting best effort in outreach activities through dissemination of weather-based agro-advisory to farmers for weather-basedcrop planning and management and demonstration of advanced production technologies in farmers' field.

Professor Swain completed his Doctoral Degree from IIT Kharagpur, India in 2002, which received the Fertilizer Association of India Silver Jubilee Award for the outstanding doctoral research in Fertilizer Usage. Before joining the IIT Kharagpur, Prof. Swain worked as Post-Doctoral Fellow at the United Nations University, Tokyo, Japan, availing Japan Society for the Promotion Science Fellowship. Prof. Swain teaches the subjects: Systems Approach in Agriculture, Soil-Plant-Water Relationships, Crop Production Systems, Organic Food Chain Management, and Management and Productivity for undergraduate and postgraduate students in Agricultural and Food Engineering. Prof. Swain is continuously supervising Undergraduate, Postgraduate, and Research Scholars in the emerging areas of sustainable agricultural production.

Dr. K. K. Singh did his post-graduation in Geophysics in 1983 and Ph.D. on Crop yield forecasting in 1988 from Banaras Hindu University, Varanasi, India. He worked more than 33 years in the government organizations viz. National Centre for Medium Range Weather Forecasting (NCMRWF) and India Meteorological Department (IMD), Ministry of Earth Sciences New Delhi, in various capacities, from Senior Scientific Officer to Head of Agromet Division & Additional Director General of Meteorology/Scientist 'G'. He immensely contributed in the Establishment and Development of operational agriculture weather forecasting and Agrometeorological Advisory Service (AAS) system at district and block level for farmers in India. He worked in the field of application of modern tools like Dynamic crop simulation models, Decision Support System (DSS) for resilient farming, Agromet DSS, Remote sensing and GIS etc. in agriculture risk management under changing climate. He worked as WMO/UNDP Fellow at Michigan State University, East Lansing, USA during Jan–Apr, 1994 on 'Crop weather simulation applied to irrigation scheduling and assessment using Medium Range Weather Forecast'. In the international project on "Extended Range Forecast System (ERFS) for Climate Risk Management in Agriculture" funded by Ministry of Agriculture, Govt. of India, Dr. Singh played key role in implementing the Application of probabilistic extended range forecasts for risk management in agriculture by establishing an end-to-end Information Generation and Application System through pilot projects in demonstration districts/sites under some of the State Agriculture Universities, and worked to develop CRM tools. In IMD, he also implemented the Agromet component of FASAL (Forecasting Agricultural output using Space, Agrometeorology and Land based observations).

He served as Coordinator of Expert Group on Agrometeorology (EG-AgM) for RA II Working Group on Climate Services (WGCS), World Meteorological Organization (WMO), Geneva during 2017–2020. He has supervised M.Sc. and Ph.D. scholars from different universities and IITs. He has ~120 research papers to his credit published in peer reviewed national and international journals. At present, Dr. Singh is the President of Association of Agrometeorologists, Anand Agricultural University, Anand (Gujarat), India.

Abbreviations

AAS	Agromet Advisory Services
ACC	Accuracy
ACCESS-S	Australia Region Using the Australian Community Climate EarthSystem Simulator Seasonal
AGCM	Atmospheric Global Circulation Model
AICRPAM	All India Coordinated Research Project on Agrometeorology
AMFUs	Agro-met Field Units
AMIP	Atmospheric Model Inter-comparison Project
APCC	APEC Climate Centre
APEC	The Asia-Pacific Economic Cooperation
APSIM	Agricultural Production Systems Simulator
AR	Autoregressive Process
ASRs	Agriculturally Significant Rains
AWS	Automatic Weather Station
BATS	Biosphere-Atmosphere Transfer Scheme
BBF	Broad Bed and Furrow
BCC	Beijing Climate Center
BMP	Best Management Practices
BOM	Bureau of Meteorology
BSS	Bright Sunshine
CanSIPS	Canadian Seasonal and Inter-annual Prediction System
CCA	Canonical Correlation Analysis
CCAFS	Climate Change, Agriculture and Food Security
CCCma	Canadian Centre for Climate Modelling and Analysis
CCM3	Community Climate Model version 3
CDF	Cumulative Distribution function
CEEMA	Centre for Extreme Events Monitoring in Asia
CEM	Climate envelope model
CERES	Crop Environment REsource Synthesis
CFL	Courant-Friedrichs-Lewy
CFS	Climate Forecast System

CGCM	Coupled Global Circulation Model
CGIAR	Consultative Group on International Agricultural Research
CIS	Commonwealth of Independent States
CLIGEN	Climate Generator
CMA	China Meteorological Administration
CMIP5	Coupled Model Intercomparison Project Phase 5
CMO	Central Meteorological Observatory
CPC	Climate Prediction Centre
CPTEC	Centro de Previseao de Tempo e Estudos ClimAaticos
CRM	Climate Risk Management
CROPGROW	CROP GROwth
CWC	Crop Weather Calendar
DJF	December-February
DMI	Danish Meteorological Institute
DSS	Decision Support Systems
DTR	Diurnal Temperature Range
EAMAC	East Asian Monsoon Activity Centre
ECMWF	European Centre for Medium-Range Weather Forecasts
ENSO	El Nino and Southern Oscillation
EOF	Empirical Orthogonal Function
ERF	Extended Range Forecast
ERFS	Extended Range Forecast System
ETL	Economic Threshold Level
ETS	Equitable Threat Score
FAO	Food and Agriculture Organization of the United Nations
FAR	False Alarm Rate
FCP	Final Consensus Product
FGP	Food Grain Production
FileW	Weather File
Filex	Experiment File
FSU	Florida State University's
FWB	Farmer's Weather Bulletin
GCFS	German Climate Forecast System
GCMs	General Circulation Models
GDD	Growing Degree-Days
GDP	Gross Domestic Product
GEFS	Global Ensemble Forecast System
GFS	Global Forecast System
GPC	Global Producing Centre
GPCLRFs	Global Producing Centres for Long-Range Forecasts
GSCU	Global Seasonal Climate Update
GTS	Global Telecommunication System
GVA	Gross Value Added
HK	Hanssen and Kuipers Discriminant
HPC	High-Performance Computing

HSS	Heidke Skill Score
HWRF	Hurricane Weather Research and Forecasting
IARI	The Indian Agricultural Research Institute
IAV	Inter-annual variability
ICAR	Indian Council of Agricultural Research
ICASA	International Consortium for Agricultural Systems Applications
ICRISAT	International Crops Research Institute for the Semi-arid Tropics
ICTP	International Centre for Theoretical Physics
IFS	Integrated Forecasting System
IITM	Indian Institute of Tropical Meteorology
IMD	India Meteorological Department
INPE	Instituto Nacional de Pesquisas Espaciais
IOA	Index of Agreement
IOD	Indian Ocean Dipole
IPCCs	Intergovernmental Panel on Climate Change's
IRI	The International Research Institute for Climate and Society
ISM	Indian summer monsoon
ISMR	Indian Summer Monsoon Rainfall
JF	January-February
JJAS	June-September
JMA	Japan Meteorological Agency
KIAPS	Korea Institute of Atmospheric Prediction System
KMA	Korea Meteorological Administration
LAMs	limited Area Models
LARS-WG	Long Ashton Research Station-Weather Generator
LC-LRFMME	Lead Centre for Long-Range Forecast Multi-Model Ensemble
LSS	Likelihood Skill Score
MAAS	Micro-level Agro-advisory Services Dissemination
MAM	March-May
MB	Mean Bias
MBRT	Mean Bias-remove Technique
MDS	Minimum Data Set
MM4	Mesoscale Model Version 4
MM5	Mesoscale Model Version 5
MMB	Modified Mean Bias
MME	Multi-model Ensemble
MMEav	Simple Averaging of all the Individual Models
MoES	Ministry of Earth Sciences
MOM	Modular Ocean Model
MOT	Ministry of Transport
MPI-ESM	Max Planck Institute for Meteorology
MRF	Medium Range Forecast
MRI	Meteorological Research Institute
MRWF	Medium-Range Weather Forecast
MSC	Meteorological Service of Canada

MST	Multiplicative Shift Technique
MulGETS	Multi-site Stochastic Weather Generator
NARP	National Agricultural Research Project
NATP	National Agricultural Technology Project
NCAR	National Centre for Atmospheric Research
NCEP	National Center for Environmental Prediction
NEACC	North EurAsia Climate Centre
NICRA	National Innovations on Climate Resilient Agriculture
NMC	Nested Grid Model
NMHS	National Meteorological and Hydrological Services
NMME	North American Multi-Model Ensemble Project
NOAA	National Oceanic and Atmospheric Administration
NSEC	Nash-Sutcliffe Efficiency Coefficient
NWP	Numerical Weather Prediction
NWS	National Weather Service
OBS	Observational Data
OLS	Ordinary Least Squares Estimation
OND	October-December
ORSS	Odds Ratio Skill Score
OSU	Oregon State University
PCR	Principal Component Regression
PCs	Principal Components
PDE	Partial Differential Equations
Po	Precipitation Occurrence
POD	Probability of Detection
POFD	Probability of False Detection
PSS	Peirce's Skill Score
PSU	Pennsylvania State University
QMM	Quantile Mapping Method
RCC	Regional Climate Centre
RCC-LRF	Regional Climate Centre node on Long Range Forecasts
RCMs	Regional Climate Models
RegCM	ICTP Regional Climate Model
RH	Relative Humidity
RMC	Regional Meteorological Centre
RMSE	Root Mean Square Error
RPN	Recherche en PrAevision NumAerique
RT	Regression Technique
SARRA	System for Regional Analysis of Agro-Climatic Risks
SAWS	South African Weather Services
SCS	Soil Conservations Service
SD	Standard Deviation
SF	Seasonal Forecast
SLP	Sea Level Pressure
SOM	Soil Organic Matter

SR	Success Ratio
SRF	Short Range Forecast
SST	Sea Surface Temperature
SUBSTOR	SUBterranean STORage
SVD	Singular Value Decomposition
SZT	Standardized Reconstruction Technique
TCC	Tokyo Climate Centre
TS	Threat Score
UBSTOR	Simulation of Underground Bulking Storage Organs
UKMO	Met Office at United Kingdom
UM	Unified Model
USDA	United States of Department of Agriculture
UTC	Universal Time
VSRF	Very Short Range Forecast
WeaGETS	Weather Generator of Ecole de Technologie Supérieuré
WGs	Weather Generators
WMC	World Meteorological Centre
WMO	World Meteorological Organization
WOTF	Without Transformation Functions
WRF	The Weather Research and Forecasting
WTF	With Transformation Functions
WWW	World Weather Watch

Chapter 1
Panorama

The agricultural sector is a major contributor to the global economy and is seen as a key factor in achieving sustainable economic growth. In 2018, the global gross domestic product (GDP) was comprised of 4% from agriculture, while in certain developing countries, this percentage was higher than 25%. Although agriculture only makes up a small portion of the global economy, it is a vital source of income for many people worldwide. In 2010, approximately 2.6 billion people worldwide relied on agriculture as a source of income, either as direct participants or as dependents of those actively engaged in the sector. In 2012, approximately 7.1 billion people, which is 19% of the world's population, were involved in farming. However, the income generated from agriculture (including hunting/fishing and forestry) only accounted for 2.8% of the total global income (World Bank 2012; Ayana et al. 2022). In the present day, the majority of people in lower and middle-income countries are employed in the agricultural sector, which contributes a larger portion of their national income and employment.

Agriculture, food, and related industries, in the USA, contribute ~\$1.109 trillion to its GDP, which is approximately 5.2% share (UDSA 2022). In 2020, the agricultural and food sectors employed 19.7 million full- and part-time workers, representing 10.3% of the total US workforce. Direct employment on farms accounted for approximately 2.6 million jobs, which is equivalent to 1.4% of the total employment. In 2020, 19.1 million jobs in the USA were supported by employment in agriculture- and food-related industries, which accounted for 10.3% of the total employment. Of the total number of jobs supported by the food industry, 10.5 million were in food service, eating, and drinking places, while 3.3 million were in food/beverage stores. An additional 3.3 million jobs were created by other agriculture-related industries. An additional 3.3 million jobs were created by other agriculture-related industries.

The Common Agricultural Policy of the European Union provides substantial funding to all its member states for the agricultural sector due to the importance of the agriculture sector. In 2019, the agriculture industry provided employment

U. C. Mohanty et al., *Climate Risk Management in Agriculture*,
https://doi.org/10.1007/978-3-031-51862-1_1

for 9,476,600 people, and 3,769,850 people in food production in 2018. It also contributed 1.3% to the European Union's Gross Domestic Product in 2020. In 2019, Romania had the highest number of people employed in agriculture among European countries, while Denmark had the most people employed in food production in 2018. The farming sector generates an additional €0.76 ($0.81) for the EU economy for every Euro invested. In 2020, the gross value added from farming in the EU's primary agricultural sector was €178.4 billion ($190.02 billion), which is the difference between the value of all products produced and the cost of services and goods used in production.

Agriculture is a vital part of the Indian economy, accounting for approximately 18% of the total GDP and providing employment to around 50% of the population in 2018 in the agricultural and related sectors. India is ranked sixth in terms of nominal GDP and third in terms of PPP (Purchasing Power Parity) among the world's economies. In 2018, the country was ranked 139th in terms of nominal per capita GDP ($2,134) and 122nd in terms of purchasing power parity (PPP) per capita GDP ($7,783) (World Bank data, https://data.worldbank.org/indicator/NY.GDP.PCAP.CD?locations=IN). The India Economic Survey (2020–21) has revealed that the contribution of agriculture to India's GDP has reached its highest level (about 20%) in 17 years, making it the only positive aspect of the country's economic performance in 2020–21.

Agriculture and its related industries are the main sources of income in India. A large majority of rural households in the area, approximately 70%, rely on agriculture as their main source of income, with the majority of farmers (82%) being small and marginal. India is the leading producer, consumer, and importer of pulses globally, accounting for 25%, 27% and 14% of the world's production, consumption, and imports respectively. In addition to this, in 2017–18, India's annual milk production was 165 million tons (MT), making it the top producer of milk globally. India is the second-largest producer of a variety of crops, including rice, wheat, sugarcane, cotton, groundnuts, fruit, and vegetables. The Indian agricultural sector has seen a positive surge in growth during the 2020–2021 fiscal year (Economic Survey 2021). The agricultural sector has seen significant growth due to the expansion of related industries such as livestock, dairy, and fisheries. The agriculture and allied sector has demonstrated resilience in the face of the COVID-19 pandemic, with growth of 3.6% in 2020–21 and further improvement to 3.9% in 2021–22.

According to the Indian Meteorological Department (IMD), country's annual average temperature increased by about 0.6 °C between 1901–10 and 2009–18 from its pre-industrial levels. This change has led to 11 warmest years as recorded in the last 15 years, and the magnitude and degree of warming have become more severe. As carbon emissions persist and those built into the climate system come into effect, temperatures worldwide are expected to rise from 3 to 5 °C by 2100 with an unequivocal rise of 1.5 °C by 2030.

The increased population pressure, the decline in net cultivable area, and the depletion of the natural resource base have questioned the sustainability of global agricultural production in general, India in particular, where agriculture is predominantly rainfed and a majority of farmers are marginal. According to World Health Organization (WHO), climate change poses the most significant risk to global health and food security in the current century. Food security is achieved when all individuals have the ability to acquire adequate, safe, and nutritious food that meets their dietary needs and food preferences for a healthy and active lifestyle. The issue of climate change has become a major global concern in order to guarantee food and nutritional security for the increasing population. The adverse effects of climate change and climatic fluctuations are becoming increasingly visible in India and many other nations. Studies have raised alarm about the potential impacts of climate change on food production and food security. It appears that climate change has encouraged the mankind to increase food-grain yield variability and market food price volatility (Brown and Kshirsagar 2015; Fuss et al. 2015) and affects the government's policies to ensure sustainable food security (Kumar et al. 2015). Climate variability on both spatial and temporal scales is a major concern for seasonal crop production, as it can lead to the frequent occurrence of extreme weather and climate events such as droughts, floods, and tropical cyclones. To ensure sustainable agricultural production in the face of climate change and variability, considerable investment is needed in short-term climate forecasting and field-level adaptation and mitigation measures.

Climate change is already having an impact on agriculture, with changes in crop phenology, physiology, biochemical pathways, and more leading to an increase in pest and disease outbreaks, water shortages, and soil fertility degradation. These effects are not evenly distributed across the globe. Hundal and Kaur (1996) conducted a study to assess the effects of climate change on the yield of wheat, rice, maize, and groundnut in Punjab, India. It was found that if all other climate factors remained constant, a temperature increase of 1, 2, and 3 °C from current levels would result in a decrease in wheat grain yield by 8.1%, 18.7%, and 25.7%; rice by 5.4%, 7.4%, and 25.1%; maize by 10.4%, 14.6%, and 21.4%; and groundnut pod yield by 8.7%, 23.2%, and 36.2%, respectively. The production of several other crops such as sorghum, pulses, oilseeds, cotton, sugarcane, fruits, vegetables, etc. is getting adversely affected by climate change.

Crop yield prediction based on seasonal climate predictions can help to reduce the adverse impacts of climate variability and its extremes in the current climate. Yield forecasting can be used as an adaptive technology to help mitigate the effects of climate change. The year-to-year fluctuations in seasonal average temperature and precipitation in many places around the world are more pronounced than the expected changes for the middle of this century, according to Deser et al. (2012). Changes in growing season climates and the resulting food production shocks can help societies become better equipped to handle the risks posed by climate change.

Climate change can have a significant impact on the availability and quality of forage, the amount of water needed for its cultivation, and the overall vegetation patterns of grazing lands. Plants and forage crops will continue to be exposed to warmer temperatures and increased carbon dioxide. The availability of water is expected to be impacted by changing precipitation patterns in the coming decades.

Climate change can adversely affect the productivity, composition, and quality of species and have potential effects on forage production and other ecological roles of grasslands (Rojas-Downing et al. 2017). The livestock sector will be significantly impacted by water availability issues, as it relies on water for animal consumption and forage crop production (Thornton et al. 2009). Livestock accounts for approximately 8% of global water usage by humans, and an increase in temperature can cause the water consumption of animals to double. As the ocean's water level increases, saltwater is increasingly infiltrating freshwater aquifers near the shoreline. Salinization can lead to chemical and biological contamination, as well as an increase in the levels of heavy metals present in various water sources, which can have a detrimental effect on animal production. Salinization caused by water can have a detrimental effect on the metabolic, reproductive, and digestive functions of animals. Chemical impurities and heavy metals can have a detrimental effect on the cardiovascular, excretory, skeletal, nervous, and respiratory systems, as well as compromising the hygienic quality of production (Nardone et al. 2010).

It is anticipated that the impacts of future climate change on food production in countries located in lower latitudes will be considerable, while the effects in higher latitudes may vary. Vulnerable populations, particularly those in tropical and subtropical countries of the developing world, may be at greater risk of food insecurity due to climate change. For example, in South America, the amount of arable land area that could be lost ranges from 1 to 21%, while in Africa it could be 1–18%, in Europe 11–17%, and in India 20–40%. In many African countries, semi-arid conditions already make agricultural production difficult, and climate change is expected to further reduce the length of the growing season and render large areas of marginal agriculture unviable.

In tropical regions, the incidence of pests and diseases is particularly severe due to the favourable climate, multiple cropping, and the presence of alternate pests throughout the year. In South Asia, crop yields are significantly impacted by the presence of pests and diseases. The effects of global warming may have an impact on the growth and development of all organisms, including insect-pests. Temperature is the most influential abiotic factor in determining the distribution and population of insects, as they are ectothermic creatures. The ambient temperature has a direct effect on the survival, growth, development and reproduction of insects, as they are unable to regulate their body temperature. Variations in precipitation, temperature and wind velocity may affect the migratory patterns of pests. Forecasting of the population of Lipaphis erysine Kalt (aphids) on mustard crops grown in the northern part of India during the winter season has been established based on the movement of western disturbances. The Mediterranean air brought in by western disturbances brings about cloudy and favourable conditions for the growth of aphids on mustard crops, with its cold and humid nature. A sharp rise in the number of aphids was observed when the mean daily temperature was between 10 and 14 °C, accompanied by a relative humidity of 67–85% and cloudiness of more than 5 octas.

Agriculture is a vital factor in the development of the nation, so any major alteration in climate on a global level will have an effect on local farming and consequently, the global food supply. It is estimated that the average air temperature in India will rise by 0.1–0.3 °C in the *kharif* season by 2050 and by 0.4–2.0 °C by 2070 when compared to the period of 1961–1990. The projected increase in temperature during the *rabi* season in 2050 is estimated to be between 0.3–0.7 °C, while in 2070 it is expected to be between 1.1–4.5 °C. At the same time, there is also a chance of more extreme weather events, such as earlier monsoon onset, more frequent and intense droughts, and floods.

The challenge of creating suitable adaptation technologies to reduce the negative effects of climate change on agricultural production has become increasingly important. Some of the field tested technologies such as reduced flooding in rice, dry seeded rice, broadbed and furrow seed bed, mulching, organic input management, etc. are recommended as resource conservation technologies for climate change adaptations/mitigation. Planning such agricultural operations based on weather information is vital for successful agricultural prediction. Forecasted weather, nevertheless, has become an essential input for the farmers to plan the activities starting from land preparation to harvest and post-harvest operation.

The weather/climate forecasts at different time scales are one of the important components to reducing the risks in various sectors. Although the use of these forecasts varies from sector to sector, the agricultural and allied sectors are benefitted most from the use of different spatiotemporal scales of weather and climate forecasts. In the agricultural sector, without prior information on weather and climate, an excess/deficit seasonal rainfall can greatly damage crop productivity which in turn may have a great impact on the nation's economy, particularly for an agro-economic country like India. However, the risk in the agricultural field associated with the climate and weather can significantly be reduced by the use of skillful weather/climate forecasts and appropriate adaptive response through advanced decision support systems.

The advancement of technology has hugely uplifted the understanding and prediction of weather and climate systems. In addition to the development of sophisticated high-resolution dynamical models, the high-performance computing systems, big data storage, and fastening internet speed facilitates the user to avail data from any corner of the world. On the other hand, since a few thousand years, farmers across the world are aware of the fact that prior information on rainfall/temperature at a daily-weekly-monthly-seasonal scale is one of the potential factors to minimise the loss and maximise crop production, which is true till today. Thus, the farming community is seeking reliable forecasts of weather/climate parameters. The emergence of complex global models and the development of high-performance computing systems facilitate in both ways—(i) generating/simulating satisfactory forecasts and (ii) storing the huge data after the simulations that can be accessed by the user through the internet if the organiser makes it freely available. In this context, many reputed organisations/institutes have developed sophisticated, complex and advanced global models for the prediction of weather and climate at monthly and seasonal scales. These predictions are updated at regular intervals and available on the website of the corresponding organisations.

Forecasting meteorological parameters on a monthly to seasonal scale is one of the most difficult tasks for the scientific community. The prediction of weather can be challenging due to the complex interactions between land, air, and sea, as well as the interplay between small-scale convective activities and large-scale flow. There are different degrees of variation of the meteorological parameter in space and also in time (sub-seasonal to season, inter-annual, intra-seasonal, etc.). For example, the Indian summer monsoon is a yearly occurrence that takes place between June and September, providing 80% of the country's annual precipitation. A deficit (excess) in the monsoon years of India does not indicate that every place in India have received a lesser amount of rainfall compared to normal. It may happen that in deficit monsoon years, some places received normal rainfall, and the remaining regions may receive normal or above-normal rainfall. Thus, for a specific year, a statistical measure of received rainfall is heterogeneous over space/ India. The changes in space and time have a significant effect on food grain production and the GDP of the nation.

Accurate and timely forecasting of meteorological parameters on a monthly to seasonal scale is essential for long-term planning, decision-making, management, and mitigation in various sectors such as agriculture and its related industries, power and hydrology, transport and tourism, government, and industry. Advance intimation of accurate sub-seasonal to seasonal is one key factor for deciding various mitigation and adaptation strategies in agriculture for increasing productivity. Similarly, deficit rainfall may create water stress in the ground water, for which a fruitful Agro-met advisory board is required. Therefore, predictions of climate/ weather parameters at monthly and seasonal scales will be very useful for society.

Several reputed organisations regularly generate the monthly to seasonal scale forecast over the globe using modern day numerical modelling (Global Climate Model) or statistical methods (linear and non-linear such as AI/ML) or both. These forecasts are used to disseminate through a web portal and make available in graphical and/or digital format that can be made useful after downscaling and bias corrections. The monthly and seasonal forecasts (after downscaling and bias corrections) have some skills that are mostly above some performance level while predictability a week to four weeks involves higher uncertainty and skill is very poor. Thus, meteorological forecasts are generated at the scale of monthly to seasonal where the confidence level is high. But, the crop model needs input on a daily scale. Therefore, the predicted monthly/seasonal scale meteorological parameters such as rainfall and temperature need to be disaggregated at a daily scale for the use of the crop model. Several standard disaggregated methods exist and appropriate methods can be used to generate daily sequences from the monthly/seasonal forecast.

As agricultural production involves several factors including land type, soil characteristic, genotype, water and nutrients management, pest control, intercultural operation, weather, etc.; the use of crop modelling as a decision support system (DSS) can test these factors in limited time and expenses for making the right decision on planning and management of the production. These simulation models must be accurately calibrated and validated to be effectively used to assess the various management strategies to counter the negative impacts of climate change/variability on agricultural production. The information generated from the DSS can be coupled with crop-weather calendar in generating climate risk management (CRM) matrices for different locations and cropping systems. The CRM matrix involving weather and crop model forecasts should be improved over time with regular feedback from the farmers based on its application and usefulness in managing climatic risk. The climate information and its associated risk are assessed and proper management policies used be developed based on the CRM tools. Therefore, the development of CRM tools are very much essential for benefit of the farming community and minimises the loss due to variability in weather/climate.

This book comprises of all the issues required to understand and model weather, climate and crop system and the steps for linking the weather information in crop models for yield forecasting. The key issues related to climatic variability and food security, different spatiotemporal scales of weather/climate forecasts and their importance in agricultural production planning and management have been well discussed. The fundamentals of weather and climate systems including prediction systems are explained in detail for the benefit of the readers. More importantly, the availability of different scales of forecast predicted by several organisations, bias estimation and correction techniques, and generation of the single deterministic forecast using multi-predictions (model) and multi-techniques (statistical) are also dealt with in detail for their use in crop models. As per the requirement of crop models, the monthly/seasonal scale weather forecast is disaggregated for crop yield forecasting and generation of CRM matrices, which is essential for preparing agro-advisory and are elaborated well for strengthening the theoretical and practical knowledge among the researchers. Finally, detailed theoretical descriptions of different components from the prediction models to the generation of CRM have been presented as hands-on practice for the application of the weather/climate information on agricultural production management decisions.

Chapter 2
Climate Variability and Food Security

2.1 Introduction

In the last century, the world has witnessed significant changes in the climatic variables affecting both terrestrial and oceanic ecosystems. By definition, climate change is a long-term alteration in the average weather patterns of a region or the world as a whole, that is attributed to an increase in global temperatures (Metz et al. 2007). This can include changes in temperature, precipitation, humidity, and other atmospheric conditions. The use of modelling approaches to assess long-term changes in climatic variables and their impacts on natural ecosystems at the regional and global scale, as well as projections of climate change, has enhanced our comprehension of the complex temporal patterns of climates. Climate projection suggests that extreme heat events, changes in rainfall patterns, increased flooding and drought events with higher intensity food and water scarcity, more diseases, and economic loss is considered to happen in the near future, and the sustainability of natural ecosystems and human health will most likely be at great risk (Gitz et al. 2016). In addition, the potential effects of increased levels of atmospheric CO_2, N_2O, and CH_4 on future climates should not be overlooked.

Food security and climate change are both associated with a range of interconnected risks and uncertainties. Climate change will intensify the challenges to fight against hunger and malnutrition (Gitz et al. 2016). The effects of the current climate extremes on food production are already visible in various parts of the world, and these impacts are expected to be more severe with future climate changes (Lobell and Gourdji 2012). Sustainable food production is fundamental for food security (Hazra and Bohra 2021). According to the current estimates, food grain production has to be increased by 60% to feed the ever-increasing population by 2050 (Mârza et al. 2015). Achieving food security in a sustainable manner is a major challenge for developing countries in the current century (Shukla et al. 2019). The potential impacts of climate change on agricultural production could

U. C. Mohanty et al., *Climate Risk Management in Agriculture*,
https://doi.org/10.1007/978-3-031-51862-1_2

have far-reaching implications for those who depend on it for their livelihoods and income, as it could bring about a range of direct and indirect risks (Gitz et al. 2016) (Fig. 2.1).

India has achieved remarkable progress in terms of food security. Despite the remarkable advances made since the green revolution, India still has a large number of undernourished people, accounting for one-fifth of the world's total. India is already facing the daunting task of sustaining agricultural production to meet the needs of its largest population (Kumar and Sharma 2013). In addition, climate change has become a further danger to the food security of the nation. Concerning this, the Government of India has given great importance to research and development in order to address climate change in the agricultural sector. In February 2011, the Indian Council of Agricultural Research (ICAR) initiated a major project called 'National Initiative on Climate Resilient Agriculture' (NICRA) during the XI Plan, with the aim of increasing the resilience of Indian agriculture (Rao et al. 2015). The alteration in the pattern of climate and its related climate variability has a major impact on the components of crop food security, such as availability, access, utilization, and stability. This chapter focuses on food security and the associated risks due to climate variability.

2.2 Climatic Variability in India—Present Trend and Future Projections

The climate conditions in India vary significantly between states, as evidenced by research conducted by Birthal et al. (2014) and Kumar et al. (2016). The historical data uncovered that the mean temperature has considerably elevated over the years (Fig. 2.2). It is estimated that the average air temperature will rise by 1.7–4.9 °C by the end of this century, according to Chaturvedi et al. (2012). A steady increase in both minimum and maximum temperatures has been observed in recent decades (Kothawale and Rupa Kumar 2005; Birthal et al. 2014) (Fig. 2.3). To make the global climate model (GCM) scenarios usable bias correction was carried out and the resultant scenarios indicate significant warming of the Indian region in future climates. Monthly variations were noted for the projected rise in temperature. Maximum temperatures are projected to be more in northern latitudes and more during the *rabi* season. Although there is no definitive proof of long-term changes in rainfall at the national level, regional studies have indicated a shift in precipitation patterns (Jain and Kumar 2012; Mallya et al. 2016). In recent years, the amount and pattern of rainfall have shifted in various agro-climatic zones, and research has suggested that these changes will lead to more frequent droughts and floods in the near future. Indian agriculture is largely dependent on rain, with the majority (70%) of farmers being marginal and smallholders who produce 40% of the country's food grains (Birthal et al. 2014).

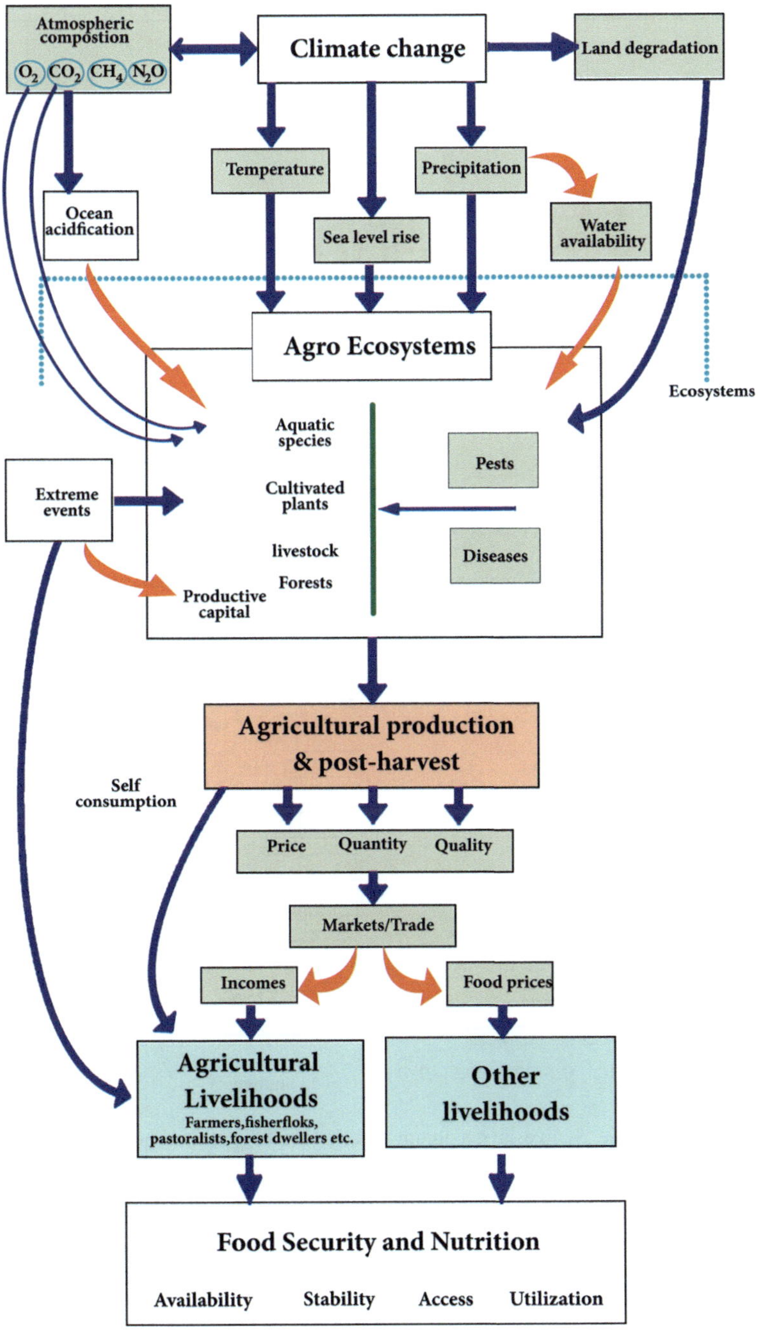

Fig. 2.1 A schematic illustration of the domino-like consequences of climate change on food security and nutrition is presented by Gitz et al. (2016)

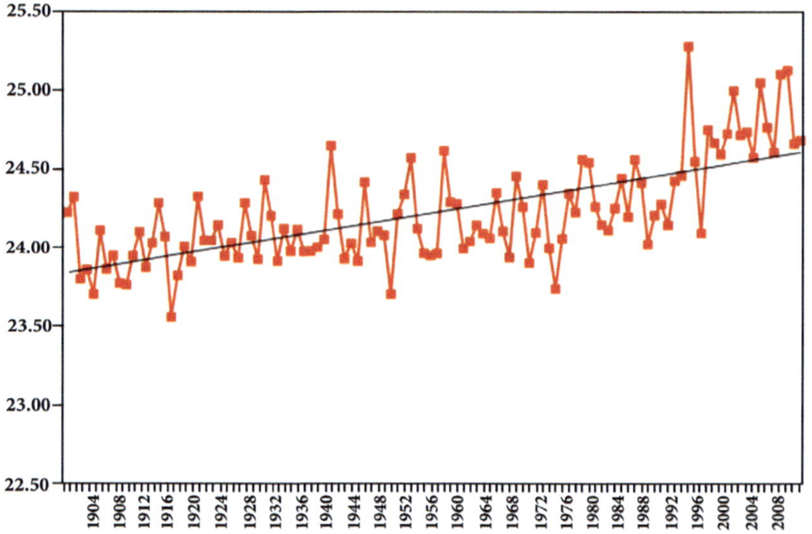

Fig. 2.2 The average annual temperature across India from 1900 to 2010 (*Source of data* India Meteorological Department, New Delhi)

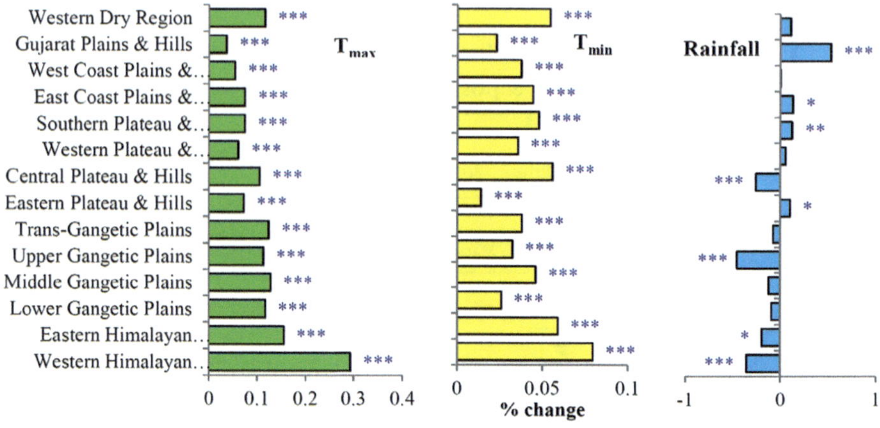

Fig. 2.3 The percentage change in maximum temperature (T_{max}), minimum temperature (T_{min}), and rainfall from 1966 to 2011 in different Agro-climatic zones of India. *Source* Singh et al. (2015): Authors Estimation. Census of India (2011), Singh (2006), NICRA District Agricultural Contingency Plans, Indian Meteorological Department, Directorate of Economics and Statistics, Ministry of Agriculture and Farmers' Welfare. *, ** & *** indicates 10, 5, and 1% level of confidence

India is already facing water scarcity in many areas, and the effects of climate change are expected to exacerbate this situation (Chakrabarty 2016). According to Cruz et al. (2006), the deficit in rainfall and droughts caused wetlands to dry up, which is observed as severe in 14 major wetland ecosystems. They also estimated

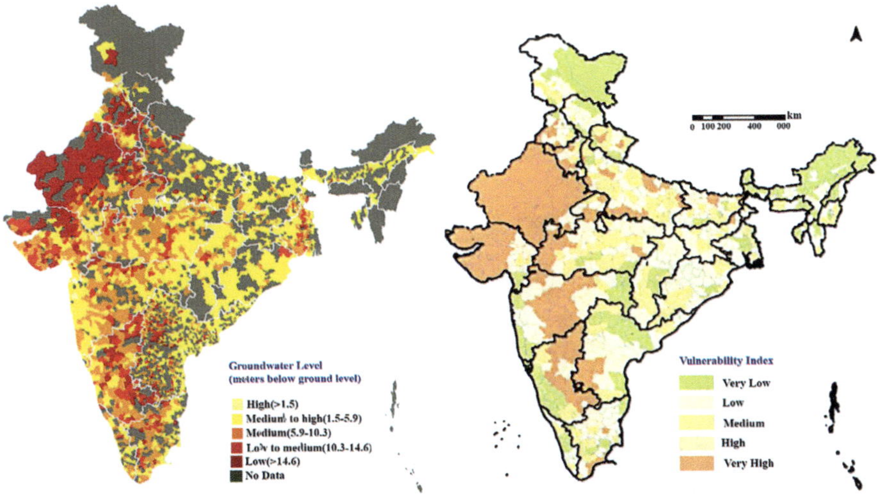

Fig. 2.4 The decreasing level of groundwater in India (*Source* World Resources Institute, http://www.wri.org/blog/2015/02/3-maps-explain-india%E2%80%99s-growing-water-risks) (**a**), vulnerability scale of Indian agriculture to climate change (**b**) (*Source* Rama Rao et al. 2018)

that ~54% of India faces extremely high water stress (Fig. 2.4). The states of Punjab and Haryana, which are responsible for the majority of India's rice and wheat production, are facing a severe water shortage in the northwestern region of the country. In India, it has been estimated that 12.6% of the arable land is prone to flooding, and approximately half of the cultivated area is affected by natural disasters annually (Kumar et al. 2015a). India's food production is highly susceptible to the effects of climate change, as the agricultural sector is still heavily reliant on the monsoon season. For example, in the western Indo-Gangetic Plain (IGP), food systems are particularly susceptible to issues such as over-irrigation and rising water tables, which lead to soil salinity, whereas, in the eastern IGP, farmers are faced with the challenge of rising sea levels and increased risk of flooding, which they have limited capacity to address.

A majority of Indian agriculture (around two-thirds of the cropped area) is rainfed and highly dependent on the unpredictable monsoon (Mall et al. 2006). Around half of the Indian population relies on climate-dependent industries such as farming, forestry, and fishing for their daily sustenance (Kumar et al. 2016). Figure 2.5 shows the Stressors Vulnerability Scale of 121 Krishi Vigyan Kendras (KVKs) districts of the National Initiative on Climate Resilient Agriculture (NICRA).

Farmers in the region were increasingly worried about the fact that monsoon was arriving later in recent years. This perception was particularly evident in the northern and southern regions. In these two areas, the majority of farmers experienced an early end to the monsoon season. This implies a decrease in the length of the crop growing season, which has major implications for crop planning and

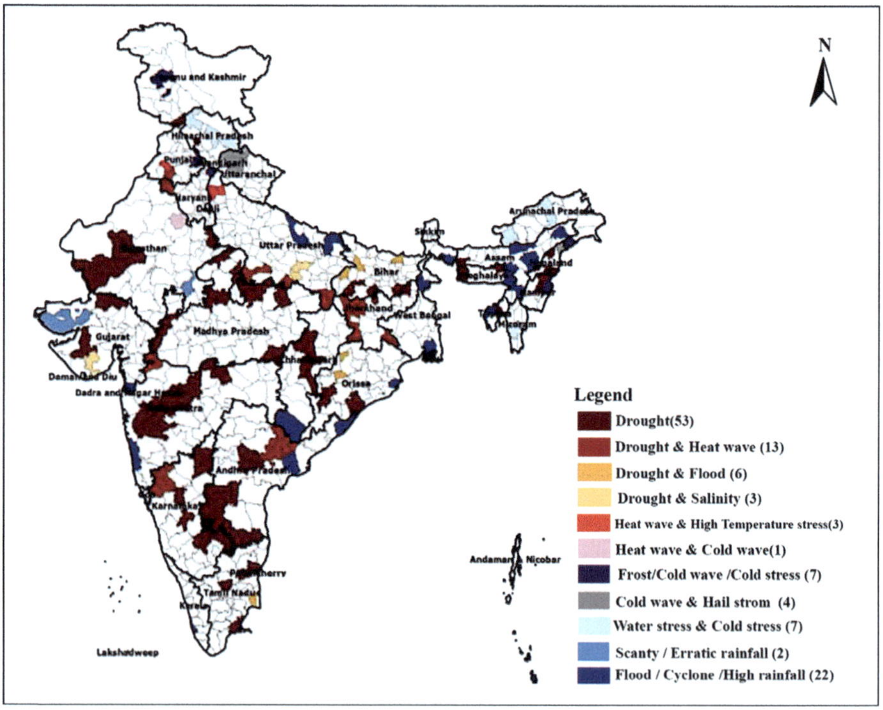

Fig. 2.5 Map of 121 districts along with the climate vulnerabilities (*Source* NICRA)

management. The majority of farmers in all the zones reported a decrease in the amount of rainfall they received each year. A significant number of farmers in the eastern and hilly regions did not observe any pattern in the yearly precipitation. A majority of farmers in all the zones have noticed that summers and winters are becoming increasingly hotter, which is a sign of changing climate (Rama Rao et al. 2018).

2.3 Food Production in India

Agriculture is a major contributor to India's economy, accounting for approximately 17–18% of the country's Gross Domestic Product (GDP) and providing employment to over half of the population. India has the highest net-cropped area in the world, with the US and China coming in second and third respectively, as per the Indian Economic Survey 2018 (https://www.indiabudget.gov.in/budget2019-20/economicsurvey/). According to the Intergovernmental Panel on Climate Change's (IPCC) fifth assessment report (AR5) and special report (SR15), the potential consequences of climate change, such as rising temperatures, altered precipitation patterns, extreme weather events, and reduced water availability, could lead to a decrease in agricultural productivity (Pachauri et al. 2014).

Smallholder farmers around the globe are seen as particularly susceptible to the effects of climate change, due to alterations in temperature, precipitation, and the occurrence or severity of extreme weather events that can have a direct impact on their crop and livestock yields, as well as their family's food security, income, and overall welfare. It is clear that global warming is causing the global mean sea level to rise in two ways: melting glaciers and ice sheets worldwide are adding water to the ocean, and the ocean's volume is increasing as the water warms. The increasing sea level poses a risk of saltwater intrusion into agricultural lands in small islands and coastal areas, thus threatening their existing agricultural production. Changes in precipitation and sea level rise can lead to soil salinization, resulting in decreased food and water security due to reduced agricultural production. The sustainability of global food systems may be jeopardized by climate change, as it can have a significant impact on future price movements and the short-term volatility of prices (Wheeler and von Braun 2013).

India is particularly vulnerable to the impacts of global warming due to its tropical climate, low economic status, and extensive coastal areas. High temperatures and extreme precipitation can have a detrimental effect on productivity by causing physiological changes. The IPCC (AR6) has warned that climate change, combined with the increasing demand from a growing population, could put food security at risk in the future. Furthermore, it can also affect soil fertility, the occurrence of pest infestations, and water availability, which can have an impact on food grains, animal husbandry, and fishing.

The following section examines the effects of climate change and its associated variability on the major crops, animal husbandry, and fisheries that form the daily diet of the majority of people in Southeast Asia, particularly India.

Food Grain

One of the major issues confronting the Indian agricultural system is its low productivity, which is exacerbated by higher temperatures and more erratic precipitation patterns that reduce the yields of staple crops such as maize, wheat, and rice (Sage et al. 2015). The yield of cereals in developed countries has been very low since the 1980s, with no significant improvement in recent years (Fig. 2.6).

Rice is a dietary staple for a large portion of the world's population, particularly in Asia, and is essential for food security (Yearbook [2013], http://www.fao.org/docrep/018/i3107e/i3107e00.htm). India is the second-largest producer of rice in the world, with a significant portion of it being cultivated in rainfed conditions during the *kharif* season (Joshi 2015). This crop accounts for 23.3% of the total gross cropped area and 43% of India's total food grain production (Singh 2009). India is responsible for around 67% of the total rice production in South Asia. Figure 2.7 illustrates the area, production, and yield of rice in India from 1995 to 2016. The rice fields had an average yearly increase of 0.59%. The production and yield growth were 2.51% and 1.53%, respectively, from 2000–01 to 2010–11.

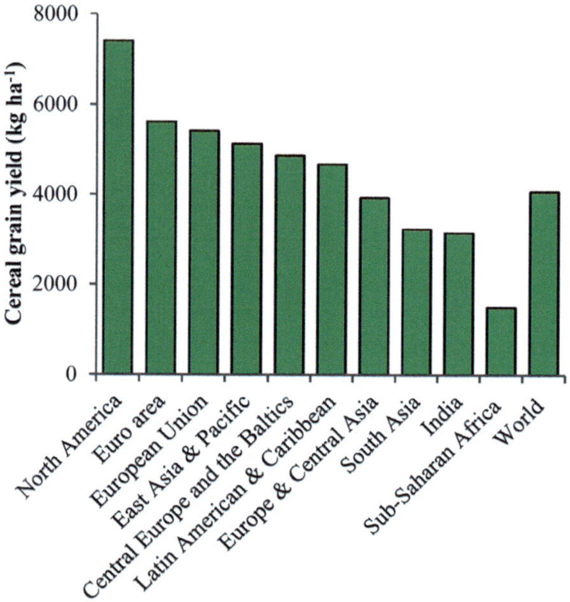

Fig. 2.6 Country-wise yields of cereal grains (kg per hectare, 2017). *Source* World Bank Database (https://data.worldbank.org/indicator/AG.YLD.CREL.KG)

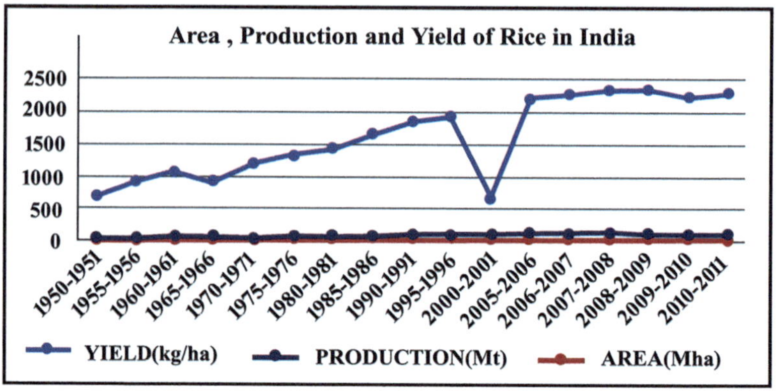

Fig. 2.7 Area under food grains (in Mha) in India in Million hectares from the period 1995–96 to 2015–16 (*Source* Singh et al (2020))

Unfortunately, the area's growth rate became negative (–0.73%) due to the monsoon failure that hindered production growth. The green revolution, which saw the introduction of high-yielding production technologies and high-yielding rice varieties, resulted in a peak in productivity growth of 3.19% between 1980–81 and 1989–90. Recent years have seen a decrease in productivity growth (Singh et al. 2013).

Rice cultivation in tropical regions is highly susceptible to climatic conditions such as temperature, precipitation, and solar radiation, which can have varying effects on the crop during its various growth stages (Yoshida 1979).

Rice is the most important *kharif* crop in India, and it is highly sensitive to low temperatures and rainfall. Eastern India accounts for 60% of rice production in India's production of which 80% is rainfed. Rice cultivation in rainfed is more susceptible to temperature and rainfall due to its dependency on climate. A slight change in these climate variables drastically affects the rice yield. The intense heat and radiation of the monsoon season in Eastern India drastically reduce the amount of rice that can be harvested (Garrity et al. 1986). The onset of cold during anthesis has a significant impact on rice production. It is majorly observed in the late-transplanted rice in eastern India.

High temperatures in the range of 20–30 °C and high precipitation levels of 1500–2000 mm are beneficial for the growth of rice plants and the yield they produce (Yoshida 1981). However, temperatures exceeding 35 °C can cause heat stress, resulting in sterility of the spikelets, unproductive pollen, and a decrease in grain quality (Nguyen et al. 2014).

In Eastern India, recurrent floods and droughts are a common occurrence in the majority of rain-fed rice areas in Assam, North Bihar, Orissa, and West Bengal. According to Prasad et al. (1986), approximately 10 Mha of lowlands in Bihar, Orissa, and West Bengal are subject to flash floods and water saturation. Research has demonstrated that floods and droughts are the primary impediments to highland rice production in Bihar, leading to a significant decrease in yield (Thakur 1994). In Bihar alone, almost 95–100 thousand tons of highland rice is lost from the plains every year. The losses caused by flooding in the rainy areas and the lowlands of Bihar account for around 12–27% of actual production per year. Various studies show that drought is the main contributor to the rice yield gap in the highlands, where floods such as flash floods represent the main constraint for highland rice in the highlands (Herdt 1996). In India, maize is particularly vulnerable to drought, with yield reductions being higher than for other crops (Leng and Hall 2019). Drought can cause a decrease in transpiration rates, leading to leaf rolling and drying, reduced leaf expansion rates and plant biomass, immobilization of solutes, and heat stress. Consequently, the projected climatic changes, particularly those involving increased variability in precipitation, could render some regions unsuitable for rainfed rice cultivation. In a study conducted by Singh et al. (2017), the use of statistical models such as the Climate Envelope Model (CEM) and Boosted Regression Trees (BRT) was explored to map the climate suitability for rice cultivation in India at a regional level. Their research revealed that between 15–40% of the current rainfed rice farming could experience a decrease in yield or may not be suitable for cultivation. Northern and eastern parts of India are more at risk than some parts of the western and on the contrary, central India may get benefit from the increased precipitation levels. In a study conducted by Gupta et al. (2015), district-level yield data for rice, sorghum, and pearl millet in India were analyzed. The results showed that rice yields are sensitive to rainfall extremes, while sorghum yields are increasing with extremely high temperatures. However, there was no evidence of any effect of climate variables on pearl millet yields.

The Indian Agricultural Research Institute (IARI) employed the INFOCROP model (outlined in Chap. 8) to evaluate the potential impacts of climate change on India's main crops, such as rice and wheat, by taking into account alterations in climate variables like carbon dioxide, temperature, and precipitation. The effect of these variables may increase or decrease crop yields. They reported that rice yields would likely be decreased by increasing mean temperature from 2 to 4 °C. The eastern regions of India are expected to be hardest hit by rising temperatures and decreasing radiation, resulting in relatively lesser grain count and shorter duration for grain filling. In contrast, the yield reductions are due to increased temperatures and higher radiation in Northern India. Although the additional CO_2 could benefit the plants, this effect was offset by increasing temperature.

In many parts of India, a rise in temperature (around 2 °C) leads to a decrease in wheat grain yield. The effects of climate change are more pronounced in the lower productivity regions than in the higher potential productivity regions, such as northern India. They found the boundary changes for the irrigated wheat production in India, which is shown in Fig. 2.8. The yield loss prediction is more in the rainfed regions than in irrigated ones due to the rainfall variability. The yield reduction in the sub-tropical areas ranges from 1.5 to 5.8%. It is comparatively more in tropical regions, which indicates that greater crop loss can be expected in warmer regions.

Wheat is a major food crop in India, contributing 13.64% of the world's total production and 14% of the global area (Ramadas et al. 2019). The production has increased from 12.3 to 70.8 Mt (1 Mt = 1000 kg), from 1965 to 1999 and the increment of land area and the green revolution played a major role (Kumar and Singh 2014). The production reached 106.21 Mt in the year 2019–20. The demand for wheat is expected to exceed 130.0 Mt by 2050 (Attri and Rathore 2003). The increasing demand for wheat production under adverse environmental change scenarios would be a difficult task.

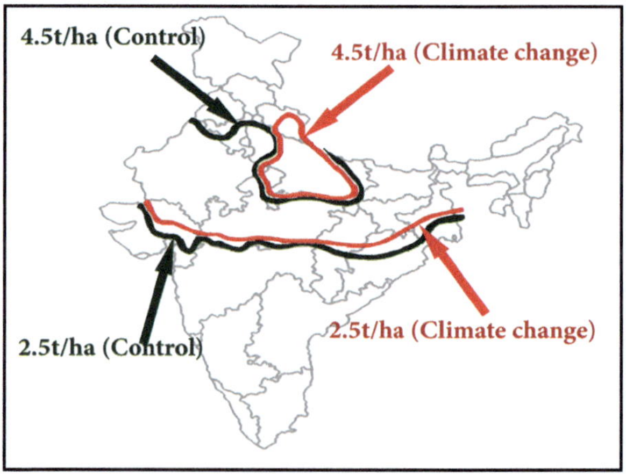

Fig. 2.8 Boundary shift of irrigated wheat cultivation (*Source* IARI Report)

The potential yields of many tropical and subtropical regions are likely to be reduced due to the projected temperature increases, as reported by Houghton et al. (2001). Climate change has a considerable effect on the growth, development, and productivity of plants, as it affects the stomatal opening and the rate of physiological processes through its influence on temperature and CO_2. Higher temperatures accelerate biochemical reactions and also increase transpiration losses. Stomatal conductivity decreases in plants at C3 and C4 with increasing CO_2 concentration. C3 plants appear to gain more from higher CO_2 concentrations in terms of dry matter production, as they experience increased leaf expansion, higher photosynthesis rates per unit area, improved water use efficiency, and a rise in photorespiration rates. The process of nitrogen (N) uptake is influenced by changes in the rate of physiological processes resulting from increased temperature and increased CO_2. Plants tend to keep the C: N ratio constant for better growth and development. Due to the increased levels of CO_2 in the atmosphere, plants absorb more nitrogen, which helps them to grow and develop better, thus counteracting the negative effects of rising temperatures. Hazra et al. (2019) did an experimental study on wheat crops in eastern India with an elevated CO_2 level and they found that the increment of N dose of fertilizers (50% higher than the recommended) has a significant increment in wheat yield. Attri and Rathore (2003) did a study on different wheat cultivars of enhanced temperature and CO_2 scenarios and found that the wheat yield is increasing with the temperature rise of 1.0–1.5 °C and 466 ppmv of CO_2. But, further, an increase in the temperature resulted in the yield reduction. Lal et al. (1998) also concluded the same i.e., an increase in temperature affects the wheat yield irrespective of the enhanced carbon dioxide level in the atmosphere. Mor (2017) demonstrated that the yield of wheat is positively impacted by variations in winter rainfall, yet it is likely to be detrimentally affected by increasing temperatures in future climate change scenarios. It is projected that the yield will decrease by 8.3%, 6.6%, and 12.9% in 2020, 2050, and 2100, respectively. This decline will most likely affect smallholder farmers more than the medium and larger holders farmers in India. Research conducted by the All India Coordinated Research Programme on Agro-meteorology revealed a negative correlation between wheat grain yields in major wheat-growing districts and minimum temperatures (Tn) during the post-anthesis period (Rao et al. 2015) (Fig. 2.9).

Adaptation of timely sowing and heat tolerance varieties plays a significant role in climate-changing scenarios. Kumar et al. (2020) did an empirical study on the wheat yield in Uttar Pradesh and Haryana using the Ricardian model. They found that the yield reduction is due to the increase in temperature and also, and they found that climate change is non-linearly related to crop production. Jha and Tripathi (2011) also proposed that in order to counter the impacts of climate change, late and early-sown varieties, and heat-tolerant varieties should be made available in the market, and farmers should be encouraged to make use of these facilities to maximize crop production. Consequently, non-climatic factors also play a role in agricultural growth. On the other hand, the wheat yield will increase 13.9–15.4%, 23.6–25.6%, and 25.2–27.9% for the periods 2020, 2050, and 2100

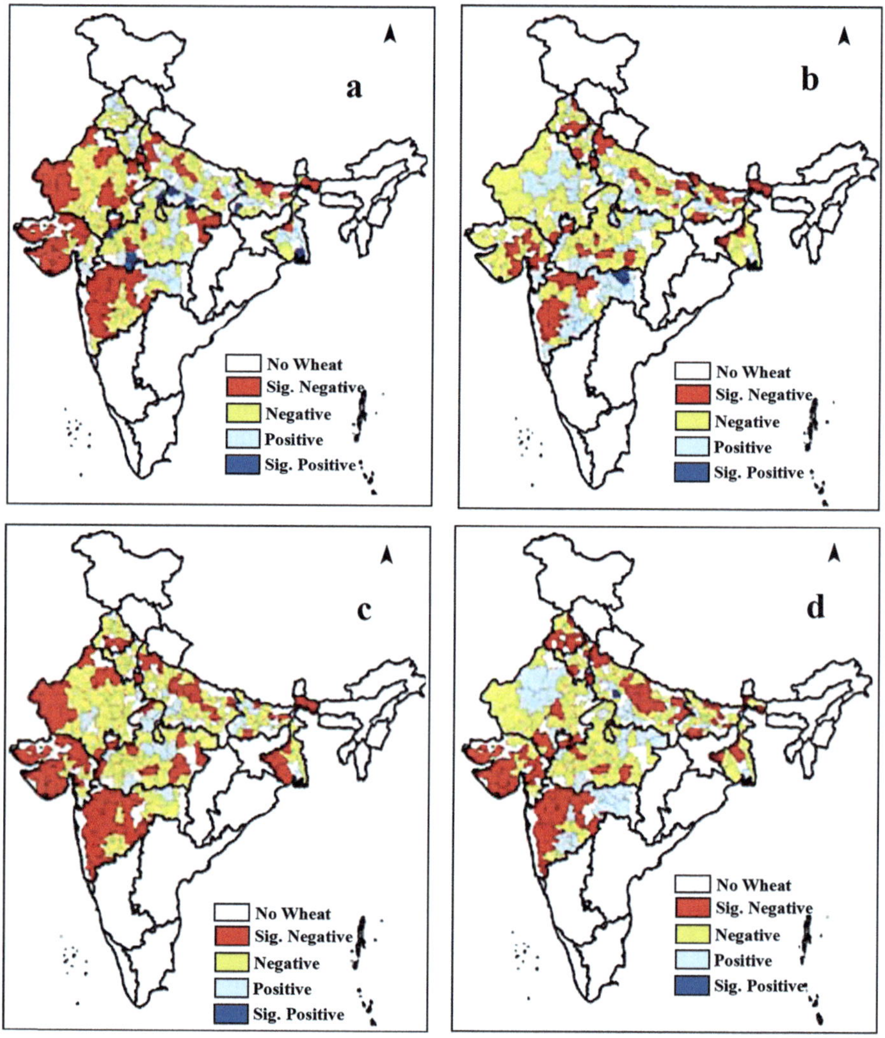

Fig. 2.9 Relationship of district wheat yields with **a** seasonal Tx, **b** February Tx, **c** seasonal Tn, **d** February Tn (Tx, and Tn represents Maximum and minimum temperatures respectively)

with the increment of irrigation water facilities due to the reduction in the rainfall in the *rabi* season (Abeysingha et al. 2016). Kashyap and Agarwal (2021) demonstrated that irrigation can help to counteract the impacts of fluctuating precipitation on wheat yields in a changing climate.

Maize is India's third-largest staple food crop and accounts for 9% of the national food supply. It is one of the promising agricultural diversification opportunities in the upland regions of India. Maize is a highly adaptable cereal crop that can be grown in a variety of environmental conditions, and it needs a moderate climate for optimal growth (Kogbe and Adediran 2003). In India, the area under maize cultivation has been steadily increasing over the years, reaching 9.2 million hectares in 2014–15 and yielding 24.2 Mt (Patidar et al. 2020). Climate change in precipitation variability and temperature fluctuations is expected to significantly impact agricultural production. The research conducted by Allen et al. (2003) and Patidar et al. (2020) has shown that an increase in mean annual temperature of 1 °C can have a detrimental effect on maize production in South Asian regions, with a 17% decrease in yield and a 10% and 8% decrease in grain and biomass yield, respectively, in Madhya Pradesh state. The precipitation will likely increase in future RCP scenarios, and it resulted in a decrease in the maize yield. They found that the delay of sowing dates will likely reduce the effects of temperature and rainfall variability. Guntukula and Goyari (2020) conducted a study on maize production in Telangana and found that the average minimum temperature had a significant negative effect on maize yield. Additionally, maximum temperature and precipitation were positively correlated with maize yield, although the maximum temperature was not statistically significant. Furthermore, precipitation and minimum temperature were identified as risk-mitigating factors, while maximum temperature was identified as a risk factor for maize yields. Ahmed et al. (2020) also studied the impact of temperature on maize production and grain yield, finding that these are inversely proportional to the amount of nitrogen fertilizer applied in a changing climate. However, the temperature positively impacts the maize yield in the upper Gangetic plains. The temperature is low in those areas in winter, which is not favourable for maize production. The projected temperature will most likely rise due to climate change which could result in an increase in maize production.

2.4 Fruits and Vegetables

India produces 169.1 Mt of vegetables and 90.2 Mt of fruits in an area of 10.1 Mha and 6.3 Mha, respectively. Between 1996 and 2010, the Ministry of Agriculture reported a 13.5% increase in the area of vegetable production and a 13.4% increase in production. India is the second largest producer of vegetables globally, with a 14% share of the world vegetable market, behind China. India is a major producer of vegetables, being the second-largest producer in the world. It is the leading producer of ginger and okra and ranks second in the production of potatoes, onions, cauliflower, eggplant, and cabbage. Additionally, India is the top producer of bananas (25.7%), papayas (43.6%), and mangoes (40.4%) among fruits.

In the last 25 years, the production of vegetables around the world has doubled, and the value of the global vegetable trade is now higher than that of staple food grains. This has been beneficial to small farmers, as it provides them with higher incomes and more employment opportunities per hectare than staple food. Vegetables are also an excellent source of micronutrients, making them a valuable resource for overcoming deficiencies. The yield of most vegetables is significantly impacted by a variety of environmental conditions, such as temperature, solar radiation, humidity, precipitation, wind, drought, and salinity. These factors are further exacerbated by climate change, leading to reduced yields due to extreme temperatures and limited soil moisture.

Environmental stress is a major contributor to crop losses globally, reducing the average yields of major crops by over half. The cultivation of tropical vegetables is subject to a range of conditions that vary depending on the season and region. Furthermore, climate change is exacerbating the effects of environmental pollution on vegetable production. High temperatures, reduced irrigation water, flooding, and salinity are all factors that limit crop productivity. Additionally, extreme weather conditions can lead to soil degradation and erosion, further reducing crop yields. To counter these effects, it is important to improve nutrient use efficiency in order to increase plant growth due to increased atmospheric carbon dioxide.

High temperatures can cause disruption to the biochemical processes that are necessary for the proper functioning of plant cells, particularly those related to photosynthesis in higher plants. This can lead to a decrease in tomato productivity, with fewer fruit sets and smaller, lower-quality fruit. For pepper, high temperatures during the pre-anthesis stage did not have an effect on the pistil or stamen, but post-pollination temperatures could inhibit fruiting, indicating that fertilisation is sensitive to high-temperature stress.

Climate change is anticipated to have a major impact on water availability, with water stress likely to have a detrimental effect on crop production, particularly of vegetables. As temperatures rise, a decrease in rainfall and an increase in evapotranspiration can reduce the amount of irrigation water available, leading to extreme water stress for plants. Vegetables are typically composed of 90% water (Midmore and Poudel 1996), making water a major factor in the yield and quality of these crops. Drought can have a drastic impact on vegetable production, as it reduces the amount of water available to the plants.

The productivity of many crops, particularly vegetables, is being adversely affected by the rising salinity of the soil, especially in irrigated areas that provide 40% of the world's food. This is a major concern as it can have a detrimental effect on crop production. Onions are particularly sensitive to saline soils, while cucumbers, eggplants, peppers, and tomatoes are moderately sensitive. This information is according to the United States Department of Agriculture (USDA). In hot and arid climates, high levels of evapotranspiration can lead to a significant loss of water, leaving salt deposits around the roots of plants. This salt stress can cause a decrease in turgor, wilting, leaf curling, reduced photosynthesis, cell damage, and even death of the plant. In coastal regions affected by poor quality and high saline irrigation water due to groundwater contamination and saltwater intrusion from natural or human-induced events, the salinity of the water can have a detrimental effect on agriculture. Traditional wells are often used for irrigation in inland areas, but the salty deposits in these wells can lead to increasingly saline water, which can reduce the yield of irrigated crops.

In the tropics, vegetable production is usually carried out during the dry and wet seasons, but the high humidity caused by heavy rains during the rainy season often restricts it. Floods can be particularly damaging to vegetables, as they reduce oxygen levels in the root zone, thus hindering aerobic processes. Tomatoes are especially vulnerable to flooding due to their limited genetic variation in this regard. The flooding of tomato plants leads to an increase in the body's ethylene, which can be damaging to the plants. This is due to the low oxygen levels, which cause a rise in the production of 1-aminocyclopropane-1-carboxylic acid (ACC) in the roots. The rapid growth of epinastic leaves is a common response of tomatoes to humid conditions, and the role of ethylene accumulation has been studied. As temperatures rise, the effects of flooding on tomato plants become more severe, often leading to rapid wilting and death after a brief period of inundation.

Potato cultivation in India is mainly concentrated in the Indo-Gangetic plains, where it is grown during the mild, cool winters. The fall/winter crops of the northern Indian states of Uttar Pradesh, West Bengal, Bihar, Punjab, and Haryana account for a large proportion (84%) of the total potato production in India. The sensitivity of the potato harvest to climate variability implies that climate change and global warming will have a major effect on potato growth in India. Even a slight rise in temperature can drastically reduce the tuber yield without significantly impacting photosynthesis and overall biomass production. Additionally, high temperatures can also affect the quality of the tubers by causing thermal germination and internal necrosis. The presence of a high concentration of carbon dioxide has been found to have a generally positive effect on the growth and yield of crops, although some negative effects have been observed. For example, potatoes and vegetables have been known to ripen prematurely, and a heatwave can lead to significant crop losses (Prasad and Rana 2006). In a study conducted by Singh et al. (2013), the INFOCROP-POTATO model was used to assess the impact of climate change on potato production. It was found that a carbon dioxide concentration of 400 ppm and a temperature increase of 1 °C would result in an 11.1% increase in potato production by 2050. However, if the carbon dioxide concentration were to rise to 550 ppm and the temperature to 3 °C, it would lead to a 13.

The temperature rise in the coming decades is likely to cause an increase in weather anomalies such as floods, droughts, and cold and heat waves. The 1999 super cyclone off the coast of Orissa had a devastating effect on cashew plantations, with many trees being uprooted due to the sandy soils. The extreme cold weather experienced in the northern states of India from 2002 to 2003 had a detrimental effect on all fruit crops. However, temperate fruit crops such as apples, plums, and cherries benefited from the prolonged cooling, resulting in a high yield. On the other hand, the heat wave that occurred in March 2004 had a negative impact on crops such as tea, potatoes, and vegetables in the same belt. Additionally, long periods of drought in summer caused significant damage to crops such as cocoa, black pepper, coconut, coffee, tea, and cardamom along the west coast in 1982–1983 and 2003–2004 (Rao et al. 2008). The winter of 2010 saw an increase in nighttime temperatures in some areas of the country, which had a negative effect on the flowering of mangoes. To protect horticulture from the potential impacts of climate change, such as cold and heat waves, floods, droughts, and rising sea levels, proactive measures must be taken to ensure national climate resilience.

By 2050, the productivity of coconut in India is projected to rise by 10%, and by 2080, it is expected to reach 20% higher than current yields. However, on the east side, yields are predicted to decrease by 8% by 2050 and 31% by 2080 in comparison to current yields. In particular, Kerala, Maharashtra, and parts of Tamil Nadu and Karnataka are expected to experience an increase in yields, while Andhra Pradesh, Orissa, Gujarat, and other parts of Tamil Nadu and Karnataka are likely to experience a decrease. Global warming and climate change are likely to have a mixed effect on coconut production in Kerala. While increased temperatures may lead to higher yields, the increased frequency of floods and droughts could have a negative impact. Furthermore, any rise in temperature during the second phase of nut development could reduce the size of the nut and thus the crop yield and oil content. This stage is particularly sensitive to high temperatures. In the past three decades (1951–80) compared to the next three (1981–2009), a slight decrease in coconut production was observed under field conditions in Kerala, as opposed to the INFOCROP-COCONUT model. For three decades between 1951 and 1980, coconut productivity in Kerala was significantly high (5,762 nuts ha^{-1}) compared to the 1981–2009 period (5,670 nuts ha^{-1}), with a 1.6% decrease. During this time, there were notable changes in the distribution of precipitation, dryness index, number of summer droughts, humidity index, and temperature. The rise in temperature, the drought index, the number of severe summer droughts, and the decrease in the precipitation and humidity index have all been identified as contributing factors to a slight decrease or stagnation in coconut productivity over a period of time (Rao et al. 2013). As a result, it is expected that coconuts' productivity will be negatively impacted by the projected climate change scenario, as floods and droughts are likely to become more frequent and intense.

Bananas are plants that thrive in moist conditions, and in India, they are mainly cultivated in the Gangetic plains of Bihar, the Godavari River watershed in Andhra Pradesh, the Tapti River basin in Maharashtra, and the Cauvery Delta region in Tamil Nadu. Additionally, other areas with access to water sources are also suitable for banana cultivation. India is the largest producer of bananas in the world, accounting for 29.2% of global production with a total output of 29.8 million metric tons (Mt) in an area of 0.8305 million hectares (Mha) (Singh and Singh (2020)). A rise in temperature of 1–2 °C in the subtropics of the nation, as predicted by the global circulation model, could be beneficial for the growth of bananas. This could lead to an expansion of banana cultivation into Maharashtra, Gujarat, Uttar Pradesh, Bihar, Jharkhand, West Bengal, Orissa, and the northeast, thus increasing production. The regions located in the center and northern parts of the country are likely to be conducive to the growth and development of bananas, potentially leading to an increase in yield from current production levels. Nevertheless, if the temperature remains high for an extended period of time during the summer months, coinciding with the flowering and development of the bananas, it could result in a decrease in productivity. Bananas, as a C3 plant, have been observed to perform better under high CO_2 concentrations in the atmosphere. Photosynthesis and biomass accumulation have been seen to increase with an increase in atmospheric CO_2 levels (Schaffer et al. 1997). However, it is yet to be determined if the effects of high temperatures can be counteracted by increasing the concentration of CO_2. Banana cultivation can be improved and intensified if there is a reliable source of water throughout the year. As climate change alters the precipitation pattern, new cultivation methods must be adapted to meet the needs of a growing population. Ensuring a steady supply of water is essential for successful banana production.

The quality of horticultural products is largely determined by environmental factors such as soil, temperature, light, and CO_2. Unfortunately, climate change has had a negative impact on the nutritional value of many fruits and vegetables. Ascorbic acid and sugar levels in tomatoes were enhanced when exposed to elevated carbon dioxide levels (Islam et al. 1996). However, this increase in CO_2 can lead to a decrease in the protein and mineral content of the fruit. Vitamin C, sugar, acids, and carotenoids in oranges, tomatoes, and strawberries were all improved when exposed to higher CO_2 concentrations. When exposed to elevated levels of CO_2, the total sugar and acidity of grapes increased during the middle stages of ripening, but not in the later stages (Bindi et al. 1996). Additionally, sour oranges grown in open chambers with 400–700 μmol mol^{-1} CO_2 had a 7% higher vitamin C content than those grown in normal conditions (Idso et al. 2002). Field studies conducted by Wang et al. (2003) revealed that the increase in total biomass and yield of strawberries was not accompanied by a significant increase in vitamin C. Furthermore, Njoku et al. (2011) found that high-temperature stress can lead to a decrease in vitamin C, starches, sugars, anthocyanins, and volatile substances, as well as aromatic compounds of fruits. Irrigation can be beneficial for the quality of fruits and vegetables, as it can lead to an increase in sugar, anthocyanins, and volatiles. However, too much water stress or a combination of high temperatures and water stress can have a negative effect on the quality of fruits and vegetables, reducing their levels of vitamins, antioxidants, and minerals (Shivashankara et al. 2013).

2.5 Animal Products

In India, livestock is a major contributor to the agricultural sector, with an estimated 20.5 million people relying on it for their livelihood. It is estimated that small agricultural households derive 16% of their income from livestock, compared to the average of 14% for all rural households. Livestock plays a vital role in the livelihood of two-thirds of rural communities in India, providing employment to 8.8% of the population. It contributes 4.11% to the country's GDP and 25.6% to the total agricultural GDP. India is the largest producer of milk in the world, accounting for 20.2% of the global milk production. It also has the largest egg production in the world, with 5.7% of the global share, and 3% of the global meat production. According to the 2019 livestock census, India ranks first in the buffalo population, second in goat farming, second largest in the poultry market, third in the sheep population, and tenth in the camel population in the world. Uttar Pradesh shares the largest amount of milk (16.3%) and meat (15.1%), Andhra Pradesh shares the largest amount in egg production (18.2%), and Rajasthan shares the largest wool production (35.9%) in the country.

Adapting to and mitigating the impacts of climate change on livestock systems is essential for their sustainability. Sejian et al. (2015) highlighted the importance of this, as extreme weather conditions can have a detrimental effect on farm animals. Therefore, strategies must be implemented to reduce the risks posed by climate change and ensure the long-term viability of livestock production. The impacts of extreme weather conditions on cattle are significant, with climatic extremes and seasonal variations in the availability and quality of grasses having a detrimental effect on animal welfare, productivity, and reproductive performance (Sejian 2013). Climate change has a significant impact on livestock production, as evidenced by the effects of extreme weather conditions such as heat waves, floods, and droughts. These events can lead to production losses and even the death of farm animals (Gaughan and Cawsell-Smith 2015). Animals may be able to adapt to hot climates, but the coping mechanisms used for survival can have a negative effect on their performance. Figure 2.10 illustrates the impact of climate change on livestock production.

Climate change has a significant impact on livestock production, leading to financial losses due to reduced milk and meat production, reproductive efficiency,

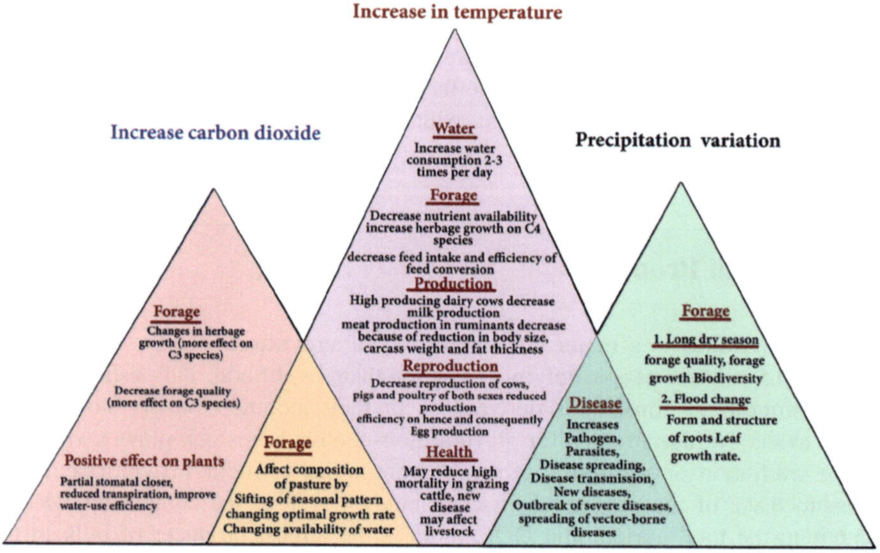

Fig. 2.10 Impacts of climate change on livestock (*Source* Rojas-Dowing et al. 2017)

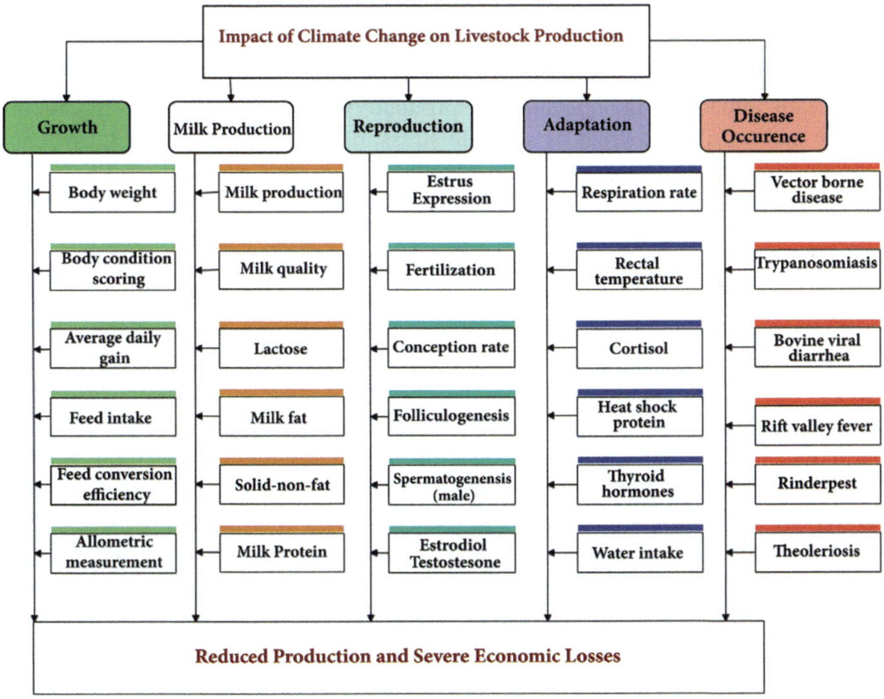

Fig. 2.11 Impacts of climate change on livestock production (*Source* Seijian et al. 2016)

and cattle health. Heat stress is one of the major factors contributing to these losses. Figure 2.11 illustrates the various effects of climate change on animal production, with most of the losses resulting from indirect effects such as reduced or unavailable food and water resources.

The quantity and quality of forage are significantly impacted by changes in atmospheric CO_2 content and temperature. C3 crops are more sensitive to increased CO_2 levels than C4 crops, which have higher water use efficiency (Thornton et al. 2009). Temperature and CO_2 levels can also alter the dynamics of species competition due to changes in optimal growth rates. High temperatures and arid conditions can lead to a decrease in the availability of nutrients for farm animals, as increases in temperature can increase lignin and cell wall components in plants, decreasing digestibility and degradation rates (Polley et al. 2013). C3 plants have a greater potential to improve their forage quality than C4 crops when exposed to higher levels of carbon dioxide. Additionally, C3 plants have a higher crude protein content and digestibility than C4 plants.

In many areas, the unpredictable nature of rainfall during the growing season has had a detrimental effect on the production of animal feed. As climate change continues to take its toll, livestock production systems are likely to experience more negative than positive impacts. Due to the complexity of managing, analyzing, and interpreting the effects of multiple stressors on farm animals, most studies examining the impact of environmental stress on animal production, reproduction, and immune status have focused on one stressor at a time. Animals raised in tropical environments are particularly exposed to multiple stressors, which can have a significant impact on their health and productivity. When animals were exposed to heat and nutritional stress as separate stressors, the impact on their growth and reproductive performance was not as severe as when they were exposed to both at the same time. Furthermore, the adaptation strategies of the animals to the individual stressors were different from those used when both stressors were combined (Sejian et al. 2010). When two stressors occur simultaneously, the physiological effects on the body can be severe. Heat stress in animals can lead to a decrease in feed intake and an increase in water intake. This can have a detrimental impact on the body's ability to adapt and maintain itself during the stressful period. Climate change has a significant impact on the endocrine state of livestock, leading to increased maintenance requirements and decreased performance. Heat stress, in particular, has a detrimental effect on milk production and composition, particularly in animals with high genetic potential. Furthermore, environmental stressors can reduce body weight, average daily gain, and body condition of livestock. Milk yield is significantly reduced when animals are exposed to heat stress, with the most productive animals being the most affected (Seijin et al. 2016). This decrease in milk production is accompanied by a decrease in fat content, low-chain fatty acids, solids non-fat, and lactose content; and an increase in palmitic and stearic acid (Wheelock et al. 2010). Additionally, the feed intake of milch animals is also reduced due to heat stress, leading to a further decrease in milk production (Sheikh et al. 2017). Heat stress can have a detrimental effect on dairy animals' post-absorptive nutrition metabolism (Baumgard and Rhoads Jr. 2013). Additionally, it has been linked to impaired embryonic development and increased embryonic mortality in cattle. Furthermore, heat stress during pregnancy can slow the growth of the fetus and increase the risk of fetal loss. Moreover, it can also interfere with the secretion of hormones and enzymes that regulate the functioning of the reproductive system, leading to a decrease in the conception rate of up to 27% (Sheikh et al. 2017).

In a study conducted by Upadhyay et al. (2007) on the daily milk production data from 1994 to 2004 in Karnal, Haryana, it was found that an increase in maximum temperature and a decrease in minimum temperature (3 °C) could lead to a reduction of 30% in milk production due to the shortened lactation period. Kumar et al. (2015b) conducted a study to assess the impact of climate change on livestock production in rainfed areas of India. By analyzing the milk productivity data from 1992 to 2007 and the climate variables of the same years, they found that meteorological variables such as precipitation and the maximum temperature had a significant effect on the milk productivity of animals in these areas. Their findings suggest that an increase in climate variability, particularly in terms of precipitation (both in deficit and in excess) and temperature, is likely to have a detrimental effect on animal productivity.

2.6 Monthly to Seasonal Scale Weather/Climate Variability and Associated Food Risk

In 2010/2011, a drought and heat wave caused Russia to issue a wheat embargo, leading to food price spikes due to reduced crop production and export bans (Welton 2011). National governments and commercial entities in countries that are dependent on food imports are highly attentive to changes in food production and export prices in major food-exporting countries, as well as their own domestic production (Iizumi et al. 2013). This is due to the fact that consumers in many countries are more reliant on food imports than before (http://www.fao.org/docrep/014/i2330e/i2381e00.pdf). The need for monitoring and predicting climate-related disruptions to agricultural production, as well as providing timely warnings of such events, is growing globally.

Agriculture is the foundation of the Indian economy, with 65% of the population living in rural areas and relying on agricultural-related activities for their livelihood. Approximately 70% of the working population is dependent on agriculture for their income. India has two main rainy seasons, namely the southwest summer monsoon (June–September; JJAS) and the northeast monsoon (October–December; OND). The Southwest summer monsoon is a major contributor to the annual rainfall in India, providing 80% of the total. This has a significant impact on the country's agricultural production, as it supplies water for the two main crop-growing seasons, *kharif* (summer) and *rabi* (winter). Variations in the monsoon rainfall on a monthly and seasonal basis can have a major effect on the total food grain yield of India and its economy, which is largely dependent on agriculture (Krishna Kumar et al. 2004). During the period of 1979–92, India had the highest growth rate in per capita food production in the world, at 1.6% per annum. Agriculture is a major contributor to the country's GDP, accounting for nearly 21%, and 43% of India's total land area is used for agricultural purposes. The inter-annual variation in JJAS (June–July–August) rainfall is the most prominent factor influencing food production in India. A decrease of 10% of the long-term mean JJAS rainfall leads to a significant decrease in rice production (Abrol et al. 1999). Therefore, year-to-year fluctuations in JJAS rainfall can have a major impact on food production in India. The interannual variability of the Annual Indian Summer Monsoon Rainfall (AISMR) has a significant effect on the Gross Domestic Product (GDP) and agricultural production of India and Sri Lanka (Webster et al. 1998). The October–November–December (OND) season is the main rainfall season in the southern peninsular region, which is essential for agricultural production in this area. However, compared to the Southwest Monsoon Rainfall (ISM), the OND rainfall has received very little attention in terms of studies on its interannual variability and teleconnections. The southeastern tip of the Indian peninsula and Sri Lanka are particularly dependent on the OND season for their annual rainfall, as the Western Ghats create a rain shadow region during the summer season. Approximately 50% of the total annual precipitation in this region is received during the OND season (Prasanna and Yasunari 2008). In comparison, other parts of India receive most of their annual rainfall during the ISM season.

At global, national and sub-national scales, the impacts of monthly and seasonal climate on crops vary significantly. This is particularly true for the Indian sub-continent, where rainfall and temperature patterns are not uniform across the region. Rainfall and temperature variability have been identified as the primary determinants of crop yields in numerous studies (Fishman 2016; Bhatt et al. 2019), with other climate-agriculture-environment interactions such as irrigation and fertilisation playing a role depending on the geographical location, crop type, and cropping season.

Research conducted by Preethi and Revadekar (2013), Revadekar and Preethi (2012), and Prasanna (2014) have demonstrated the influence of various rainfall levels on crop yield. It was found that JJAS rainfall surpassing the 85th, 75th, and 65th percentiles was associated with *kharif* food grain production. However, the correlation between extreme rainfall and food grain production was found to be insignificant, while moderate rainfall events showed a significant correlation. Prasanna (2014) conducted a study to examine the variations of JJAS seasonal rainfall and its effect on *kharif* and *rabi* food grain yield in India. The results showed that the total food grain yield during the *kharif* (summer) season is strongly correlated with the JJAS rainfall (Fig. 2.12). Moreover, a highly significant relationship was observed between the month of July and the JJAS rainfall, while a very weak correlation was seen in August.

An increase or decrease in rainfall can have an effect on food grain yield, particularly during the *rabi* (winter) season, although the correlation is not always clear (Fig. 2.11). Variations in OND rainfall can have an indirect effect on the *rabi* crop, as the JJAS rainfall influences water and soil moisture availability in many parts of India. If the JJAS season experiences a decrease in rainfall activity, this can lead to a decrease in crop yields. However, prolonged rainfall breaks can also have a negative impact on crop growth, resulting in reduced yields. Therefore, it is important to be able to predict monthly and seasonal rainfall categories for agricultural planning purposes.

Global yield forecasting has an advantage over regional forecasting in that it provides insight into potential yield variations in both food-exporting and food-importing countries, which can be beneficial for food agencies that strive to maintain national food security. This is in contrast to regional yield forecasting services, which are limited to specific areas such as Europe, Africa, the US, and China. A study conducted by Iizumi et al. (2013) found that global yield forecasts for rice and wheat can be accurately predicted three months prior to harvest, with losses of up to 19% of the harvested area. However, reliable predictions for maize and soybean yields are not possible due to their greater dependence on moisture conditions, which are less predictable than temperature. The previous study had two main drawbacks: the low accuracy of maize and soybean yield predictions, and the limited area where reliable predictions could be made (about one-fifth of the global harvested area). These issues are of great importance, as these crops are international commodities and the limited area where accurate predictions can

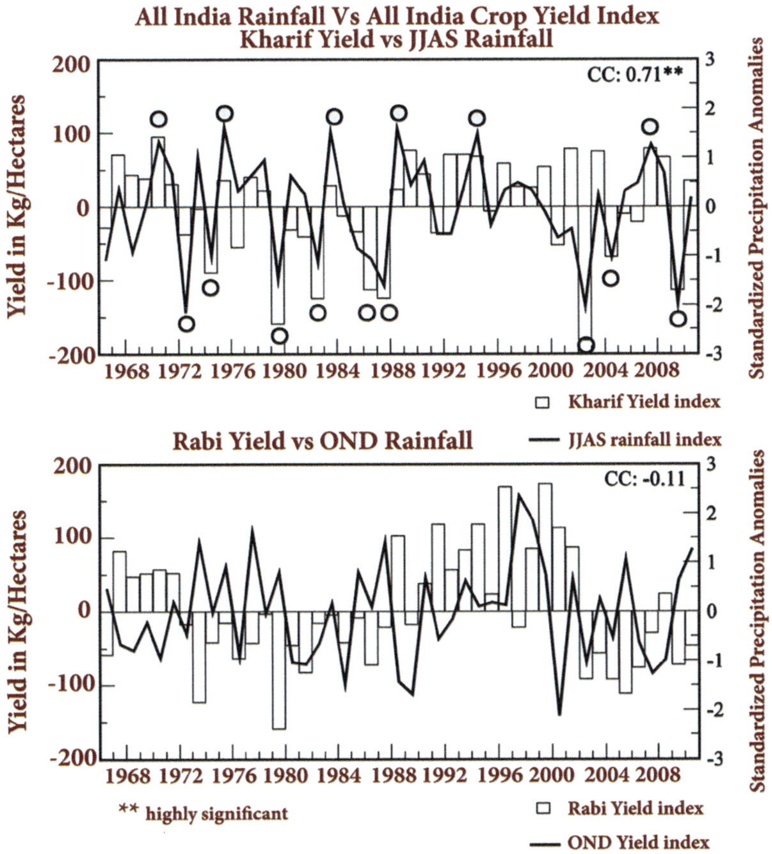

Fig. 2.12 All-India rainfall index versus all-India crop yield index. **a** SW monsoon and **b** NE monsoon (*Source* Prasanna 2014)

be produced is insufficient for food agencies in many countries. To address these issues, it is necessary to improve the methodology used in global yield forecasting, particularly in terms of the accuracy of meteorological variables (such as rainfall and temperature) at seasonal and sub-seasonal scales. This will enable more reliable predictions of crop yields worldwide.

Crop yields, vegetables, and fruits are largely dependent on climatic conditions, such as rainfall and air temperature, as well as their spatiotemporal distributions over the life cycle of the crop. Additionally, extreme weather events, such as heat waves, heavy rainfall, floods, and drought, can have a detrimental effect on agricultural production over a period of time ranging from one day to several weeks. Extreme weather events can have a devastating impact on agricultural production, which is the livelihood of a quarter of the world's population. These events can lead to crop failure, loss of livestock, and destruction of infrastructure, all of which can have a significant economic and social impact. Furthermore, these extreme events can reduce animal production, further impacting the livelihoods of those who depend on agriculture and related sectors. Therefore, accurate predictions at different time scales are essential in order to minimize losses and maximize profits in the agricultural and related sectors. However, the direct use of Global Climate Model (GCM) products is not suitable for regional scale predictions due to the highly heterogeneous nature of important climate variables such as rainfall and relative humidity. The accuracy of meteorological forecasts using advanced mesoscale modelling systems for up to a few days is very high and can be applied in the agricultural sector. However, forecasts on a monthly to seasonal scale have lower accuracy and involve more uncertainty due to the use of General Circulation Models (GCMs).

The MME technique is a popular approach in the climate science field for enhancing the accuracy of weather and climate predictions at different time frames (Mohanty et al. 2013, 2019a, b). This method combines the results of multiple General Circulation Models (GCMs) to calculate the statistics of a given climatic variable. The MME technique has the potential to enhance global yield forecasting, as it is based on the outputs of multiple GCMs, such as the Scale Interaction Experiment–Frontier Research Center for Global Change coupled GCM (SINTEX-F1) (Luo et al. 2005). The Indian Institute of Technology Bhubaneswar (IITBBS) has developed a reliable and robust forecasting system for monthly to seasonal predictions at a regional level (Mohanty et al. 2013, 2019a, b). The model and methodology used for this purpose are discussed in detail in Chap. 6. The use of monthly to seasonal scale forecasts in agriculture for generating climate risk matrices is discussed in Chaps. 7–9. The potential of monthly and seasonal forecasts, along with climate variability at these scales, to provide real-time information for agricultural purposes is explored in Chap. 10, with some observational case studies. The need for continuous weather forecasts and their applications in agriculture is further discussed in the following chapter.

Chapter 3
Hierarchy of Agromet Forecast

3.1 Introduction

Food is necessary for us to keep alive, and also for the sustainability and development of society. Thus, it is imperative to make the ratio equal to one or higher between the supply and demand of food required for survival. Since a long time, farmers are trying to understand which crops and their variety are suitable to specific patterns of weather parameters, particularly rainfall and surface air temperature. Thus, the farmers are acquainted with traditional knowledge of the weather pattern and plant's behaviour/response to the particular weather/climatic conditions within their life cycle.

In order to have better crop production, it is necessary to understand the variability and dynamics of the weather phenomena, particularly rainfall events and surface temperature, which was inheritably realised by the Indians more than a few thousand years back. In early date around 5000 years ago from now, the farmers in India mostly followed the predictions of the meteorological variables (particularly rainfall) provided by the Hindu Almanac for agricultural practices and to date the same traditional farm-related experience is applied in practice in many places in India (Gadgil et al. 2002; Lobo et al. 2017). On the other hand, the need for weather information has been realised in the USA a few centuries ago, and daily weather information is recorded in a systematic and organised way since 1776 when the third president of the United States, Mr. Thomas Jefferson reinforced the farming practices based on the local climate (Kelmm and McPherson 2017). During the end of the eighteenth century, at least from 1793, Farmer's Almanac, containing the descriptive weather forecast based mainly on the reoccurring weather patterns and proxy data, is published to facilitate the farm-practices in the USA (Hale 1992; Fiebrich 2009).

Around the third century of BC (more than 2000 years ago), Kautilya, also known as Chanakya, believed to be a contemporary of Aristotle, predicted that

U. C. Mohanty et al., *Climate Risk Management in Agriculture*,
https://doi.org/10.1007/978-3-031-51862-1_3

there is a monthly distribution of rainfall over the Indo-Gangetic plains during the summer monsoon (June–September), which is almost close to the climatological evidence in the present scenario. He suggested that the first and fourth month (June and September) of the year contributes approximately one-third, while the middle two months (July and August) contributes two-thirds of the seasonal total rainfall, and therefore, advised to adopt the agricultural practices accordingly for that region (Shamasastry, 1929). In his famous book 'Kautilya's Arthasastra' he illustrated that the positions of the planets could be useful for the prediction of rainfall distribution when correlated with time. It may be interesting to note that some of the crops, such as sorghum and pigeon pea, have been cultivated in India using almost similar farm practices for the past 2000 years (Achaya 1994). On the other hand, the farmers have acquired traditional knowledge from their experience based on weather conditions and adopted strategies to make farm related decisions for reducing the loss. In India, several proverbs exist since ancient times, which indicate 'if and then' conditions based on the knowledge of the nature of rainfall variability. For instance, a proverb that Kautilya referred to for the Indo-Gangetic Plain is:

> Awatadarnahidiyo Jat no diyo hasta
> Bina khane dona gaye Pahunaaur Grihasta

Meaning if there is no rain in Ardra (which is considered as the time period of 22 June–5 July) after the onset of monsoon and no rain in Hasta (which is considered as the time period of 27 September–9 October) during the retreat of the monsoon, then both guest and host have to go without food (Gadgil et al. 2002). The use of proverbs by the farmers clearly indicates their keen interest to understand the positive or negative impact of weather and climate variability on crop production. Before the digital network era, most farmers used the proverb as a thumb rule in rural areas for farm practices. However, with the advancement of observational networks and the increase of spatio-temporal data density, synthesising of data and analysing with sophisticated technology and advanced scientific methodology to understand the dynamics and behaviour of the weather and climate events have improved. Present-day meteorological forecasts using dynamical and statistical methods are increasingly in demand in different sectors, specifically the agricultural sector, due to their satisfactory skill since the studies by many researchers showed that anticipation of weather in many proverbs does not have any scientific basis (Basu 1953).

The traditional farmers have segmented the one year time frame by units having almost two weeks (13–14 days) time periods based on the solar calendar and each unit has been named differently known as 'Nakshatras' (Table 3.1). Based on the weather and climate conditions for a particular 'Nakshatras', the farmers used to take farm-related decisions. Thus, in traditional farming, the farmers do not use a weekly or monthly scale of time; instead, they use a bi-weekly time frame for planning resources and farm operations.

It is indeed interesting to note that the farmers used to make farm-related decisions based on some specific natural events. For example, traditional farmers and

Table 3.1 Time period of each Nakshatra that Indian farmers used for farm-related planning (Gadgil et al., 2002)

Nakshatra	*Period*[a]	*Nakshatra*[a]	*Period*[a]
Ashwini	13 April–26 April	Swathi	24 October–5 November
Bharani	27 April–10 May	Vishakha	6 November–18 November
Krittika	11 May–24 May	Anuradha	19 November–2 December
Rohini	25 May–7 June	Jyeshtha	3 December–15 December
Mrigashira	8 June–21 June	Moola	16 December–28 December
Ardra	22 June–5 July	Purvashadha	29 December–10 January
Punarvasu	6 July–19 July	Uttarashadha	11 January–23 January
Pushya	20 July–2 August	Sravana	24 January–5 February
Aslesha	3 August–16 August	Dhanishtha	6 February–18 February
Makha	17 August–30 August	Shatabhisha	19 February–3 March
Pubbha	31 August–12 September	Poorvabhadra	4 March–17 March
Uttara	13 September–26 September	Uttarabhadra	18 March–30 March
Hasta	27 September–9 October	Revati	31 March–12 April
Chitta	10 October–23 October		

[a] After India meteorological department 1999

villagers of India believe that if there are flying termites in the evening and/or ants moving their eggs to higher grounds then there is a high probability of rainfall occurrence, and they also believe that if there are dark clouds in winter and continuous drizzling then there is a high chance of pests and diseases outbreak (Anandaraja et al. 2008).

With the evolution of farm practices, it is strongly felt that decisions based on a time frame of the 'Nakshatra' concept may not be sufficient to have better crop yield production. In the past hundred years, many instruments have been set up in different places, known as meteorological observation centre/stations, to take observational data. The analysis of the observed station data at different time scales and its implementation with the agricultural practices have emerged as modern agricultural practices. Modern crop practices mostly have a bit different time frame from traditional crop practices (Gadgil et al. 2002). Figure 3.1 shows a nice graphical representation of traditional and modern cropping systems for Anantapur, India.

The crop life cycle requires climatic/weather conditions in an optimum range, which varies from crop to crop. It is noteworthy to understand and quantify the crop-weather relationship in the different growth stages. Therefore, it is revealed that the advance intimation of weather and climate conditions at different time scales will be very much helpful for making farm-related decisions. The India Meteorological Department (IMD), established in 1875, is an Indian government organisation, which is responsible for issuing data on weather and climate forecast. India is an agro-economic country, and in the mid of twentieth century, most of the agricultural practices were dependent on rainfall. In 1945, IMD started to provide weather services to farmers by informing the weather conditions and

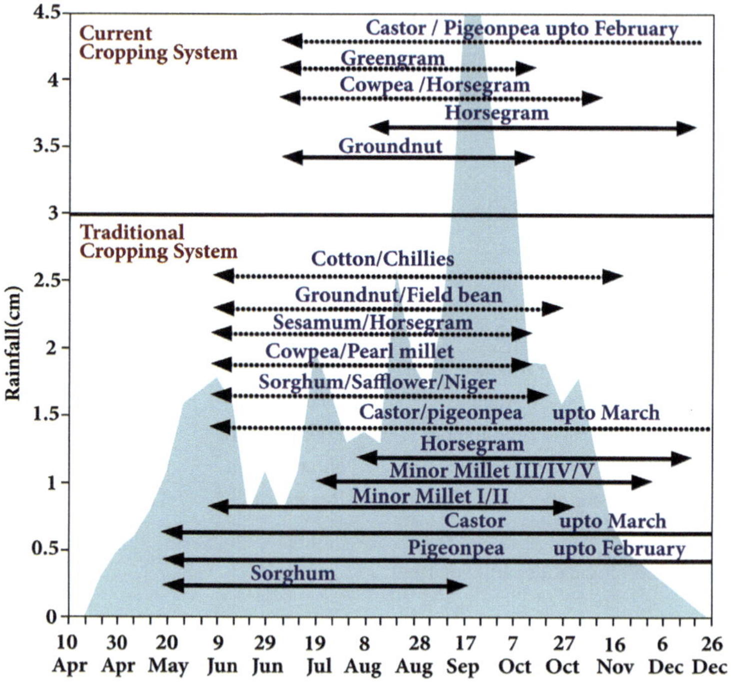

Fig. 3.1 Traditional and current cropping systems in the Anantapur region. Average weekly rainfall is shown in the background. The solid (dashed) lines indicate major sole crops (intercrops) (Gadgil et al. 2002)

broadcasting through All India Radio. The broadcast is made every day evening in regional languages throughout the year and a second bulletin is also provided in the morning during the summer monsoon season to facilitate the farmers by informing about the next two days' weather prediction at the district level and an outlook for the following 2 days. However, only a 2-days forecast with the next 2-days outlook is insufficient for the phenology of the climate-plant decision support system. There is a need to have a seamless forecast system starting from a few hours to a seasonal time scale for farm-related decisions and policy planning.

3.2 Seamless Weather Forecast for Agriculture

Weather and climatic conditions are one of the most important factors for agricultural production. The atmospheric state i.e. weather plays a key role in different stages of crop growth, influences the development of crop yields, varying degrees of impact due to pests and diseases, measurement of water stress, etc. which altogether has a massive effect on crop production. Therefore, understanding the phenology of the crop-climate system is very much essential and for a particular location and specific crop, weather forecast at different time scales is inevitable for the betterment of the crop yield production. The different time scale forecasts are useful for taking immediate action by making agriculture to policy planning, as most of the farming operations are tied with crop phenology. In this context, a reliable seasonal forecast could be helpful not only to the farmers by changing the policy of the crop varieties but also for minimising the loss to the farmers, which in turn strengthens the economy of the country. In the early date after the establishment of the meteorological department dedicated to weather and climate study, the forecast primarily was issued using synoptic meteorology and was mostly subjective. The agricultural weather forecasts have some unique features to realise better farm practices. For example, some major crop needs torrential rainfall during the preparation of the field for sowing and sufficient soil moisture availability to the seed zone of the sowing crop. Thus, reliable forecasts on agriculturally significant rains (ASRs), which is the minimum need for soil water is very much helpful for the farmers to start the farming of targeted crop. Therefore, a reliable forecast of the monsoon onset at the beginning of the summer monsoon, or skilful forecasts of the thunderstorms at the end of the pre-monsoon season may fulfil the need of the farmers and they can make the forecasts usable. In many regions, the climatology of dependable precipitation at different probability percentages and potential evapotranspiration is used to consider the start and end of crop growing period (Cocheme and Franquin 1967; Brown and Cocheme 1973; Venkataraman 2002). It is well established that advanced intimation of satisfactory weather forecasts can be usable by the farmers and have a positive impact on agriculture, except for some adverse weather events. The response of the same crop varies with different stages of its growth cycle even if a similar type and intense weather phenomena occur. In addition, the response also varies at the same stage of crop growth for varieties of the crop. Because the way the crops respond to the weather varies and different degrees of impact are observed with the crop varieties, the farming practices vary region-to-region and also in the same season.

The weather forecast simply can be defined as the future state of the atmospheric conditions for a given period of time issued for a point or region. As the weather warning at different stages of crop growth is very useful, there is a demand for different time scale forecasts. With the advancement of understanding the dynamics and physics of the atmosphere-ocean-land-ice-vegetation complex interactions, dense and reliable observations, sophisticated physical/dynamical and statistical models, and weather/climate forecasts at different time scales are issued by several operational agencies worldwide. The weather forecasts are made to make usable by agricultural and allied sectors and the other sectors. Primarily based on the time and space scales, from the agricultural viewpoint, the available meteorological forecasts that are useful for a given crop cycle can be categorised as follows:

a. Now-casting
b. Very Short Range Forecast[a]
c. Short Range Forecast
d. Medium Range Forecast
e. Extended Range Forecast
f. Long Range Forecast/Seasonal Forecast

[a]In many organizations, the now-casting forecast includes the VSRF time scales and no VSRF is separately provided

Thus, the forecasts are classified into various categories. These categories are mainly defined on the basis of temporal coverage of the forecasts as classified above. The method of different scale forecasts and the average skill of these forecasts varies. Therefore, it may be useful to understand the skill of different temporal scales of forecasts.

The forecasting of the weather for the next few hours is generally referred to as now-casting. The forecasts delivered for a lead time of a few hours to a day and 1–3 days are called very short range and short-range forecasts, respectively. Medium-range forecasts are made for periods covering 4 to 10 days ahead. The extended range forecasts are covering the time period of 1 to 4 weeks, and seasonal scale forecasts are made for periods of a month to a season.

It is easier to forecast in the temporal range of now-casting than the other forecasts because the possible occurrence of local weather features such as showers, thunderstorms, snowfall, etc. can be guessed up to convinced accuracy by analysing the latest manual observations as well as the latest radar and satellite data sets. Present-day numerical weather prediction (NWP) models are able to forecast very short, short, and medium range atmospheric events with satisfactory accuracy. This success is achieved since very short, short, and medium range atmospheric events are highly dependent on their initial values and also with the improvements of observational data sets in higher density as well as improvement of model physics and dynamics and availability of high performance reliable computing systems. Extended-range and seasonal forecasts are different from weather forecasts since it involves the prediction of deviations from the seasonal climate. The seasonal scale events are mainly governed by seasonal scale features specifically by some large-scale features, slowly varying land surface parameters and some semi-permanent features. The skills of different scales of forecasts may be represented in a schematic diagram (based on inputs from various available published work on forecast skill over different temporal scales, researches and forecast application scientists) as shown in Fig. 3.2. It is seen in Fig. 3.2 that though there is a good degree of predictability skill in now-casting, very short range, and short-range weather forecast, the skill is waning with time. The seasonal scale forecast which is mainly based on dynamical-statistical methods has some predictability skills. The sub-seasonal range i.e. extended range prediction is one of the toughest tasks due to the inherent complexity in understanding and modelling different scales of

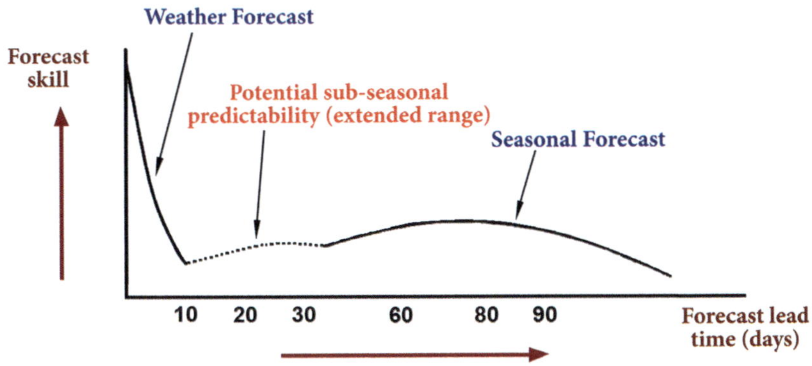

Fig. 3.2 Schematic diagram for skill of different scale forecasts with forecast lead time

atmospheric interaction processes and the confidence is lesser than the other scale forecasts and thus shown as a dotted line in the diagram.

A brief on each category of weather and climate forecasts along with its utilization particularly in the field of agriculture is described in the next.

3.2.1 Now-Casting

Definition: The forecast of the atmospheric state for the next few hours (mostly 2–3 h) over a particular region is known as now-casting.

Methods: Various methods have been adopted for now-casting, and most of these methods have a potential skill in predicting the atmospheric state within a few hours range. It is well in practice to provide now-casting using the synoptic view of the atmosphere. Synoptic view, also known as 'Bird's Eye View' is the instantaneous image of the atmosphere seen from the sky, i.e. the state of the atmosphere at a given time over a large area. The forecaster is used to anticipate now-casting using their past experience and synoptic view. The atmospheric state at a given time can be understood through a synoptic chart which is a concise portrait of the current weather describing the pressure patterns, horizontal (zonal and meridional) wind vector (direction and speed), the existence of fronts, temperature, and relative humidity. Analysis and extrapolation of surface and upper air observational data such as Automatic Weather Stations (AWS) and radiosonde data is one the useful techniques to provide a forecast for the next few hours. Now-a-days, satellite and radar imageries are used worldwide for the now-casting. Some empirical methods are also used for forecasting the weather at now-casting scale.

This technique is mostly used to predict several weather phenomena like convective storms, extra-tropical, and tropical storms, fog, and low clouds, locally forced precipitation events, sand and dust storms, wintertime weather (snow, ice, glazed frost, blizzards, avalanches), wildfires, and contaminated areas by air pollution (due to chemical or radioactive accidents) and the other mesoscale weather extremes. However, this technique is widely used to forecast severe weather or climate extremes. Thus, the spatial scale depends on the weather events and is typically in the range of a few hundred meters to a few ten kilometers.

Forecast Outputs: In the now-casting forecast, a number of atmospheric parameters such as precipitation intensity and distribution, surface air temperature, wind speed and direction, relative humidity, cloud types and amount, solar radiation, the occurrence of thunderstorms and associated lightning, etc. can be predicted.

Skill: The skill of the now-casting forecast is very high, which means that the accuracy is very high. Thus, the forecasters are very confident about this type of forecast, and users can use this forecast and act accordingly.

Utilisation: Now-casting is important to understand whether a station or region will experience any extreme weather events or not in the next few hours, and in that case, prediction of precipitation, wind, temperature, and occurrence of thunderstorms with lightening at now-casting scale is the most important to save the lives and damages of properties. However, it has a very limited application in the agricultural field.

Limitation: The now-casting forecast is confined to a few hours; thus, it is difficult to disseminate the forecast using a broadcasting system. As the time scale of now-casting is only about 2–3 h, which may not give sufficient scope for use as the routine farming operation starts early in the morning on regular basis.

3.2.2 Very Short Range Forecast (VSRF)

Definition: In this very short-range forecast (VSRF), the forecast of the atmospheric state is made for the time scale of 2–3 to 12 h. In many organizations, the now-casting forecast includes the VSRF time scales, i.e., now-casting forecasts range up to 12 h.

Methods: Like now-casting, empirical methods and analysis of the synoptic chart are used for the prediction of the VSRF. The forecaster is used to predict VSRF using their past experience and synoptic view. Analysis and extrapolation of AWS and radiosonde data, satellite, and radar imagery are effectively used for VSRF. Besides the methods stated earlier, numerical weather prediction (NWP) models of mesoscale categories are widely used for predicting weather events at a very short-range scale.

The VSRF is used to predict all the weather events as mentioned in now-casting such as convective storms, extra-tropical, and tropical storms, fog and low clouds, the occurrence of thunderstorms, etc. The spatial scale of VSRF depends on the weather events and is typically in the range of a few ten kilometres to a few hundred kilometers.

Forecast outputs: The forecasting techniques used for VSRF provide mostly all important atmospheric parameters (precipitation intensity and distribution, surface air temperature, wind speed and direction, relative humidity, cloud types and amount, solar radiation, etc.).

Skill: The skill and accuracy of the VSRF are very high. The ability of NWP models to predict the weather at the VSRF range is significant, with higher predictability and lower errors both in the time and space of the weather events. The forecasters are confident about this type of forecast, and users can use this forecast and act accordingly.

Utilisation: VSRF is important to understand whether a station or region will experience any extreme weather events, such as thunderstorms, heavy to extreme rainfall events, strong wind, etc., in the next few hours. This may be used for specific operations such as fertiliser application, pesticide spray, intercultural operations, etc. to avoid the loss of inputs and reduce the production cost. The broadcasting of such weather information reaching the farmers is, however, a challenge as the range is only for 2–12 h.

Limitation: The VSRF forecast is in the range of ~2–12 h, thus, the VSRF is not suitable for broadcasting, and hence has limited use in an agricultural operation.

3.2.3 Short-Range Forecast (SRF)

Definition: The Short-range Forecast (SRF) usually predicts weather events in a time scale from 12 h to 3 days.

Methods: The mesoscale NWP models driven by general circulation models (GCMs) are the most important tools in predicting weather events at the SRF scale. The outputs from mesoscale NWP models are analyzed and mapped to generate the forecast. Products from the limited area models (LAMs) and GCMs are also analysed, mapped, and interpreted to generate SRF. Empirical models and the forecaster's past experience are also used to provide this type of forecast. The basic information on the weather events can be generated by using AWS data, maps from meteorological radars, images from meteorological satellites, and analysis of NWP models, which are used to provide a robust forecast with significant accuracy.

The SRF is used to predict mesoscale weather events such as convective storms, extra-tropical and tropical storms, fog and low clouds, etc. The spatial scale of SRF typically depends on the weather events and is typically in the range from a few ten kilometres to a few thousand kilometres.

Forecast outputs: The forecasting techniques used for SRF provide mostly all important atmospheric parameters (precipitation intensity and distribution, surface air temperature, wind speed and direction, relative humidity, cloud types and amount, solar radiation, etc.). The forecasts are generally updated typically at 6–12 h intervals.

Skill: The skill of the SRF is high, and the accuracy is satisfactory. With the advancement of technology, a better understanding of model physics and dynamics, the use of dense reliable observational data, and modern data assimilation techniques altogether enhance the forecast skill at the SRF range. The ability of mesoscale NWP models to predict the weather at the SRF range is significant, with higher predictability and lower errors both in time and space of the weather events. This type of forecast has a wide application in all sectors.

Utilisation: SRF of weather events is one of the most important issues because of its huge usability in different sectors. This kind of forecast can be effectively broadcasted through several media such as newspapers, television, web, etc. This type of forecast has potential use in the agricultural fields for intercultural operations such as fertilizer and pesticide application, irrigation water management, hoeing, weeding, etc.

Limitation: Skillfull adaptive strategies are needed to be developed for the farmers to make more use of the SRF of weather events.

3.2.4 Medium Range Forecast (MRF)

Definition: The Medium Range Forecast (MRF) usually predicts weather events in a time scale of 4–10 days.

Methods: The NWP models (LAMs, GCMs) are the key tools to providing medium range forecast. In the NWP run, instead of a single forecast, ensemble forecasts used through probabilistic methods emerge as a robust method for this scale. In an ensemble forecast, the same numerical model is integrated several times and in each time integration some "perturb conditions" are applied to keep the rest of the model configuration the same and each run produces slightly different forecasts. Considering all the "perturbed" forecasts from a particular model is known as "ensemble" forecasts. The perturbations can be generated by changing the initial time run, incorporating observations or changes slightly the initial value data over a particular region which has a great influence, perturbing the physics of the model. The ensemble forecast is generated in such a way that the observations can lie within the ensemble range, thus, the ensemble mean will likely have more skill because it will closely follow the observed mean and the chaotic nature of climate will essentially be smoothed out. The ensemble forecast can then be used to generate the deterministic or probabilistic forecast. Nowadays, the usefulness of the probabilistic forecast in terms of cost-benefit ratio is higher.

A number of ways are used to generate MRF of weather and climate. Interpretation of the outputs from NWP models mapped to generate the forecast. Products from the limited area models (LAMs), and GCMs are also analysed, mapped, and interpreted to generate MRF forecast. Empirical models and the forecaster's past experience are also used to provide this type of forecast.

The MRF is used to predict synoptic scale weather and climate events such as extra-tropical and tropical cyclonic storms, high and low-pressure systems, etc. The spatial scale of SRF typically depends on the weather events and is typically in the range from a few hundred kilometers to few thousand kilometers.

Forecast outputs: The forecasting techniques used for MRF provide mostly all important atmospheric parameters (precipitation intensity and distribution, surface air temperature, wind speed and direction, relative humidity, cloud types and amount, solar radiation, surface pressure, etc.). The forecasts are generally updated typically at 12–24 h intervals.

Skill: The skill of the MRF is high or moderate till 5 days forecast, however, the forecast skills deteriorate faster after 5 days.

Utilization: Medium range forecast is an essential requirement for farming operations as this is useful for deciding sowing time to ensure seed germination, soil moisture and the right seeding technique in general, for uniform stand establishment. More specifically, the MRF is used in intercultural operations as SRF for efficient use of resources such as fertiliser and water by the crop and increasing farmers' economic grain while reducing the production cost. Hence the MRF has a potential usage in the farming sectors.

Limitation: It is necessary to develop useful adaptive strategies through farmers' seminars and capacity building for wider and intensive use of the MRF forecast by the farmers.

3.2.5 Extended Range Forecast (ERF)

Definition: The Extended Range ERF usually predicts the average weather events in a time scale of 10–30 days. The extended range forecast is a time scale that lies between the medium range forecast and seasonal scale forecast, with an upper time scale of about a month. It may be noted that this ERF forecast generally provides a weekly average of atmospheric state, particularly some important surface meteorological parameters such as precipitation/rainfall and surface air temperature (also known as 2 m temperature) for the first week to the fourth week.

Methods: The statistical approaches (linear and nonlinear) and dynamical physical models i.e., NWP models both are used to generate extended-range forecasts. In statistical approaches, a number of predictor fields are considered as inputs. The predictor fields and the meteorological variables are significantly and statistically correlated with the predictand variables (or target variables that will be predicted) and also have a physical relationship with the predictand and play dominating role to influence the state of that predictand field. The predictor field may be from the outside of the domain of the predictand field. Single predictor fields or multiple predictor fields are used as inputs in the statistical models to generate the extended range forecast. In many operational agencies, the bias-corrected (details are given in Chap. 6) NWP model products are used as predictor fields for ERF scale prediction. The NWP (regional climate models (RCMs), LAMs, GCMs) are used for predicting the atmospheric state and forecasters use it to compute the aggregate weekly situation of the atmospheric state. In the ERF, ensemble forecast is widely used, and the probabilistic approaches show higher confidence in the application sectors. It may be noted that the ERF scale forecast is one of the most difficult among all scales since the forecast time scale is between~one week to four weeks or more specifically 10 to 30 days and predictability reduces with higher uncertainty in the NWP since it is neither a complete initial value problem nor a dominating boundary value problem. The scale is beyond influence of initial value i.e. the forecast time scale is sufficiently long so that much of the atmospheric memory of the initial conditions is lost. On the other hand, time is too short to establish a signal by the surface boundary conditions, which is a slowly varying process with less time to capture the surface variability. However, this ERF scale forecast is the most important of all scale forecasts for the agriculture sector since this scale forecast covers the main time scale of crop phenology and is helpful for farmers to make decisions.

The ensemble forecast of NWP models, bias-corrected NWP models, and statistical techniques, along with the forecaster's past experience altogether, generate the deterministic and probabilistic forecast with uncertainty in a forecast.

The ERF is used to predict generally regional scale, typically in the range from few hundred kilometers to few thousand kilometers. Since this scale of forecasts is the most important for the agricultural sectors, thus immense efforts by scientists are made to generate robust and skillful forecasts at a shorter spatial scale, maybe at the district or county/state level.

Forecast outputs: The ERF forecasting techniques provide mostly two important surface meteorological parameters such as precipitation (precipitation intensity and distribution and surface air temperature) in the form of mean and departure from the mean (may be a magnitude or in percent of departure, see details in Sect. 5.4). The forecasts are generally updated typically on a weekly basis.

Skill: In the context of coupled models forecast by the India Meteorological Department (IMD), it is seen that the forecast of monsoonal epochs such as onset, active phase, break phase, and withdrawal is potentially useful. It is also found that the transition between the active and break phases, and vice-versa are forecasted with reasonable skill. At all-India scale, the seasonal scale forecast is significant for up to about three weeks and at homogeneous rainfall region (summer monsoon) in India, the skill is significant for up to two to three weeks over the central region, which is known as the monsoon core region, northwest India. However, at the smaller spatial scale, such as the meteorological sub-division scale, the forecast is significant for up to two weeks.

The ERF skill is the most uncertain compared to the other scale forecast. The skill is lower, and the uncertainty is higher. It is found that a probabilistic forecast is more useful than a deterministic forecast and usable to the farmer's community.

Utilization: Extended range forecast is mostly used for deciding the right sowing time of crops so that the crop should not suffer from excess water stress (flood) or water deficit (drought) during seedling stages, which is highly sensitive to all extreme weather conditions. Any weather-related damage to the crop at the seedling stage makes a complete failure of the crop, cropping season, and huge economic loss to the farmers. The ERF also helps farmers' preparedness for intra-seasonal water management to avoid yield loss due to drought at critical growth stages of crops.

Limitation: It is necessary to develop useful adaptive strategies and educate the farmers through training for using the ERF forecast in their farming operations.

3.2.6 Long Range Forecast (LRF)/Seasonal Forecast (SF)

Definition: The seasonal forecast mostly provides the expected outlook of the average weather in a time scale from one month to season.

Methods: The statistical approaches (linear and nonlinear) and complex coupled ocean–atmosphere general circulation models are both used to generate seasonal scale forecasts. The statistical approaches using potential predictors are widely used worldwide to predict the seasonal scale forecast. The statistical approaches, including downscaling that consider the linear and nonlinear relationship between the predictor and predictand fields are proven reliable methods for generating seasonal scale forecasts. The NWP models (RCMs, coupled GCMs) with a number of ensemble realizations are used for predicting the atmospheric state, particularly precipitation and surface air temperature (SST) statistically averaged over a month/season (next three months). The bias-corrected of NWP models output and ensemble mean is also another useful technique for generating seasonal scale forecasts. The seasonal scale forecast is influenced mainly by the large-scale upper air fields, slowly varying surface processes, and heterogeneity of land surface, and all these are slowly varying processes. The forecast scale is beyond one month and it is quite reasonable to capture the variation of a slowly varying process such as sea surface temperature and thus have a satisfactory skill.

The ensemble mean forecast of NWP models, bias-corrected NWP models, and statistical downscaling using linear and nonlinear relationships between predictor and target variables along with the forecaster's past experience altogether generate the deterministic and probabilistic forecast, which is mostly expected situation averaged over a month/season and predictions are made at least one month in advance.

The seasonal scale forecast provides forecasts generally on a regional to continental scale, typically in the range from few hundred kilometers to few thousand kilometers.

Forecast outputs: The SF forecasting techniques provide mostly two important surface meteorological parameters such as precipitation (precipitation intensity and distribution), surface air temperature) in the form of mean and departure from mean (may be a magnitude or in percent of departure, see details in Sect. 5.4). The average state of the other meteorological variables (such as wind, SST, soil moisture, etc.) can also be predicted using the coupled GCMs and/or statistical approaches. The forecasts are generally updated each month typically.

Skill: The comparison between the operational forecast by the IMD and actual rainfall has shown an average error in the range of ~6–10% when computed for the all-India scale and the variation in the error range due to the variation in the period for analysis. This is seen on an all-India scale during the summer monsoon. It is found that using the advanced coupled GCMs and sophisticated statistical approaches one could be able to reduce the errors. The evaluation of skill of seasonal forecast is made using hit score considering individual met-subdivision for a period of the recent 9 years for each season. The hit score is computed using the tercile category (below, normal, and above normal categories) of rainfall and surface air temperature (Mohanty et al. 2019a). It is found that seasonal rainfall has a similar signal with an actual category for 60% of total met-subdivisions during the summer monsoon season. The skill is higher with more than 70% met-subdivisions for the other seasons, such as post-monsoon, winter, and pre-monsoon seasons (Mohanty et al. 2019a, 2019b). It is also observed that the skill of forecast is higher for the surface air temperature than the rainfall/precipitation.

Utilization: The seasonal forecast is mainly used for policy planning and management of different sectors such as agriculture, water management, hydropower, etc. This is used for the selection of crops, varieties, and sowing windows as adaptive strategies to minimize the adverse effect of weather abnormality on agricultural production.

Limitation: Useful adaptive strategy and effective policy is required to make this forecast useful.

3.3 Agricultural Practices Depend on Weather Forecast

In the above discussion, we have described different types of weather/climate forecasts based on different temporal and spatial scales, including their usefulness in different sectors. Here, we will discuss the need for different scales of weather and climate forecast for agricultural operations.

3.3.1 *Selection of Crops and Varieties (Seasonal Forecast)*

The seasonal forecast is either deterministic or probabilistic and mainly provides an outlook on the average behaviour of the weather and climate for the next season. In the deterministic seasonal forecast, the departure from the long-term mean of any variables (particularly rainfall and surface air temperature) is provided.

Thus, a farmer will have the knowledge in advance whether the farm practice will be made as traditional or need to plan for some different crop suitable with the predicted weather/climate forecast. For example, if a forecast is made for a specific region in India that the summer monsoon rainfall will likely be lesser by an amount of 20–30% from the long-term mean with a 75% probability of occurrence below normal over that region, the farmer may plan an alternate variety/crop to get maximum return from the forecasted weather variability. Moreover, the seasonal forecast will support the farmers' decision on the selection of suitable crops and variety for specific land ecosystems (upland, medium land, low land, and irrigated/rainfed farming) to avoid yield loss due to weather abnormalities, if any.

3.3.2 Land Preparation and Sowing/Planting Time (Seasonal, Medium Range and Short Range Forecast)

The decision of land preparation and sowing/planting operation are the crucial phases in the farming practice since this involves maximum expenditure and is influenced by all scales of the weather forecast, starting from the short range to the seasonal scale. Sowing or planting at right time is the most priority operation as the crop gets the advantage of seasonal weather variability as per the need of the specific growth stages starting from seed germination to maturity to attain maximum production. Every crop needs specific soil moisture, temperature, and solar radiation for germination and proper establishment. Heavy rainfall during the establishment phase, irrespective of the irrigated and rainfed ecosystem, will cause severe damage to the crop, hence huge loss of economy in general and the agricultural season in particular. Under this emergency situation, re-sowing/planting of the second crop with land preparation is time-consuming, and the crop may not have optimum weather conditions (temperature and solar radiation) during its growth phases to yield a good harvest. Hence the weather-based decision on sowing/planting time is essential for uniform crop establishment, farmers' economic gain, and maximisation of food production.

3.3.3 Intercultural Operations

3.3.3.1 Fertiliser Application (Short and Medium Range Forecast)

The decision of time of fertiliser application at various stages of the crop is one of the most crucial parts of the farming practice since it involves almost all scales of weather forecast starting from the now-casting to seasonal scale. The fertiliser application rate and form of use can also be decided based on conditions of sunshine hours, soil moisture, and soil temperature. The advanced information on weather and climate for all the scales is highly useful. A scheduled fertiliser application followed by inclement weather can wipe away the entire field's profits. Therefore, the weather and climate forecasts at all temporal scales can be used to ensure that fertiliser is applied under the right conditions. A suitable example for fertilizing is when a dry spell continues so that it doesn't get washed, but soil should be moist enough so that it gets worked into the soil.

3.3.3.2 Irrigation Scheduling (Short and Medium Range Forecast)

Each crop requires appropriate amounts of moisture, light, and temperature at various stages of the crop, and the requirement varies with various stages of the crop. Detailed and accurate historical, real-time, and forecast weather and climate information can help farmers better understand and track the growth status/stage to make informed decisions. Therefore, accurate advanced information on weather and climate at different time scales can guide the farmers in making significant and potentially costly decisions, such as whether, when, and how much to irrigate. If a farmer knows well in advance about the occurrence of a good amount of rainfall, he can save the cost of unnecessary irrigation. And by having a good idea of the expected amount of rain over a period with different time scales from the seamless forecasts, farmers can irrigate just enough to allow crops to thrive, and they can maximise the yield with less investment.

3.3.3.3 Pesticide Application (Short and Medium Range Forecast)

The decision of pesticide application time at various stages of the crop is one of the most important in the farming practice since it involves almost all scales of weather forecasts and particularly now-casting to medium range forecasts plays a crucial role. Certain weather and climate conditions are favourable for the development and growth of pests and diseases and destroy crops. The incorporation of advanced information on weather and climate conditions into pest and disease modelling can help in determining the suitable time to apply pest or disease control measures. Wind and relative humidity forecasts also play a role in this decision, as crop dusters, aircraft that spray fungicidal or insecticidal chemicals on plants from above, must be utilised when wind conditions are not apt to cause sprayed chemicals to miss their targets.

3.3.3.4 Field Workability (Short and Medium Range Forecast)

Field workability refers to the availability of days that are suitable for fieldwork. It is primarily dependent upon soil moisture and soil temperature. Accurate field-level weather information can help farmers assess the workability of their fields and become more efficient in their day-to-day operations.

3.3.4 *Harvesting Time (Short and Medium Range Forecast)*

The advanced information of meteorological parameters on short and medium range time scales plays a vital role in the harvesting of crop production and its immediate processing before storage (post-harvesting) than any other field operations because a few days of inconsistent weather (extreme weather events) at the end of the crop season can be disastrous. Therefore, short and medium-range weather forecasts significantly help in harvesting and post-harvesting operations of any crop to save the yield from unfavourable weather conditions. The primary weather factors for crop harvest are rainfall and atmospheric temperature, while for post-harvest operations, the additional information of sunshine, wind, relative humidity, and dew is also important along with rainfall and temperatures. For example, the occurrence of rainfall/precipitation may increase the moisture content in the straw of rice crops, which may delay harvest operations, and the low temperature may also cause a delay in the harvesting process. Therefore, the accurate weather forecast on short and medium range potentially help farmers for making a decision for harvesting and post-harvesting agricultural operations. The above in general are highlighted in this section; however, the farmers' decision for each operation needs a specific type of weather forecast. This needs to be swiftly focused to draw the importance of the scale of the weather forecast for a specific farm operation.

3.4 Important Hydro-Meteorological Parameters Useful for Agricultural Sectors

In the weather forecasts, there are a number of parameters that are essential components for agricultural weather forecasts. These weather components on a daily/sub-daily scale have a great impact on crop phenology, growth, and production. These components or meteorological parameters are:

(i) Precipitation (rainfall, snow, fog, and dew)
(ii) Maximum, minimum, and dew point temperatures at the surface
(iii) Relative humidity
(iv) Wind speed and direction
(v) Amount and type of cloud
(vi) Solar radiation
(vii) Bright hours of sunshine
(viii) Soil moisture conditions (stress)
(ix) Evaporation and evapotranspiration
(x) Seamless forecast to understand the evolution of meteorological variables at canopy levels
(xi) Extreme weather events and warnings.

Precipitation and surface air temperature are the two most important meteorological parameters that have a major control on the production of crop yields. The accurate forecast of precipitation and surface air temperature will be greatly helpful to the farming community. The other meteorological parameters mentioned above can be approximately estimated using these two parameters in a crop model and hence these two parameters are playing a key role in crop model simulations. For example, the bright hours of sunshine can be approximately determined using minimum and maximum temperatures for a particular day over a particular region. On the other hand, the monthly and seasonal scale mean predictions have a satisfactory skill; thus, meteorologists/scientists provide monthly and seasonal scale forecasts (details in Chap. 6). Since the crop model needs a daily sequence of the meteorological variables, thus, it is necessary to disaggregate the monthly/seasonal forecasts in a daily sequence. A detailed description of the disaggregation methods is provided in Chap. 7.

3.5 Institutional Support in Climate Forecast

A number of operational agencies provide forecasts from now-casting to long-range scale. As stated earlier that now-casting to medium range forecast is region specific, and most of the regions use to get weather forecasts at these scales from corresponding agencies focussed on their region. Besides these, some government and private agencies used to provide now-casting to medium-range forecasts at a local scale covering the entire world. However, a limited number of agencies provide extended-range and long-range forecasts at continental and global scales.

As far as India is concerned, the IMD is the responsible organisation to provide weather forecasts at all time scales.

Details of the seasonal scale forecast products available at different organizations are presented in Chap. 5.

3.6 Progress in AGROMET Forecasts in India

The IMD is the pioneer to provide seasonal forecasts and issued the first forecast in 1886. The forecast was generated using empirical statistical techniques and provided a seasonal forecast for all India during the summer monsoon season. Before 1880, the necessity of the seasonal forecast had emerged due to several famine owing to severe drought conditions and a huge crisis of food production, and a great loss of lives. In 1945, the IMD started to provide service to the farmers through Farmer's Weather Bulletin (FWB) and broadcast the forecast and usability through All India Radio. In 1976, the IMD established an Agriculture Meteorology Division at Pune with its Agro-meteorological Advisory Service (AAS) units in all state meteorological centres and dedicated to providing service to the farming community. Since then the AAS Units have been issued FWB through the state meteorological centre of the IMD in collaboration with the Agriculture Departments of the respective State Governments. However, the meteorological forecast was statistically regression-based and issued for the two homogeneous regions of India at a seasonal scale. With the advancement of the forecasting methodology and a better understanding of the weather and climate, the seasonal forecasts are generated comparatively at a higher spatial scale (three homogeneous regions) and all India scale. However, this coarse resolution seasonal forecast could not able to meet the need of farmer's requirements. In a more practical sense, the farmer needs to forecast from a few days to a seasonal scale over a comparatively smaller domain to make it useful in farming practices. In order to meet the requirement of the farmers, short and medium-range forecasts have been started to issue by meteorologists.

3.7 Need for Monthly and Seasonal Forecast for Agricultural Practices

Reliable high-resolution weather forecasts are in increasing demand by the government, industry, traffic, media, farming community, and risk management departments of most countries worldwide. The seasonal scale forecast of meteorological parameters for one month to a season in advance is one of the most challenging issues to the scientific community due to complex interactions between land-air-sea and also small-scale convective activities with the large-scale flow. The variation of the meteorological parameters from year to year is also not coherent over any region. For example, while some regions experience above-average rainfall, others suffer from a deficit. Thus, the anomalies (differences from the long-term average) of the variables are positive for some of the regions and negative for others. There is also a large sub-seasonal variation in a region within a season. For example, in 2002, there was an unprecedented deficit of 49% in the all-India average rainfall in July, while the rainfall was close to the average during all the other months. Similarly, in 2009, the all-India rainfall was close to the average in July but well below the average in all the other months (June: 53%, August: 73%, and September: 79%). A quantitative assessment of the impact of monsoon on food grain production (FGP) in the country and the Gross Domestic Product (GDP) by Gadgil and Gadgil (2006) has shown that despite a substantial decrease in the contribution of agriculture to GDP from about 50% to 20% during 1951–2004, the impact of droughts has remained 2–5% of the GDP throughout. They have also shown that droughts in the current decade had as much impact on FGP as they had in the 1960s.

Therefore, the timely forecast at month to the seasonal scale of meteorological parameters (such as precipitation and surface air temperature) over a region and its various sub-regions is not only necessary but crucial for deciding various mitigation and adaptation strategies in agricultural, power and hydrological sectors. For example, the weather forecast (seasonal or monthly) for agriculture is a specialised forecast issued to serve an important segment of the county's economy. The development of this service is necessary since modern-day agricultural operations are becoming increasingly dependent upon detailed and accurate predictions of meteorological elements. Meteorological information is important for making many operational decisions. The seasonal scale prediction of rainfall and temperature may be used to reduce climate risk in agriculture management. Prediction of weather (rainfall and temperature) in advance may be used to reduce risk. For example, farmers planning to prepare land for sowing winter crops might not do so if they were given an adverse forecast in autumn. A farmer might decide to feed a "failing" crop to livestock in the spring on the basis of an adverse forecast. Therefore, the advance intimation of the likely behaviour of monthly and seasonal rainfall helps the farmer to avail opportunities and to make decisions that could enhance agricultural productivity and maximise returns or minimise the loss. Among various types of forecasts made for different temporal scales viz. short range, medium range, and long or extended range, the extended range forecasts are highly valuable to the farming community, government, and industry for long-term planning, decision making, management, and mitigation. The extended range forecasts could provide indications of monsoon rainfall variability.

The life cycle of the agricultural crops that are essential for daily life may span from a few weeks to a few months. On the other hand, several weather/climate events significantly affect crops in different stages in their life cycle. These climate and weather events have a typical life span and typical size; thus, it is very important to know the events and their spatiotemporal spread. In addition, the fundamentals of prediction systems and their different components in association with the forecast all are described in the next chapter.

Chapter 4
Weather and Climate Forecast

4.1 Introduction

Weather and climate both describe the state of the atmosphere for a particular region. Weather indicates the instantaneous condition of the atmosphere (e.g., pressure, temperature, relative humidity, wind etc.), while climate is defined as a statistical average of the weather for a longer period of time. As per the World Meteorological Organization (WMO), climate for a particular region can be prescribed if the statistical average of the weather is computed for a period of 30 years or more. Thus, the climate is a quasi-steady statistical ensemble average of weather conditions for a few decades.

The weather and climate directly or indirectly impact the social and cultural system in many ways. The weather or climate conditions or events that have the potential to injure or cause loss of life or damage to property are known as weather or climate hazards. Thus, hazards are threats to society, economy, and environment. The disasters, on the other hand, are the widespread destruction of lives and damage of properties. There are a number of natural hazards, such as geophysical, hydrological, meteorological, biological, biophysical, and extra-terrestrial hazards, in which weather and climate events are the cause of hydrological, meteorological, and biophysical hazards that accounts more than 90% of the total of natural hazards in terms of loss of lives and damages of property worldwide. According to the International Disaster Database, the economic loss caused by weather and climate extreme events in India alone itself is about $3 billion per year, which is 10% of the global financial losses of $30 billion per year. Therefore, there is a demand for understanding, monitoring, and predicting weather and climate extremes in this present global warming era (Vitart and Molteni 2009). In the next, we will discuss the different events of the weather and climate system that are of interest to understand since reliable forecasts of these events are in great demand.

U. C. Mohanty et al., *Climate Risk Management in Agriculture*,
https://doi.org/10.1007/978-3-031-51862-1_4

4.2 Weather and Climate Events/Systems

There are several weather and climate events that are considered as natural hazards and are triggered mostly by natural disasters. A brief on the major natural hazards and or disasters under the umbrella of weather and climate is provided below.

4.2.1 Drought

The drought is commonly known for the prolonged deficient situation as compared to the normal condition. There are three types of droughts that affect the agriculture sectors, these are meteorological drought, hydrological drought, and agricultural drought. The existence of meteorological drought for a longer time has greatly influenced the occurrence of hydrological drought. In other words, the meteorological drought is the major cause for triggering and sustaining hydrological drought. Similarly, the hydrological drought is the one of most prevailing conditions for the manifestation of agricultural drought. Meteorological drought occurs due to primarily prolonged dry conditions or break in precipitation. However, there are several other ways in the atmospheric conditions that may contribute to the drought. For example, a warmer temperature will increase the evaporation rate from the ground/soil, which in turn extends the low precipitation period compared to the cool condition. This drier soil will tend to reduce plant cover and, as a result, will suppress the rainfall occurrence. Thus, a drier area will further increase the drying conditions. Therefore, drought acts as a positive feedback mechanism by persisting its own condition. In the warming era, it is highly likely that the intensity and frequency of drought will increase.

Drought can affect the agricultural and allied sectors, water management and resources, hydro-power, transportation through rivers. Prolonged drought is highly favourable for deadly wildfires, which can affect the lives, damage property, and can pollute the environment.

4.2.2 Floods

The floods are defined as when water is inundated over the land. The floods may occur due to several reasons such as heavy or extreme rainfall events, cloudbursts, glacial lake burst, storm surges, etc. There are different categories of floods based on the spatial and or temporal scales and the cause behind them, such as flash floods, fluvial floods, rural ponding, urban flood, coastal flooding, etc.

The heavy to extreme rainfall events may cause water logging over the land area. If the combined effect of infiltration through soil and runoff is lesser than the amount of rainfall falling from the sky, then that place will most likely be filled up with water and cause floods. In the context of global warming, there is an increase in global mean surface temperature, which in turn enhances the temperature of the atmosphere. It is established that the holding capacity of moisture in the air will be increased with the increase of temperature. Theoretically, more moisture in the atmosphere will lead to amplification of the amount of rainfall/precipitation during the convective activities. Consequently, an increase of intense rainfall events has been observed in the last few decades worldwide. Thus, global warming will very likely increase the extreme rainfall events and thus a threat of more number of flood events. A cloud burst is a sudden, very heavy rainfall mostly associated with thunderstorms. This kind of event generally occurs for a short duration and over a small region. It is a local scale phenomenon; however, the occurrence of the cloud burst over the hilly region may greatly affect the downstream area through a flash flood. The glacial lake burst often occurs due to the melting of the ice and the chances of the glacial lake burst is becoming higher with global warming. The downstream will be affected highly due to flash floods. The prolonged rainfall over upstream may often cause fluvial flood in the downstream area. On the other hand, tropical cyclones/hurricanes are one of the main reasons to have storm surges and cause coastal flooding. Besides the direct impact of coastal flooding due to storm surges, the infiltration of salt water in the coastal agricultural land greatly affects agricultural production adversely.

4.2.3 Storms

By definition, storms are low-pressure system. There are several categories of storms based on their characteristics.

4.2.3.1 Thunderstorm

Thunderstorm is a fast-developing vertical cloud system which usually develop when warm and moist air rise over the comparatively cooler regions. Thunderstorms are associated with gusty winds, heavy rainfall, lightning, and thunder. The life time of the thunderstorm is short (a few hours) and more frequent over regions with high temperature and moisture, i.e. over the tropical belt. However, it is seen over most of the world except in the polar region. Thunderstorms are dangerous mostly to the crops. Lightning can cause loss of lives if it strikes on a human or animal and can damage the properties (houses, electrical circuits and electronic instrumentation) severely.

4.2.3.2 Hailstorm (Ice Storm)

Hailstorms are often accompanied by thunderstorms and larger hail falls from the cloud to the ground. It mostly occurs over the region when the temperature is close to freezing point and condensation takes place slowly, then often, the air comes out from the water droplet and makes clear ice. The ice then starts to fall due to gravitation attraction from colder to warmer environment. At the same time, the evaporation from the ice surface takes place and makes it lighter, which then again moves upward due to strong updraft motion. The ice is formed again over that ice particles during its updraft motion and becomes larger and heavier, which again starts to fall. This may repeat several times and finally it comes out from the cloud and falls on the ground. A hail may have many layers of ice due to the upward and downward motion of the ice particles (the same number of layers that ice particles follow the upward motion), and the size may be from 5 to 15 mm in diameter. The hailstorm mostly occurred over the mid-latitude regions; however, a few events also occurred over the tropical region. The lifetime of the hail storms may be, in general less than 10 to 20 min. The hailstorms are extremely dangerous to crops, fruits, and other agricultural products. It is destructive to the buildings, and hailstorms of the larger size of hail may be dangerous to the animals exposed to it.

4.2.3.3 Snowstorm (Blizzard)

Snowstorm or blizzard is one of the severe weather hazards associated with the strong wind and falling or blowing snow. It is seen mostly over the mid-latitudes and Polar Regions and bringing the temperature lower. During the snowstorm, the wind speed with low temperature may be 50 to 80 km per hour in the mid-latitude, while winds may spill at the ice edge with a velocity of 160 km per hour in the Polar regions.

4.2.3.4 Dust Storm (Sandstorm)

The dust storm or sandstorm occurs over the semi-arid or arid regions where the surface is dry and, loose sand and dirt exist over the surface. During the warmer conditions, the area exhibits a low-pressure which causes the gustwind, and the strong wind uplifts and blows the sand and dirt which is known as a dust storm or sandstorm.

Dust storms can have several impacts on the dry land/semiarid/arid regions. It may cause the loss of soil, and may damage/move the organic matter and the nutrient-rich particles, which leads to loss of agricultural productivity. It reduces visibility which hinders the normal transport systems on the road and aircraft. The dust storms may transport the diseases from one place to another, which may increase the pollution level and immediate impact on the lungs.

4.2.3.5 Tornado

A tornado is a violent whirling motion of an air column having small diameter and developed within a convective cloud and the ground. Most of the violent tornadoes are spawned by supercell storms that exhibit mid-level rotation. This mid-level rotation can be narrowed at the ground due to a strong updraft, which in turn intensifies the rotation due to conservation of angular momentum. This process is known as vertical stretching, and if it continues then a narrow whirling air in funnel shape can occur which is known as tornado. The tornadoes are mostly seen with thunderstorm clouds. The tornadoes mostly occur in the mid-latitudes; however, a few tornado events are also observed over the tropical regions. The maximum wind speed of a tornado may reach up to 500 km per hour. It has a life time usually a few ten minutes, and affects locally/ small regions.

The massive wind speed of a tornado may cause a huge destruction of the area (mainly buildings), including great loss of life if it strikes the populated area. All the agricultural sectors, including crops, trees, fruits, etc., can be destroyed on the path of the tornado.

4.2.3.6 Tropical Cyclone

Tropical cyclones are also rotating motion of air column like tornadoes, but the size of the tropical cyclone is very large compared to the tornadoes. It generates over the warmer oceans, and the whirling motion intensifies if the atmospheric states are favourable for its formation. Once it intensifies over the ocean, it draws energy from the sea surface through evaporation from the surface, condensation of moisture inside the cyclone, and release of latent heat. The cyclones are associated with massive windspeed and torrential rain. The diameter of a tropical cyclone may be 500 km. The wind speed increases linearly from the outer periphery of the cyclone up to the 'eye wall', which is 15–50 km away from the cyclone centre. This increase is noticed towards the axis of rotation or cyclone centre. The speed of wind then reduces, and the centre of the cyclone is often calm, warmer, and having low pressure, called the 'eye'. The intensity of a cyclone can be measured by calculating the dropping pressure at its center compared to the mean sea level pressure. The wind speed is also directly proportional to the difference of pressure between mean sea level and cyclone centre. In a tropical cyclone, the maximum wind speed may reach 250 kmph, and the gusty wind may be more than 300 kmph. Tropical cyclones are known by various names in different parts of the world, such as Hurricanes, Typhoons, Willy willies, Tropical cyclones, Cyclones, etc. Although all these names indicate the similar characteristics of the storm, however, the intensity to reach at that naming stage varies. A tropical cyclone intensifies mostly over the ocean still it reaches near the coastal region, and the dissipation of the cyclone mostly starts after that, particularly after the landfall, the strength of the storm decays faster owing to lack of the available energy underneath over the land. The coastal region is affected by the storm surge associated with the cyclone. The frictional force between the strong wind of the cyclone and sea surface water is the major cause to pile the sea water and move to the same direction of the wind. The height of the surge also depends on the intensity of the cyclone during its landfall.

The massive wind speed over a larger area for a longer duration can drastically affect the agricultural sectors, forests, constructions (building), and other properties. The spatial coverage as well as strength of the wind and the duration of stay of cyclone over the land depends mostly on the intensity of the cyclone. In addition to the massive wind speed that itself damages the properties and causes the loss of lives, torrential rain and coastal flooding due to storm surges combine together to make extreme disaster conditions affecting severely agriculture, societal infrastructure, lives (loss or injuries) of animals, birds and the other sectors in many ways. The landfall of an extremely severe cyclonic storm (the maximum wind speed at eye wall of the cyclone attains in the range of 167–221 km per hour) may damage the property, equivalent to a few billion dollars. The loss of lives is higher in a tropical cyclone in comparison to other disasters worldwide. A tropical cyclone has an immediate, direct, and long-term impact on society.

4.2.3.7 Heat Wave

The heat wave is a situation when the day's maximum temperature exceeds some critical values along with a positive departure of a predefined magnitude from the long-term average. The critical value and predefined magnitude of positive departure vary from region to region. For example, based on IMD definition, the critical values are 40, and 30 °C for the plain regions and hilly regions, respectively in India, and the departure from normal is from 4.5 to 6.4 °C, while for the coastal regions, the critical value is 37 °C. This warming condition of above normal temperature may sustain for a few days to a few weeks over a smaller region to a larger region. The heat wave is mostly seen in the tropical and mid-latitude regions. Climate change in association with global warming is a potential factor in increasing the frequency and intensity of the heat wave events, and the observational evidence supports it. The heat wave may affect the health and its severity may exacerbate the mortality. The heat wave highly damages agricultural production, such as crop yields, fruits, and its allied farming sectors through the sickness/death of cattle, pets, etc.

4.2.3.8 Cold Wave

In contrast to the heat wave, the cold wave situation occurs when the day's minimum temperature drops down to a certain critical value from its long-term average by a predefined magnitude for a particular region. The cold wave often occurs over the polar and mid-latitude regions; however, it is barely seen in the tropics when the cold front from the mid-latitude propagates through the tropical region during the winter seasons. Likewise, heat and, cold waves are also dangerous to the health which may cause mortality. It may also greatly damage agricultural production.

Different events associated with the weather/climate and their typical life span and size are schematically represented in Fig. 4.1. Description of Seamless Forecast (Nowcasting to climate) is provided in detail in Chap. 3.

Therefore, the atmospheric hazards are various types based on space and time scales and the forecast systems of those hazards are also developed for different spatiotemporal scales. The most important weather/climate hazards that greatly impact in society are briefed in Table 4.1. It may be noted that the impact of the weather/climate hazards may be short time scales (few hours to few days) or may impact for a longer period of time (few weeks to few months or more), and also the intensity of impact varies events to events. The weather/climate hazards affect the different sectors of society, and most of the impact is negative, however, in a few cases, it may be neutral or positive.

The different degrees of negative/neutral/positive impact of weather/climate hazards are summarized in Table 4.2 with three classifications of intensities such as low-, moderate-, and high-impacts.

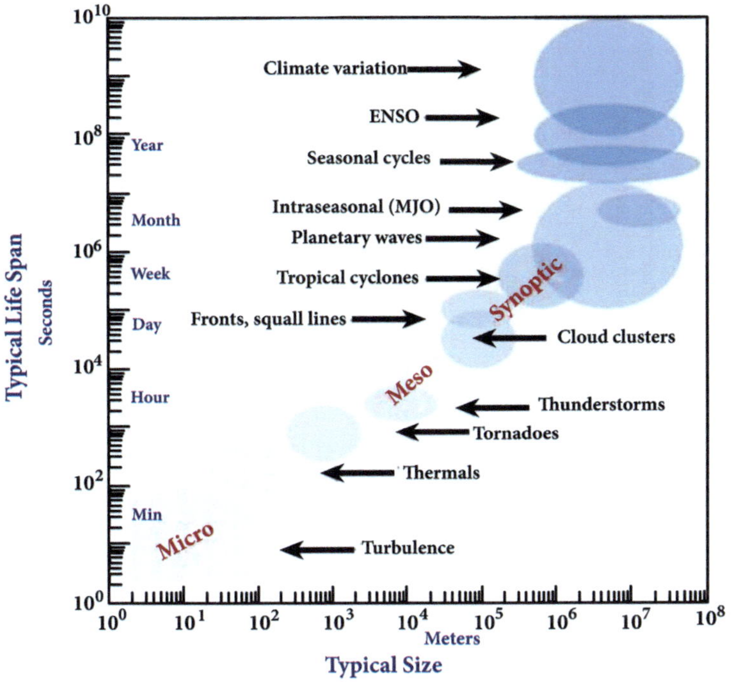

Fig. 4.1 Schematic view of various weather and climate events with their typical life span and size (Adopted from Laing and Evans 2011)

Table 4.1 Different weather hazards with their spatiotemporal scales and forecast categories adopted from Gill and Malamud (2014)

Event name	Definition	Component hazards (where applicable)	Spatial scale	Temporal scale	Forecast category
Drought (meteorological)	A prolonged period with substantially less rainfall than normal (climatology), resulting in a serious deficit of surface and groundwater resulting agricultural stress	Meteorological drought, that leads to hydrological drought and then agricultural drought	Few hundred to few thousand	Several weeks to year or more	Extended range-long range
Floods	The inundation of typically dry land with water due to large excess of rainfall than normal leading to standing water	Flash flood, fluvial flood, rural ponding, urban flood, coastal flooding, storm surge, glacial lake bursts	A few ten kms to a few hundred kms	Hours to Days to a week or so	Nowcasting-short-medium
Cyclone	A significant perturbation of the atmospheric system in association with strong convective activities and whirling motion involving heavy precipitation and violent winds, thunder, and lighting	Tropical cyclone, hurricane, typhoon	~100–1000 kms or more	~A few days to a week	Short-medium
Thunderstorms	Strong convective activities in association with strong winds, gusts, precipitation, lighting, and thunder	Multi-cell thunderstorm Nor'wester	~10–100 kms	Few hours	Now casting-very short
Hailstorm	A part of thunderstorms where super-cooled water becomes bigger size ice and fell down, mostly associated with thunderstorms, responsible for loss of lives (outside the home) and damages of trees and properties				Now casting-very short
Lightning	The atmospheric discharge of static electricity, occurs mostly cloud to cloud or cloud to ground, the second one dangerous can be responsible for loss of lives (outside home) and damages of trees and properties				Now casting

(continued)

Table 4.1 (continued)

Event name	Definition	Component hazards (where applicable)	Spatial scale	Temporal scale	Forecast category
Tornado	Extremely violent rotating wind from the cumulonimbus cloud to the surface, often associated with precipitation, thunder, and lighting		~Few meters to a few ten kms	Few seconds minutes to an hours	Now casting-very short
Dust storm (sandstorm)	A severe weather condition is having strong winds filled with dust/sand occurs, these particles of dust/sand uplifted and blown with the strong winds/gusts. The phenomena is like a thunderstorm but a dry condition occurs over mostly arid regions	Sandstorm, dust storm, Andhi	~10–100 kms	Few hours	Now casting-very short
Dust devil	Dust devils a very small whirling air over a small place, normally seen in a day time		A few cms to a few meters	A few seconds	Now casting
Snowstorms	A significant perturbation of the atmospheric system, with heavy precipitation of snow		~10–100 kms	Few hours-day-week	Short-medium
Heat wave	A prolonged period of temperatures above the normal average for a designated period of time (either short or long-term, local, regional, or global)	Heat waves, climatic change	~10–100 kms or more	~Days	Short to medium
Cold wave	A prolonged period of temperatures below the normal average for a a designated period of time (either short or long-term, local, regional, or global)	Cold waves, climatic change	~10–100 kms or more	~Days	Short to medium
Heavy to extreme rainfall	Strong convective activities associated with enormous rainfall within short time and localized	Flood/flash flood/ land slide	~Few ten kms to few 100 kms	~Few hours to days	Short to medium

Table 4.2 Potential impacts (low, moderate, and high) of different weather events in various sectors for immediate/short (hours to day) and long (weeks or more) time periods. Minus (–) and plus (+) signs indicate negative and positive impacts; single, double, and triple signs indicate low, moderate, and high impacts, respectively. Zero (0) sign for neutral or negligible impacts (Courtesy: Jahn 2015 and modified according to tropical region)

Sector	Flood		Drought		Heat		Cold		Storms	
	Short	Long	Short	Long	Short	Long	Short	Long	Short	Long
Agriculture	––––	––	–––	––	–––	0	––	0	–––	––
Industry	–––	++	––	0	––	0	––	0	––	–
Tourism	––	0	–	––	––	0	–––	0	––	–
Transport	––	–––	–	0	––	0	–––	0	–––	–
Energy	–––	––	––	0	––	0	––	0	––	–
Insurance	–––	––	––	–	––	0	––	0	–––	––
Housing	––	++	–	–	–	0	––	0	––	+
Health	––	0	–	–	–––	0	––	0	––	0
Ecosystem	–––	0	––	––	–	0	0	0	––	0

4.3 Principle of the Dynamical/Physical Forecast System

Reliable predictions of the weather and climate hazards, particularly those which can be destructive in nature in terms of loss of lives and aggregate damage of properties are in great demand for risk management and policy planning. In addition to that, advance intimation of the weather and climate prediction can enormously be helpful to the different sectors such as agricultural, water management, farming sectors, hydro-power sectors, etc. In order to develop weather and climate prediction systems, it is indeed necessary to understand the dynamics, complex physical processes, and linear and non-linear interactions between different spatiotemporal large-scale circulations (~10^4–10^7 m or more) with small physical processes (~$0.4–0.75 \times 10^{-6}$ m), interactions among land–air–sea, and all the processes that are directly or indirectly involved in governing and modulating the weather and climate systems.

India was the pioneer for the forecast of summer monsoon when Blanford provided an empirical-based monsoon forecast in 1884 using the relationship between the Himalayan snow cover and Indian summer monsoon rainfall. In 1910, G. Walker provided the Indian summer monsoon rainfall forecast using the regression technique. These forecasts are for a seasonal mean over the whole India. On the other hand, V. Bjerknes in 1904 proposed that weather and climate systems are initial value problems, and reliable forecasts can be made by representing the atmospheric variables in mathematical equations of prognostic variables. He also mentioned that as many equations could be formulated to represent all kinds of dynamical, physical, mechanical, and the other processes that play a role in the weather and climate systems. Consequently, he also realized that there were no exact mathematical solutions to solve the partial differential equations. In 1922, L. F. Richardson had made an attempt to solve those complex equations through numerical methods and tried to calculate the next 6-h forecast of the pressure field over a small region. That forecast went drastically in the wrong direction, however, the cause of the failure was not known at that time. In 1928, three mathematicians namely Courant, Friedrichs, and Lewy had found out the probable reasons of the failure. They reported that the ratio ($\Delta x/\Delta t$) of space (Δx) and the time steps (Δt) must be equal to or less than the phase velocity during the solutions of these equations using numerical methods else, the solution would be exponentially amplified which is familiar now as a computational instability. This criterion, i.e. $\Delta x/\Delta t \leq c$ (where, Δx is the grid spacing, Δt is the time steps, and c is the phase velocity) is known as the CFL criteria. During the 1930s and 1940s decades, Carl-Gustav Rossby, Charney, Fjortoft and Neumann, Gilchrist and Cressman had significantly contributed in the theoretical and experimental studies for the development and improvement of the Numerical Weather Prediction (NWP) systems that can be used for the prediction of the weather and climate. In 1955, Berghorsson and Doors had first provided the weather forecast using the NWP systems. In the 1960s, the development of more complex general circulation models (GCMs) had been accomplished; however, those were simple models as

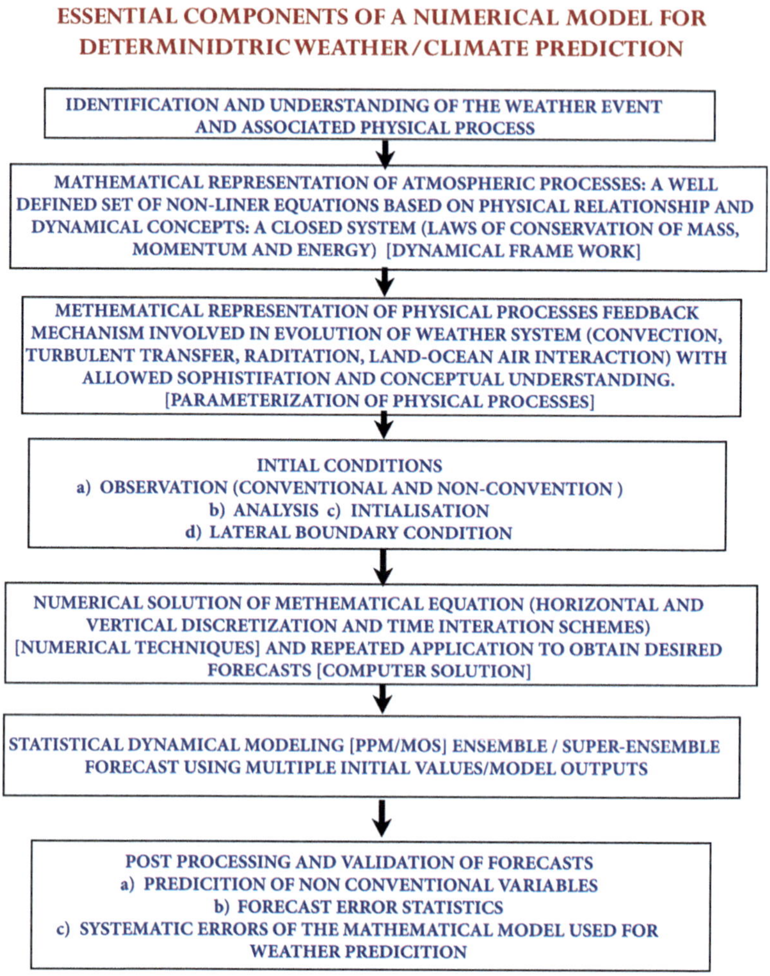

Fig. 4.2 Flow chart of the essential components of an NWP for a deterministic forecast

compared to present-day advanced, sophisticated, very complex GCMs. It may be noted that the more complex GCMs require more computational facilities than the comparatively lesser complex GCMs to complete the same job. Thus, the advancement of high-performance computing (HPC) systems has accelerated the progress of the development of NWP systems.

The forecasting system has basically three components, such as (i) data (model initialization and validation), (ii) model (dynamical or statistical), and (iii) post-processing or analysis of the model outputs. The essential components of an NWP for the deterministic forecast are represented as a flow chart in Fig. 4.2. In the following sections, the details of the dynamical or physical forecasting systems, that involve mainly data, model, and post-processing are described in brief. The forecasting systems using statistical approaches have been provided in detail afterward (Chap. 6). The schematic flowchart of the components of weather forecast is provided below.

4.4 Data (Model Initialization and Validation)

In order to understand the data required for the model initialization and validation, this subsection has been fragmented into several components.

4.4.1 Observational Data (OBS)

These are the prime sources to understand the weather and climate systems and their evolution with time. The OBS are the key factors to design, develop, verify, and validate the modelling system that can be used efficiently for the prediction of different weather and climate systems. The OBS is also required to provide the model's initial and boundary conditions for its integrations and prediction.

There are several types through which the observations are made and data are collected. During most parts of the nineteenth century and prior to it, the observational data are collected manually using the appropriate instrumentations. In the recent few decades, automated surface observing systems, RADAR, satellites, etc., are playing an important role in providing better quality data with higher temporal frequency. Manual and automated station observations, upper air observations (radiosonde, pilot balloon, radio winds), buoys, weather RADAR, satellites are in use to take the observations over the land, ocean and upper air.

The observational data sets can be categorized into two broad types: (a) conventional data in which the observations are routinely measured at the standard time defined by the World Meteorological Organization (WMO) that are 00, 06, 12, and 18 Coordinated Universal Time (UTC) over the fixed stations/platforms. The second type is (b) unconventional in which data can be obtained at any time of the day. Satellite, aircraft, drifting buoys, etc. are used to provide the non-conventional data, which are very crucial both for model initialization and validation at data sparse regions.

4.4.2 Organizing of OBS Data

The OBS data are obtained from the various places in the world and the observation points are not uniformly located over the space. Therefore, to make all the observational data usable to any user located in any point in the world, all the observations taken at various places are transmitted through a well-defined Global Telecommunication System (GTS) to store the data in a single place. This transmission is made through a high-speed networking system, and any user can avail the observational data within 4–6 h after the time of observations. The WMO has specified the standard format, code, and decode to transmit and receive the data under the World Weather Watch (WWW) program.

4.4.3 Quality Control

The observational data that have been decoded at the central places may contain errors which may be introduced through observational, and or instrumental, and or transmission processes. Therefore, it is essential to check the quality of the observed raw data and minimize the error. The process to check the quality of the data is known as a quality control check. There are several ways to check the quality of the data. It is highly recommended to check the duplicate data (a variable may have two or more values at a particular location for a particular time) and units of the data before processing for the quality control. These are:

i. Basic quality control: This is to check the format of the code and syntax of the code.
ii. Internal consistency (within one observation): The measured parameters are often physically interrelated with one another. For example, surface temperature will be low if snowfall occurs, or the measurements of variables using radiosonde at different layers.
iii. Impossible values: Sometimes, the observational data present the values which are impossible. For example, wind direction outside the range from 0 to 360°, or rainfall amount in negative.
iv. Outliers: This process detects the wild value in the observations.
v. Functionality related check: Some meteorological parameters are derived from the basic parameters using mathematical formulations. It is good to check the value of derived parameters from basic parameters using mathematical equations. For example, relative humidity, vapour pressure, and dew point temperatures can be estimated using the wet-bulb and dry-bulb temperatures.
vi. Temporal consistency: At a specific point/location, it is necessary to check if the measured parameter is consistent with time or not.
vii. Gross check: Closeness to the climatology and the deviation in extremes within the range of the climate variability for a particular time over a fixed location; or how much the parameter has deviated from the mean may be useful.
viii. Buddy check: Checking the consistency of the observations from its neighbouring observations. A denser observational network is better in this process.

4.4.4 Analysis and Reanalysis

Both the analysis and reanalysis processes use a modelling framework to produce grid points/nested data uniformly distributed over the space using randomly spaced observations keeping the physical consistency in space and time. In both processes, the 3-dimensional model is initialized with the available observations, and the atmospheric state is defined at grid points for $t=0$ time. In the analysis, the upgraded model with time is being used for generating the analysis field, while in reanalysis, one frozen model is being used to produce the reanalysis field. Thus, a reanalysis product will be invariant with time if no additional observations are incorporated. While besides the additional observation incorporation, the analysis products may also vary with the use of upgraded models. In general, the horizontal resolutions are finer in the analysis products compared to the reanalysis products.

The analysis or reanalysis fields are used to initialize the model. These procedures are used to take care of such that no imbalance between mass and velocity fields occurs. The imbalance between the mass and velocity fields can generate fictitious gravity waves of greater amplitude than the meteorological wave. The present-day model consists of damping terms that tend to dissipate the fictitious waves till the propagation of these kinds of waves exists for a short time; thus, short-range forecasts can be noisy. However, quality control and assimilation system can reduce the imbalance at initialization, hence lesser noise in the forecast. It may be noted that the forecast skill of the short-range period has notably improved after the use of the assimilation technique.

4.4.5 Data Assimilation

In the NWP system, data assimilation is a technique that combines observational data with the numerical model products in order to produce an optimal estimation of the evolving state of atmospheric or oceanic systems. Simply, data assimilation technique can be regarded as a process to keep the numerical model in track with a constant update of fresh observational data to the model. It combines observational data and the underlying dynamical principles that govern the system. It is widely used as a method to investigate various atmosphere, ocean, and land surface processes.

In the atmospheric sciences, the data assimilation technique is conceived from the fact that NWP is an initial value problem. Thus, presenting the initial state with its truest estimation improves the future prediction remarkably. Since the birth of NWP processes, there have been attempts to combine present and past observational data of the atmospheric state with the model results. As the technique involved is tedious, an automatic objective analysis method was developed by Panofsky (1949), Gilchrist and Cressman (1954), Cressman (1959), and Barnes (1964). These efforts can be considered as the primary steps to obtaining data assimilation technique followed by the comprehensive work by Gandin (1966).

The data assimilation technique consists of a set of statistical interpolation methods to improve the knowledge of past, present, and future system states by combining the experimental data from a model and the theoretical knowledge on the system. It can determine an evolving probability density function that specifies the range of possible states and the probabilities representing reality. Statistical interpolation is one of the simplest and elementary techniques that can be implemented on complex high-dimensional systems. There are various data assimilation methods; the Cressman analysis method, the optimal interpolation method, three-/four-dimensional variational analysis, and the Kalman filter. All these methods are designed on least-square methods in which the final state is estimated with minimized uncertainty. However, the difference in these methods lies in the selection of the metric used to quantify the uncertainty and the corresponding weight provided to the observations and the prior estimate.

Data assimilation is addressed in many names depending on the field of application (e.g., *state estimation*, *history matching*, *filtering*, *smoothing*), and it is often combined with *inverse methods* to extract maximum information from observations. This technique can be applied to all types of weather models, including global and regional.

4.5 Model

The dynamical or physical model is based on the principle that accounts for all kinds of governing equations, interactions between land–air–sea, and all dynamical and physical processes, including their complex interactions that plays a key role in determining the state of the weather and climate. To develop the mathematical model, the following three points are important:

(a) Complete understanding of the physical system to be modeled.
(b) Segregate the systems into different components/dynamical or physical processes which can be stated using mathematical relation.
(c) Develop a set of mathematical equations to represent the dynamical or physical processes/components of the atmospheric system.

The flow in the atmosphere is continuous; however, the atmosphere consists of a wide variety of events ranging from micro turbulence to planetary waves. Based on the characteristics and spatiotemporal extension, the spectrum of atmospheric flow has been divided into a number of parts. Next, a brief description in association with the dynamical model framework is provided in the following sub-sections.

4.5.1 *Dynamical Model*

There are seven atmospheric variables through which one can express the atmospheric state using mathematical formulations. These parameters are considered as the basic parameters, and the seven equations are known as the fundamental equations or governing equations of the atmosphere. These atmospheric parameters are pressure (p), temperature (τ), air density (ρ), specific humidity (q), and three components (u, v, w) of wind. The governing equations consist of both fundamental forces (surface forces and body forces) and apparent forces, and the other factors such as energy balance, mass conservation, equation of state, and moisture conservation. These equations are presented in the next.

4.5.1.1 Equation of Motion (Law of Conservation of Momentum)

The earth has orbital as well as rotational motion. The equation of motion for the atmosphere has been derived using Newton's second law of motion for the non-inertial frame of reference. The forces that act on a particle in the atmosphere are

(1) Pressure gradient force (surface force, fundamental force): This force arises due to the difference in air pressure at two places. It is expressed by $-\frac{1}{\rho}\ \Delta p$ where ρ is the density of the air and Δp is the variation of pressure.
(2) Gravitational force (body force, fundamental force): This force arises due to the attraction of two bodies having mass, and the center of mass of those are in different locations. Mathematically,

$$\vec{F} = G\frac{mM}{r^2}\hat{r} \tag{4.1}$$

M = Mass of the earth, m = mass of the object, G = universal gravitational constant, r = distance between the centre of masses of the two bodies. $\vec{F}$ and $\hat{r}$ are the force and unit vector of the distance r. The force on a particle due to gravitational attraction can be expressed as $\vec{F} = m\vec{g}_0$, where, $\vec{g}_0$ is the gravitational acceleration, thus $\vec{g}_0 = G\frac{M}{r^2}\hat{r}$ and can be considered as a gravitational force acting on a unit mass. This gravitational force acts towards the centre of the earth
(3) Centrifugal force (apparent force, Pseudo force): This force arises due to rotation of the earth. For unit mass, this force will be mathematically expressed as, acting radially outward from the axis of rotation. R is the distance between the particle and axis of rotation, v is the angular velocity. The force arises when a particle is on the reference frame which has a rotational motion.
Effective gravity: In the atmospheric science, the resultant force of the gravitational and centrifugal forces is considered and known as effective gravity which acts in a direction a bit shifted from the center of the earth (Fig. 4.3).

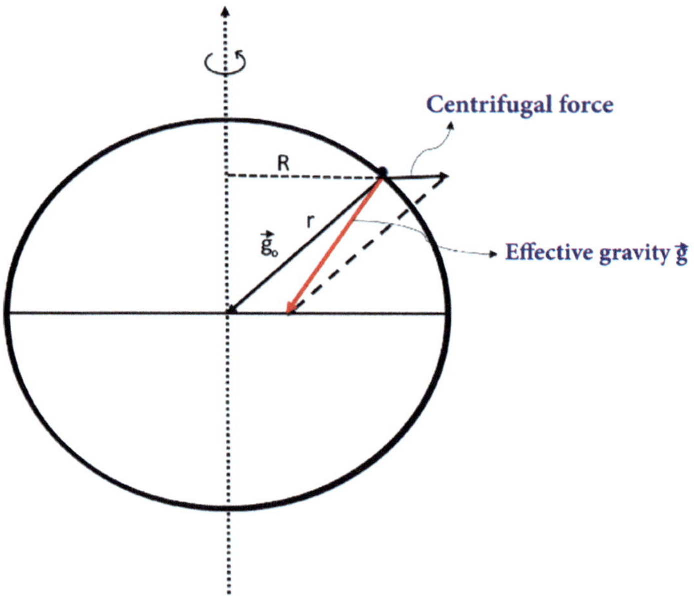

Fig. 4.3 The direction and magnitude of the effective gravity which is a resultant of gravitational and centrifugal force

(4) Coriolis force (apparent and fictitious force): This force arises due to the rotation of the earth, and it is an apparent deflecting force. The magnitude of this deflecting force depends on the rate of rotation of the earth and latitude of the particle location. Mathematically,

$$F_{Co} = -2m \times \left(\vec{\Omega} \times \vec{V}\right) \tag{4.2}$$

where, $\vec{\Omega}$ and $\vec{V}$ are the earth's angular velocity and velocity of the particle w.r.t. the earth.

(5) Frictional force: It is the resultant due to friction between moving fluid and earth surface or between two layers of a fluid having different velocity (viscous force due to internal friction). Mathematically,

$$F_{fr} = \nu \nabla^2 V \tag{4.3}$$

where, ν is the kinematic viscosity coefficient, is the dynamic viscosity coefficient, $\nu = \frac{\mu}{\rho}$, μ is the density of the fluid, and V is the relative velocity of the fluid compared to the two layers.

The equation of motion of a particle relative to the earth as a reference frame can be expressed as

$$\frac{d\vec{v}}{dt} = -\frac{1}{\rho}\Delta p - 2\vec{\Omega} \times \vec{V} + \vec{g} + \vec{F}_{fr} \tag{4.4}$$

Here, $\vec{v}$ is the velocity of the particle in x or y or z directions. Thus, the above equation represents the three equations of motions in three directions. In the atmosphere, large-scale flow is quasi-horizontal, and for a vertical acceleration is relatively very less than the horizontal velocity and the Eq. 4.4 for vertical component becomes hydrostatic equation where the pressure gradient force is balanced by the gravity.

4.5.1.2 Equations of Mass Continuity (Law of Conservation of Mass)

In a large-scale motion, the approximation is that mass flux is zero. Thus,

$$\vec{\nabla} \cdot \left(\rho \vec{V}\right) = 0 \tag{4.5}$$

And for an incompressible fluid, the density is constant following the motion, thus $\frac{d\rho}{dt} = 0$ and velocity divergence is zero, therefore, $\vec{\nabla} \cdot \vec{V} = 0$ (since the local rate of change of density is almost zero and the atmosphere can be treated as an incompressible fluid).

The expansion of the Eq. 4.5 and using the above for an incompressible fluid, the mathematical expression for the equation of mass continuity becomes,

$$\frac{\partial \rho}{\partial t} + \left[\frac{\partial(\rho u)}{\partial x} + \frac{\partial(\rho v)}{\partial y} + \frac{\partial(\rho w)}{\partial z}\right] \tag{4.6}$$

As under the average conditions, the air density does not locally vary. Therefore, the atmosphere may be treated as incompressible reducing the above equation to

$$\vec{\nabla} \cdot \vec{V} = 0 \text{ or } \frac{\partial u}{\partial x} + \frac{\partial v}{\partial y} + \frac{\partial w}{\partial z} = 0 \tag{4.7}$$

4.5.1.3 Equation of State (Ideal Gas Equation for Dry Atmosphere)

The major composition of the atmosphere is the gases (more than 99%), and these mixtures of gases follow mostly the ideal gas law which is

$$p = \rho RT \tag{4.8}$$

where, p = pressure, ρ is the density of the air, T = temperature, and R is the gas constant.

4.5.1.4 First Law of Thermodynamics (Principle of Conservation of Thermodynamic Energy)

The energy added (dQ) to a closed system is equal to the sum of increase in internal energy (dU) and work done in expanding the volume against external pressure (pda). Thus

$$
\begin{aligned}
dQ &= dU + pd\alpha \\
&= c_v dT + pd\alpha \\
&= c_p dT - \alpha dp
\end{aligned}
$$

Thus,

$$\frac{dQ}{dt} = C_p \frac{dT}{dt} - \alpha \frac{dp}{dt}$$

or

$$\dot{Q} = C_p \frac{dT}{dt} - \alpha\omega \tag{4.9}$$

where, $\dot{Q}$ is the rate of heating.

4.5.1.5 Moisture Continuity Equation

$$\frac{\partial q}{\partial t} + V\Delta q + w\frac{\partial g}{\partial z} = E - P \tag{4.10}$$

where, E and P are evaporation and precipitation, respectively.

It can be noted that all the above equations are partial differential equations having inherent non-linearity because of advection terms that contain a product of a dependent variable with the derivative of another dependent variable, except the equation of state, which is algebraic in nature. The equations represented in 4.4–4.7 are the hydro-dynamical equations, where the equations expressed in 4.8 and 4.9 are thermodynamic equations.

4.5.2 *Coordinate Systems*

4.5.2.1 Vertical Coordinate Systems in the Model

The above governing equations presented in the above section are needed to express in a fixed vertical coordinate system. The equations can be represented in the Cartesian and spherical coordinate systems. In the Cartesian coordinate system, the vertical coordinate is the height (z). The vertical coordinate can also be expressed in pressure (p), or by σ the ratio (normalization) of p by surface pressure p_s ($\sigma = p/p_s$), eta (p – pt)/(p_s – p_t) where p is the pressure at any point, p_t is the pressure at the model top, and the user can fix this according to requirement and p_s is the surface pressure and potential temperature (θ) instead of height (z), and one needs to transform the governing equations stated above accordingly. It is seen that using p or σ or θ vertical coordinates has advantages over height for the development of a model.

(1) In the pressure coordinate system, the density didn't appear in the pressure gradient force and simplified the equations. It is one major advantage of the pressure coordinate system over height.
(2) Since the sigma (σ) coordinate system is related to the pressure, it simplifies the equations and easily handles the lower boundary conditions without complicating the mathematical formulation.
(3) The sigma coordinate system naturally following the terrain (surface topography) and the terrain-following sigma coordinates better represent the continuous field.
(4) The sigma coordinate system has a tendency to increase the vertical resolution near the surface and a consistent tendency up to the top of the model. This enhances the model skill through better representation of boundary layer processes and features.
(5) One unique advantage of sigma coordinates over the height, pressure, and isentropic coordinates is that it eliminates the problem of intersecting of the ground of the vertical coordinate.
(6) In the eta coordinate system, it is not required to perform vertical interpolations, which is a must in sigma coordinates to calculate the pressure gradient force.
(7) The representation of low-level convergence is better represented in eta coordinates, particularly in the steep orographic region.
(8) Many atmospheric events such as cold air outbreaks, lee side cyclogenesis are better resolved in the eta coordinate system; hence forecast skill is improved.
(9) The isentropic coordinate system (θ) allows better resolution for baroclinic regions such as fronts.
(10) Isentropic coordinate has several advantages, such as it conserves potential vorticity, vertical motion is caused exclusively by diabatic heating (almost), etc.

4.5.2.2 Horizontal Coordinate Systems in the Model

The horizontal coordinate systems are represented based on the choice of projection used in the model. It is necessary to project the grid points of the model to the real map of the earth. This involves several types of projection, such as Marcetor projection (for the tropical region, where the distance of 1-degree spacing is almost the same in x and y directions), Lambert conformal projection (mid-latitudes region), polar stereographic projection (for the polar region) to make the grid spacing in the model close to the actual geographical spacing.

There are several ways that determine at what points in the grid the required variables of the equations will be calculated. The simplest one is that all the variables can be estimated at each grid point (Fig. 4.4a). This is known as an unstaggered system or A-type grid. In the modelling framework, there are several other ways in which some variables are used to compute at different locations within the

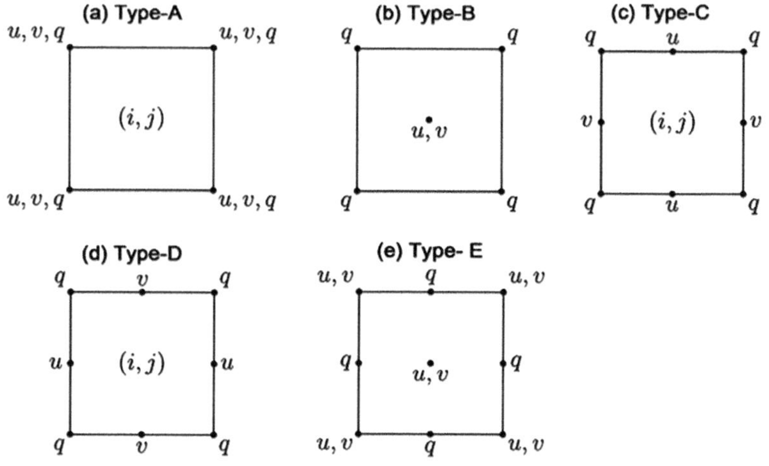

Fig. 4.4 Different types of grid systems (non-staggered and staggered)

grid, and the other variables are estimated at the intersection points (grid points); this is known as a staggered grid. There are B-type, C-type, D-type, and E-type staggered grid (Fig. 4.4).

The system of equations in the horizontal coordinate system is represented as per the projections in use. However, as the earth is nearly spherical, spherical coordinates are a natural choice.

The staggered grid has several advantages over the unstaggered grid, particularly the efficiency of the computing of the variables is higher, requires less storage system during the computation, etc.

4.5.3 Derived Equations and Variables

Besides the governing equations and basic parameters that are used in the atmospheric models, some derived equations are also included in the atmospheric model. For example, equations for vorticity, divergence, surface pressure tendency, etc., are derived and incorporated into the modelling framework. Several other important parameters like perturbed temperature in the thermodynamic equation and mixing ratio are used to compute. Regarding the derived parameters and equations, a detailed description can be found in the book by Coiffier (2011).

4.5.4 Parameterization of Physical Processes

The development of NWP models has been greatly progressed owing to the increased quality data network, a better understanding of the atmospheric processes and land–air–sea interactions, and enormous advancement in technology and computing system, still present-day NWP models are not efficient to accommodate explicitly all the small-scale physical processes that play an important role to governing the atmospheric system. For example, some physical processes such as cumulus convection, planetary boundary layer, land surface processes, radiation and radiative effects of the clouds, the role of aerosols and gases, air–sea interactions, etc. play a significant role in the atmospheric process, and it is needed to account for the effect of these processes.

In order to account for all the physical processes that are directly or indirectly modulating the atmospheric system, one needs to consider all the processes that have a great spectrum in spatiotemporal scales. Due to mainly computational efficiency, the grid spacing in a dynamical modelling framework used to be considered at a horizontal scale of a minimum length may be a few hundred meters, and lesser than that grid spacing may not be permeable from the computational and stability point of view. At the same time, it is required to represent the effect of the small-scale physical processes in the modelling framework. The process in which the aggregate effect of the small-scale physical processes (or sub-grid scale processes) for the grid box that cannot be explicitly resolved that is implicitly represented through the large-scale flow parameters is known as parameterization. As stated above, there are a few major issues for incorporating the parameterization method or scheme in the dynamical model.

(i) The computing efficiency is limited to considering the grid spacing small enough that it is able to resolve the small-scale physical processes. For example, a typical high-resolution dynamical model may have grid spacing of a few hundred meters or a few kilometers which is very large to account for the radiation processes that act in the length scale in the order of 10^{-5} m or less.
(ii) There are many physical processes that are too complex to represent in the modelling framework. Besides this, all the physical processes are not fully understood to represent in the mathematical formulation.

In order to include the physical processes that may be in resolved scale or unresolved sub-grid scale, physical parameterization concepts that represent physical forcing parametrically on the resolved flow and constants are based on a number of conceptual hypotheses and observational analyses. Thus, a significant degree of uncertainties is involved in the parameterizations concept because of the assumptions, accuracy of the spatiotemporal observational data (that may be coarse both in space and time), and unequivocal errors in the analytic results. On the other hand, the model skill highly depends on both the appropriate representation of physical parameterization and the incorporation of correct feedback mechanisms (both positive and negative) within the physical processes. In addition, if the model grid spacing is very small (high-resolution), then the model may directly resolve many physical processes that may not be resolved by the model having a coarser resolution model. In the high-resolution model, the treatment of the physical processes and interactions among them with feedback mechanisms are better represented; thus, the skill of the model theoretically be higher. However, a small noise in the physical

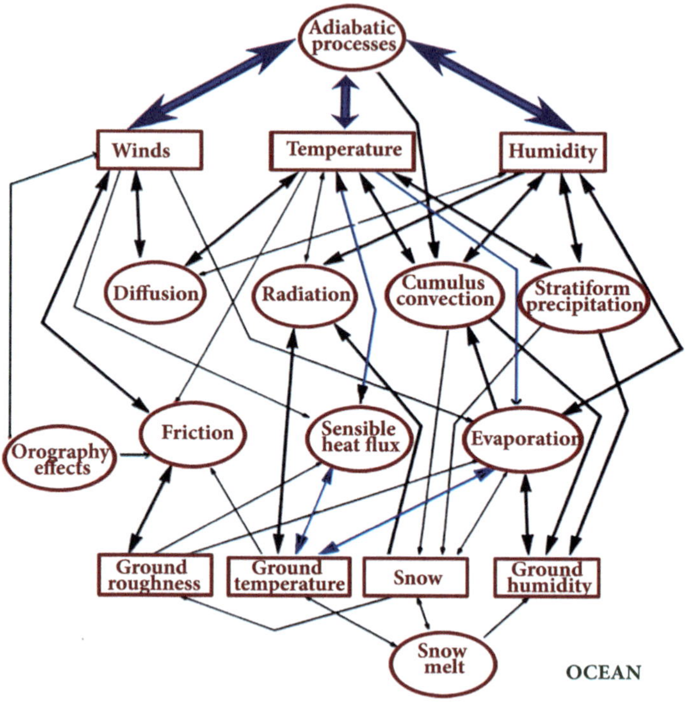

Fig. 4.5 Schematic view of the major interaction processes among different components. The greater thickness represents the stronger interactions and the way of interactions are shown through the arrow (*Source* Simmons and Bengtsson 1984)

process may drastically amplify, and hence skill will greatly deteriorate. In addition, as indicated earlier, feedback mechanisms that may be positive or negative mechanisms that affect the physical processes and amplifying or dimming the process, respectively, are other important aspects. It is important to note that both mechanisms may co-exist within the interaction of physical processes. Figure 4.5 represents a schematic view of complex interactions between different physical processes and feedback mechanisms, in which the thickness of the interaction line characterizes the strength of the interactions. Since the interactions are highly non-linear and complex in nature, thus, representation of the physical parameterizations is one of the most important factors for the success of the model in predicting weather/climate systems.

In the numerical weather prediction model, the major physical processes that are needed to be parameterized are described briefly and presented in. the next.

4.5.4.1 Parameterization of Land–Ocean–Air Interaction

It is important to estimate the exchange of heat, moisture, and momentum between the underlying surface and atmosphere. Therefore, it is necessary to characterize the underlying surface, whether it is land or the ocean (or water body), because only then can the exchange be estimated. In any numerical model, a grid box is considered over the land or water if more than 50% of the grid box area lies over the land or ocean/water, respectively. If the grid box is over the ocean, then the surface temperature of that grid box is specified using sea surface temperature, and the lower boundary is characterized by the water. On the other hand, if the grid box lies over the land, then all available observations such as elevation (topography), roughness, soil type, soil moisture (surface wetness), vegetation type, vegetation cover, snow cover and depth, surface temperature, etc. are used to represent land–surface characteristics. In this parameterization scheme, mainly friction (ground roughness), sensible heat flux (surface temperature), evaporation (surface snow, surface humidity, sea surface), etc., are represented in the model. Bulk aerodynamic method or similarity approach is generally employed for parameterizing the surface fluxes, surface albedo, surface roughness, etc., and is represented in the numerical models.

4.5.4.2 Parameterization of Boundary Layer Process

The atmosphere is divided into several layers based on its physical properties. The lowest part of the atmosphere where the frictional effects are accountable is considered as the atmospheric boundary layer. Turbulent mixing is the most important factor in the atmospheric boundary layer. In order to include the effects of the boundary layer, the diffusion equations with the coefficients derived from the mixing length theory or through turbulent kinetic energy closure approximation are used to parameterize, and it plays a key role in the transport of fluxes from the surface to the free atmosphere.

In an ideal boundary layer, the energy fluxes are conserved; thus, the energy balance can be expressed as

$$F_{\text{rad}}^{\text{surf}} - F_{SH}^{\uparrow} - F_{LH}^{\uparrow} - F_{G}^{\uparrow} - F_{M} = 0 \tag{4.11}$$

where, $F_{\text{rad}}^{\text{surf}} \rightarrow$ net flux of radiation at the surface, $F_{SH}^{\uparrow} \rightarrow$ outgoing sensible heat flux; results from the difference in the temperature of the surface and overlying air, $F_{LH}^{\uparrow}$ outgoing latent heat flux; results from the evaporation at the surface and lateral condensation with the atmosphere, $F_{G}^{\uparrow}$ the heat flux going into subsurface layers, F_{M} is the energy budget needed to be studied and modeled on the glacier bodies, i.e., between the glacier and underlying earth and glacier and overlying atmosphere.

4.5.4.3 Radiation Parameterization

The main source of heat energy of the atmosphere is the incoming short-wave solar radiation that is incident on the earth's surface, and some parts of it are absorbed; some parts of it are scattered and remain reflected back to the free space. The long wave radiation that is due to emissions from the atmospheric gases, clouds, and the earth's surface is the major cause to heat up the atmosphere. In this parameterization, the net radiative fluxes owing to the short-wave radiation and long-wave radiation, as indicated above, are computed. The amount of energy that is received at the earth's surface due to incoming solar radiation depends on the zenith angle of that location, albedo (reflectivity) of that surface, and cloud cover. On the other hand, the long wave radiation depends on the temperature of that particle/matter and its emissivity; and the composition of the atmosphere and cloud cover also play key roles in the net long-wave radiation. Since the nature of each constituent differs from one to another, thus it is necessary to separately estimate the radiative transfer for each constituent present in the entire column of the model atmosphere, and finally, net radiative transfer with a radiative balance is computed through a parameterization scheme.

4.5.4.4 Convection/Cumulus Parameterization

The representation of the convection process in the model is one of the key factors for a skillful prediction, particularly where non-hydrostatic approaches play dominating roles. In the cumulus convection process, very rapid exchanges of mass, energy, and momentum take place between the lower and upper troposphere. The deep convection is one of the most important physical processes for the weather extremes in the tropics, and this type of convection also significantly modulates the tropical circulation. A number of cumulus parameterization schemes have been developed; however, because of the reliability and consistency in the prediction skill of the model, four cumulus schemes, namely Kuo (1974), Anthes (1977), Grell (1993), Emanuel (1991) Emanuel and Zivkovic-Rothman (1999), and Tiedtke (1989) are widely used in the modelling framework. Above all four, the Kuo-Anthes type convection scheme is the simplest one in which it is assumed that the convection can only take place at grid points where the moisture convergence exceeds a defined threshold value and vertical sounding is convectively unstable. In the moisture convergence, some parts of it moisten the column, and the remaining parts are converted into rainfall through the condensation process and release latent heat. In this scheme, the maximum amount of latent heat occurs in the upper cloud layers. The Kuo scheme removes the available buoyant energy at each time step during the model integration. Two closure assumptions, namely Arakawa and Schubert closure (A–S, Grell et al. 1994) and Fritsch and Chappell closure (F–C, Fritsch and Chappell 1980), are also used, where A–S is based on the relation between convective fluxes and rainfall with the tendencies of the atmospheric state, while F–C looks the relation between the convective fluxes to the degree of instability. In the Grell scheme, only updraft and downdraft are considered, and no direct mixing is allowed between the cloud column and surroundings, except at the cloud top and bottom. A constant mass flux with height is assumed, and updraft and downdraft take place based on the maximum and minimum moist-static energy. In the Emanuel scheme, multiple sub-cloud layers are assumed instead of one single cloud layer, and the mixing in clouds is highly episodic and inhomogeneous. Here, the convection is triggered when the neutral buoyancy is greater than the cloud base level. The condensed moisture forms either clouds or precipitation. The Tiedtke convection scheme is based on mass flux and moisture convergence, and the triggering mechanism takes place when the parcel temperature is greater than the ambient temperature and a threshold temperature. This scheme has both shallow and deep convection schemes. Further details on each of the convection schemes can be found in the corresponding references, and a comparison in the performance among those schemes is reported in several literature (Sinha et al. 2013a, 2019).

4.5.4.5 Large-Scale Precipitation (Stratiform) Parameterization

The air at any grid box may become supersaturated due to several reasons, such as large-scale ascent, horizontal advection, or radiative cooling. In most of the modelling frameworks, the precipitable water or ice is computed separately and kept as a different variable; thus, the representation of clouds becomes more realistic. When the precipitable water becomes heavier and cannot suspend in the air, it starts to fall; as it is coming down to the earth's surface, it has to pass through the warmer temperature, and evaporation of water or melting of snow takes place, and transfer of latent heat can be accounted for. Thus, lowering the freezing level in association with heavy precipitation can be predicted. In large-scale precipitation, stratified clouds like cirrus or starters mainly play the role, and the precipitation takes place when the mean state relative humidity exceeds some predefined threshold value, and this is parameterized in the model.

4.5.4.6 Gravity Wave Drag

The surface is not smooth everywhere, and when air flows over a rough surface or a complex mountainous region, gravity waves generate that propagate upward. This is one crucial point in the modelling framework to take care of when the exchanges of momentum take place between the earth's surface and atmosphere. Thus, the representation of orography and stability of the lower atmospheric layers in the model is important.

4.6 Solution of the Model Numerical Equations

The fundamental equations are involved with the partial differential equations and are non-linear in nature; thus, analytic solutions of the primitive equations cannot be possible, instead, numerical methods are applied to obtain the solutions.

Keeping in view the 3-D space and time evolution of the fundamental variables present in the primitive equations, all the complex equations need to be solved at each grid point. In the modelling framework, three aspects, namely horizontal discretization, vertical discretization, and time integration, are involved in solving/estimating the atmospheric parameters in spatiotemporal scales. A brief on these aspects are described below.

4.6.1 *Horizontal Discretization*

In the numerical model, finite difference and spectral methods are commonly used for the horizontal discretization.

4.6.1.1 Finite Difference Method

The atmospheric processes in reality and representative mathematical formulations are continuous; however, as stated earlier, these mathematical equations are mostly partial differential equations and non-linear in nature. In order to solve these partial differential equations numerically, it is necessary to rewrite these equations in a discrete form with a close approximation. In the finite difference methods, derivatives that are present in the equations are replaced by appropriate finite difference approximates. In principle, a parameter over a point (grid point) is estimated by the difference between that parameter/variables of the adjacent points. Thus, the region of interest where the numerical solutions are to be carried out is divided into an equally spaced grid/points horizontally and extended vertically to create a 3-dimensional space that covers from the surface to the top of the atmosphere over the domain of interest. In this method, the model domain may be global or regional, and for the prior case, the grid points are considered from pole to pole (global climate models or general circulation models, GCMs), while for the later, the grid points of the model spaced over the domain of interest. The distance between the two grid points is known as 'grid spacing' or resolutions of the model, where high resolution indicates the smaller spacing between two grid points and vice-versa. The regular grid spacing becomes very high resolution near the poles, which creates instability in the model integrations mainly because of the CFL criteria and a tendency to converge of the variables to the pole. Earlier, filtering methods were used to avoid instability; however, this process ultimately reduces the effective resolutions and model skills. In the present time, advanced and efficient methods are developed and currently used in numerical methods.

In the finite difference methods, normally, three types of approaches are used, namely forward difference, backward difference, and central difference method.

In Fig. 4.6, grid points are shown with a spacing Δx in one dimension. Consider a variable A(x,y,z,t), which is a function of both space and time, and it is needed to compute the derivative of variable A at x from its adjacent points.

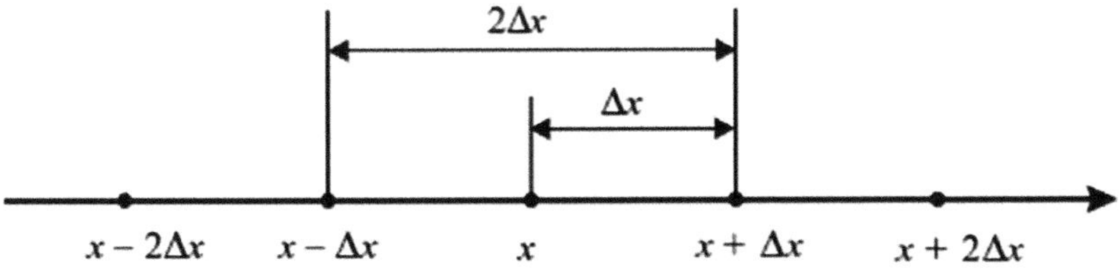

Fig. 4.6 A representative view to calculate the variables on the point x using the information from its adjacent points. Here, Δx is the grid spacing

In the forward difference method,

$$\frac{\partial A}{\partial x} \approx \frac{A(x+\Delta x) - A(x)}{\Delta x} \tag{4.12}$$

Now, the variable A at $x+\Delta x$, can be expanded using Taylor's theorem

$$A(x+\Delta x) = A(x) + \frac{\Delta x}{1!}\frac{\partial A}{\partial x} + \frac{\Delta x^2}{2!}\frac{\partial^2 A}{\partial x^2} + HO\left(\Delta x^3\right) \tag{4.13}$$

Rearranging the Eq. (4.13), one can get

$$\frac{A(x+\Delta x) - A(x)}{\Delta x} = \frac{\partial A}{\partial x} + \frac{\Delta x}{2!}\frac{\partial^2 A}{\partial x^2} + HO\left(\Delta x^2\right) \tag{4.14}$$

The above procedure is known as the forward difference method. In a similar way, instead of $A(x+\Delta x)$, if one considers $A(x - \Delta x)$ and $A(x)$, and computes the derivative of A, one will get

$$\frac{A(x) - A(x-\Delta x)}{\Delta x} = \frac{\partial A}{\partial x} - \frac{\Delta x}{2!}\frac{\partial^2 A}{\partial x^2} + HO\left(\Delta x^2\right) \tag{4.15}$$

This is known as the backward difference method. Both the forward and backward difference methods are the 1st order accuracy approximation in estimating the derivative of A.

Adding Eq. (4.14) and Eq. (4.15), one can get

$$\frac{A(x+\Delta x) - A(\Delta x - x)}{2\Delta x} = \frac{\partial A}{\partial x} + \frac{\Delta x^2}{3!}\frac{\partial^3 A}{\partial x^3} + HO\left(\Delta x^3\right) \tag{4.16}$$

The Eq. 4.16 is known as the central difference method which is 2nd order accuracy approximation in estimating the derivative of A. The term HO (Δx^n, $n>1$) is higher order terms that are ignored or truncated from the equations for simplicity. The error due to truncation is computed by the difference between the actual value and obtained value using the above equations. The truncation errors are lesser for the central difference than the other two methods. It may be noted that the higher order terms contribute less, and an approximated value that is close to the actual can also be achieved after the truncation of the higher order terms.

4.6.1.2 Galerkin Method

This is a different method than the finite difference method, and it is used to estimate the horizontal fields over space for the prediction of the weather and climate. In the Galerkin method, instead of using the variables directly to solve the partial differential equations at each point but use a series of expansions of chosen functions forming a system of ordinary differential equations, which lead to the solution of original equations with an approximation. The Galerkin method is broadly two types:

(i) **Finite element method**: In this method, the chosen function is zero except over a small domain, and the value of the field is calculated using low-order polynomials.
(ii) **Spectral method**: Here, the functions are chosen based on the Eigen functions of an operator that is defined over the whole domain. In this method, spherical harmonic functions are used to estimate the spectral solutions. These spherical harmonic functions are defined by the two integers, namely (a) wave number in the east–west direction and (b) meridional index in the north–south direction. The number of different spherical harmonic functions, each of which represents the wave longer than some specified value, represent the field variables. In the spectral method, the truncation of spherical harmonics is either triangular or rhomboidal. The numerical model based on the spectral model is commonly denoted by T382 (equivalent grid points in lat $\times$ lon $-$ 576×1152; approximate grid spacing at the Equator is 38 km or 0.34°) or R382, where the number 382 indicates that the model can resolve all the waves up to wave number 382 with triangular (T) or rhomboidal (R) truncation, respectively. The representation of the variables globally is well determined by the spectral method; however, small-scale or sub-grid scale physical processes that are very crucial in representing the local weather cannot be appropriately represented using this method because it is needed to transform each stage from spectral to grid value. The spectral model is widely used in the GCMs. The spectral method is also used in limited area models; however, noise that propagates from the boundary deteriorates the model skill, and hence forecasting for the limited time period is useful. Alternatively, boundary conditions that are prescribed from the global model enhance the skill of the limited area model for a longer time forecast, which is being used for most operational centres.

4.6.2 Vertical Discretization

The atmospheric models are three-dimensional, and along with the horizontal spacing (regionally or globally), it extends vertically from the surface to the top of the atmosphere. The vertical components of the fundamental equations are needed to express by using the vertical coordinates. Initially, the vertical coordinates are considered as simply Cartesian coordinates, e.g., z-coordinates; however, due to complexity in the solution and hydrostatic nature of large-scale motions, the Cartesian z-coordinates can be transformed into pressure coordinate system (Eliassen 1949). The advantages of the pressure coordinate are that it simplified the continuity equation; however, complexity arises over the mountain region. In order to minimize the errors in the equations due to mountain regions, Phillips (1957) introduced a transformed coordinate system, known as normalized pressure coordinate or 'sigma' (σ) coordinate, which is the ratio between the atmospheric pressure and surface pressure, mathematically,

$$\sigma = p/p_s \tag{4.17}$$

The modified normalized pressure coordinate or coordinate (Phillips 1957) is one of the better versions to clearly prescribe the atmospheric top and reducing the errors in the forecast over the mountain regions.

$$\sigma = \frac{p - p_t}{p_s - p_t} \tag{4.18}$$

where p, p_t and p_s are the atmospheric pressure at any height, atmospheric pressure at the user-defined atmospheric top, and surface pressure, respectively. In this 'sigma' coordinate system, $\sigma = 0$ at the top (since $p = pt$) of the atmosphere and $\sigma = 1$ (since $p = p_s$) at the surface. Thus σ varies from 1 to zero.

The σ coordinates are widely used in numerical models because it has the potential to smooth mountains which in turn reduces numerical errors; users may define finer vertical resolutions that help to represent boundary layers, tropopause, etc., at finer scale and improve the model skill. However, a few drawbacks are present in the σ coordinates. Some other vertical coordinates, like potential temperature, are also used in the numerical models.

4.6.3 Time Integration Schemes

The time derivative terms in the governing partial differential equations (PDE) have been calculated by time difference schemes, which are generally of the second or even first order of accuracy. Lower-order accurate time integration schemes are generally used in the PDEs as higher-order accurate schemes do not necessarily improve the results. To reduce error on the numerical simulation, accurate numerical schemes are necessary. Other than the time integration scheme, the initial condition of the model is important. The time step in the Eulerian schemes depends on the Courant–Friedrichs–Lewy (CFL) condition such that the time step is much smaller than that of the spatial scale. More accurate space differencing is one of the important things. Semi-Lagrangian schemes generally use much larger time steps than allowed by the CFL conditions.

The Leapfrog scheme is used mostly, which is an explicit three-time level scheme. Explicit schemes mean that the scheme predicts the value of the prognostic variable at each grid point for time level "$n+1$" after knowing the value of the grid point at time level n and before that. The value of the prognostic variable at a grid point at time level "n" depends on the valves at time level n, and $n+1$ is involved. Implicit schemes, on the other hand, are unconditionally stable, such that much larger time steps can be used in the numerical solution. It is better to predict the prognostic variables through simultaneous algebraic equations in the implicit scheme. Solving a simultaneous algebraic equation depends on the incursion of a matrix with a dimension equal to the number of grid points. For the model with 3-dimension and with thousands of grid points, the implicit scheme is not economical computationally and is very time-consuming.

Another one, in the semi-implicit schemes, the gravity wave terms being separated and treated implicitly. Because the inertia-gravity waves need a smaller time step. In another approach, such as the split-explicit strategy, different terms are being treated with different time steps. This scheme has very good stability properties and has been widely used in numerical prediction models. Whereas, in the split-explicit scheme, the time truncation error is higher compared to other schemes. But this scheme will not affect the results of the numerical simulation.

4.7 Post-processing

This is the final stage of the NWP modelling system. The model outputs such as rainfall/snowfall movement, the future intensity of the pressure system, and consequent weather events at desired grid points or over desired area/vertical pressure level are obtained using various interpolation schemes, including statistical techniques.

Outputs of model simulations are used to evaluate the performance of the model and also validation of the different components of the model, and to estimate the systematic errors in the NWP system. Thus, this stage comprises three steps, namely:

(a) Prediction of derived conventional and non-conventional fields at desired location/ vertical level
(b) Preparation of forecast error statistics, and
(c) Evolution of the systematic error of the model

While interpreting the model forecast, the limitations of the same must be kept in mind. Some of these have been given in the next section.

4.8 Types of Models

In this section, a brief description of different types of atmospheric models in common use for simulation/prediction of atmospheric processes is provided.

4.8.1 Simple 1D/2D Models

Simple one-dimensional (1D) or two-dimensional models (2D) are used to study and forecast specific phenomena. For example, time series of observations of a phenomenon along with the associated meteorological factors may be collected at a place. The model incorporating physical processes is developed to fit the observations.

4.8.1.1 1D Model

In one-dimensional models, an important assumption is that there is no horizontal variation of atmospheric variables such as temperature, moisture, and wind. Due to this assumption, advection processes are not being considered in the one-dimensional models. The advection in the one-dimensional models will be calculated from large-scale fields present in the model. Various parameterization schemes (column physics) can be developed, tested, and validated against special location-specific experiments over both land and ocean, which is one of the advantages of such a model. Field experiments such as MONTBLEX-90, LASPEX-97, INDOEX-99, BOBMEX-99, etc., are some of examples.

4.8.1.2 2D Model

2-Dimensional models have some additional advantages as compared to one-dimensional models, such that the two-dimensional model includes atmospheric processes either in the zonal or meridional direction. In these models, an advection of meteorological parameters (winds, temperature, and moisture) along the specified axis can be considered. Sea breeze or valley wind circulation, air-pollution dispersion, etc., can be studied by these models.

4.8.2 Numerical Weather Prediction Models

Numerical weather prediction models (NWP) are useful for the simulation of the state of the atmosphere in the short (1–3 days) and medium (4–10 days) range future. A good NWP model accurately predicts the movement and evolution of disturbances such as the frontal system and tropical cyclones. Parameterizations are empirical approximations based on large-scale (resolved) variables. Parameterization may have a theoretical justification but always needs to be tested experimentally. Different types of parameterizations such as cloud microphysics, i.e., the aggregate effect of the clouds on the resolution scale, cumulus convection, atmospheric radiation, sub-grid scale motion, chemistry, and cloud physics, are calculated. Clearly, some Parameterizations are present in GCMs, such as very slow land surface changes, or slow chemical processes, which are very specific for GCMS.

4.8.3 Climate Models

A global climate model, also known as a general circulation model, both aggravated as GCM, uses the same equations of motion as a numerical weather prediction (NWP) model. But the temporal scale for GCM is about 30 years as a result of slow changes in some boundary conditions (such as the solar constant) or physical parameters (such as the greenhouse gas concentration). Compared to NWP models, GCMs are run for a much longer time period, which needs high computational facilities and time. An error in the sea surface temperature by a few ^{0}c, or a small but systematic bias in cloudiness throughout the model, matter little to an NWP model. Whereas, for a GCM, these factors are important, because they matter over the long-term.

GCMs are coupled atmosphere–ocean models, which means that a model simulating surface and deep ocean circulation is coupled to an atmospheric GCM. The interface between the atmosphere and the ocean is the sea surface, where the transfer of water (evaporation/precipitation) and momentum occurs between the atmosphere and ocean. An accurate coupling of the atmosphere to the ocean (with long memory) is essential to simulate the El Nino-Southern Oscillation (ENSO), for instance.

A key problem in a GCM modelling system is long-term stability and sensitivity to small changes in surface conditions or radiation input. The atmosphere is neither invariant nor transitive. An almost transitive system can flip between alternative patterns. The flipping to and from ice age condition is the best example of a transitive system.

Different type of atmosphere models used for weather prediction depends upon different parameters such as resolution, boundary conditions, dynamics, physics, and input-boundary conditions required. A brief description of meso-scale, regional scale, and global models has been presented in the following sections.

4.8.3.1 Meso-Scale Models

For the simulation of sub-synoptic scale weather features (from a few km to a few hundred km) meso-scale models are used. Some of the current models that are in vogue are NCAR/Penn state mesoscale Model (MM5) or high-resolution ETA model etc. Meso-scale models may be hydrostatic/non-hydrostatic, with different horizontal and vertical grids and nesting options in the horizontal scale. Four-dimensional data assimilation may be used in the meso-scale models. Appropriate selections of time integration schemes, model physics, and initialization are important to simulate the observed small-scale features.

A huge number of scientists worldwide are using meso-scale models such as MM5, sophisticated advanced high-resolution models, i.e., WRF, HWRF, to predict the weather phenomenon from 24 to 72 h in advance with satisfactory skills and usable to the preparedness for the weather hazards like tropical cyclones, western disturbances, thunderstorms, heavy to extreme rainfall, etc.

4.8.3.2 Regional Scale Models

A regional model is also called Limited Area Model (LAM), initially used for short-range forecasts (1–3 days). However, better understanding the slowly varying processes and its incorpration in the modelling framework has made regional scale models (RCMs) a potential tool to forecast the average atmospheric state at monthly to seasonal scale with satisfactory skill. There are several regional scale models such as NCAR/Penn state limited area model, NMC the nested grid model, the French weather service limited area model, ECMWF limited area model, and FSU limited area model. The incorporation process of orography, boundary conditions, physics, and initialization in the model is different for each model. Bias present in the model is one of the limitations of the models. Model customization for a particular region is finely tuned by conducting a large number of simulations/experiments.

4.8.3.3 General Circulation Models (GCM)

A general circulation model (also known as Global Climate Model) uses the same equations of motions as any Regional Scale Model. A GCM for NWP is used to predict the weather in short (1–3 days) and medium (4–10 days) in the future. GCM climate models are run for much longer duration, such as a season to years to a decade or more. For NWP, the spatial resolution may vary between 50–300 km, and initial conditions must be very accurate.

4.9 Hierarchy of Climate Model

Classification of various types of mathematical models depends on the number of space dimensions included and the degree of model complexity emphasizing the physical foundations. For general reviews, see Schneider and Dickinson (1974), Smagorinsky (1974), Saltzman (1978), Ramanathan and Coakly (1978), North et al. (1981) and Schlesinger (2012).

Systematization of climate models can be obtained by taking into account the modes of averaging in space. Let us consider that there are three-dimensional climate fields denoted by 3-D (λ, ϕ, z). After averaging at a vertical scale, this model is now transferred to two-dimensional 2D (λ, ϕ) models or averaging in the zonal direction leading to two 2-D (ϕ, z) dimension models. If these two-dimensional models are now averaged in one of the other dimensions, one-dimensional [$1 - D$ (λ) $1 - D$ (ϕ) $1 - D(z)$] models can be obtained. The integration of one-dimensional models with respect to the remaining dimension will lead to Zero dimensional (0-D) models. Temperature is one of the best examples of a 0-D model. This temperature is the long-term average of the global mean temperature that results from the balance between absorbed solar and emitted terrestrial radiations.

Depending upon the weather/climate problem, different climate models, from zero dimensions to the most complex coupled ocean, land, and atmosphere general circulation models (GCM) can be used.

(a) **0-D Models**
In this simplest of climate models, the predicted variable is usually global mean temperature. The parameters usually included are changes in solar constant, snow albedo feedback, cloud feedback, greenhouse forcing, etc.

(b) **1-D Models**
   (i) Energy balance models: surface temperature (Ts) as a function of latitude, simple heat transport prescription is included.
   (ii) Radiative convective models: global mean temperature as a function of altitude usually has a very complex radiation scheme.

(c) **2-D Models**
   (i) Energy balance model with latitude *x* longitude variation.
   (ii) Models with latitude *x* height representations of zonal mean temperature and u-component of wind. This type of model is popular for studies of chemical processes.

(d) **3-D Models**
There are atmospheric general circulation models with prescribed sea surface temperature or with simple ocean models.

(e) **4-D Models**
The most complex of the climate models includes fully coupled atmosphere GCM, ocean GCM, and interactive land and sea-ice models.

The relative importance of various processes in different types of models can be discussed using the climate modelling pyramid (Fig. 4.7). The edges of the pyramid represent three of the basic elements of the models, and the complexity of the model is shown to increase upwards. The base of the pyramid represents simple climate models, such as one primary process.

4.10 Fundamentals in the Prediction System

Weather forecasts can be done by both subjective and objective methods. In the subjective methods, the formulation of forecasts is based on the judgments of individual forecasters. The objective methods generally involve the use of numerical

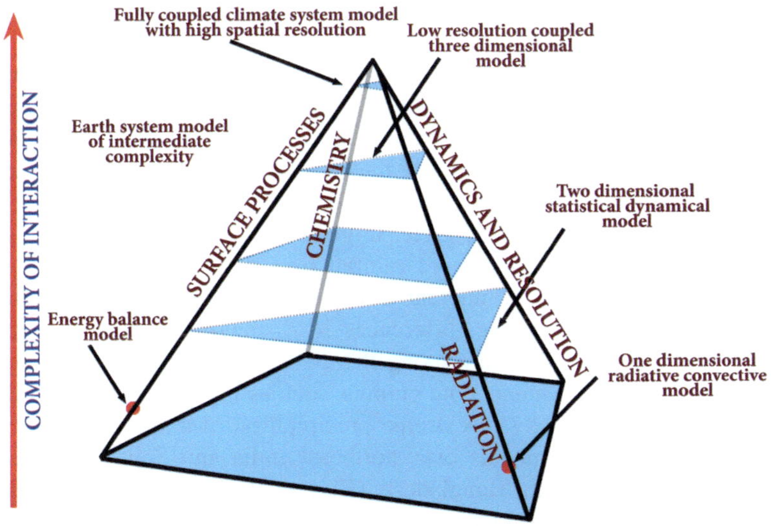

Fig. 4.7 Climate model pyramid. (*Source* McGuffie and Henderson-Sellers 2014)

and/or statistical models (Murphy and Brown 1985). Subjective forecasts of precipitation are more skillful compared to the objective forecasts for shorter lead times (i.e., 12–24 h), whereas both methods provide approximately equal skill for longer lead times (i.e., 36–48 h).

(a) **Subjective Methods**

Subjective approaches of weather and climate prediction are generally of two types such as synoptic method and the climatological method. In the synoptic method, the prediction of weather events can be done through the use of synoptic weather charts. It mainly describes the summary of the current weather situation; it means the pressure pattern, fronts, wind speed, and direction and how they change over the coming few days. These methods are generally based on empirical relationships of various meteorological parameters to heavy precipitation. It includes the relationship of heavy rainfall to low-level inflow, moisture availability, jet-stream structure, the thickness of boundary layer height, low-level equivalent potential temperature, etc. Pattern recognition is extremely important and is the foundation for forecasting heavy precipitation. Some thermodynamic variables, such as convective available potential energy and precipitable water values, are poor predictors for the prediction of rainfall intensity.

The climatological method is another simple way of producing a forecast, which involves averaging weather statistics accumulated over many years to make the forecast. Based on the previous 30 years of pattern, it mainly predicts the next day's weather. This method only works well when the weather pattern is similar to that expected for the chosen time of year. If the pattern is unusual for the given time of year, then this method will often fail to predict the weather.

(b) **Objective Methods**

Objective approaches for weather and climate prediction can be done through the dynamic method (Numerical Weather Prediction), statistical method, dynamical-statistical method, and ensemble and super-ensemble method. Several steps have been made to accurately predict the weather conditions over a region by using high-resolution Meso-scale models such as Meso-scale Model (MM5), Weather Research Forecast (WRF), etc. In general, the challenges in rainfall prediction increase with the increase of rainfall intensity (Das et al. 2008). Meso-scale models have their own inherent error for predicting heavy rainfall events (Kumar et al. 2008). There are differences in displaying extreme precipitation statistics over the Asian domain among the 20 Coupled Model Intercomparison Project Phase 5 (CMIP5) model data outputs, such as high-resolution models simulate the maximum intensity of extreme precipitation over the Indian sub-continent, medium-resolution models over northeast India and South China and the low-resolution models over Bangladesh (Kim et al. 2019). Rainfall prediction over the tropics, even at a shorter scale, is a challenging problem for the modelling community (Zhu et al. 2014).

4.11 Limitations of Models

The NWP modelling system has a number of basic limitations which, interpreting the results, should be kept in mind. Some of these are;

(i) While formulating the equations, a number of assumptions have been introduced. Thus the equations do not represent the exact nature of atmospheric flows/processes. Because of these simplifying assumptions, the set of equations represents a "model' atmosphere only.
(ii) Because of the non-linear nature of equations, the analytical solutions in general, are not known within the existing state of mathematical knowledge. Hence solutions are obtained by introducing further approximations in using perturbation techniques or using numerical methods.
(iii) Integration of these equations involves knowledge of the initial state of the atmosphere in the form of meteorological data dependent on the subjectivity of the observer, instrumental errors, etc.
(iv) Since the atmosphere parameters are highly variable in time and space, some averaging has to be done.
(v) Further, these data are observed at uneven points where observatories are located, but the numerical solution requires the value at every grid for which objective analysis is done by varying methods having in-built approximations.
(vi) During the solution stage also, approximations forcibly creep in depending on the type of computational scheme used.

(vii) In nature energy is distributed in a wide spectrum of wavelengths, including sound waves and gravity waves, which are of high frequency in comparison to the meteorologically useful waves.
(viii) These frequency waves grow much faster during integration with the passage of time, thereby showing the desired results and leading to unrealistic solutions. Therefore, before the data are put in use for integration, due care has to be taken to remove these vitiating factors by balancing the mass and velocity fields. The process is called initialization in meteorological parlance.
(ix) Because of computational constraints, the grids used for solving the process have steps long enough to take care of the description of many atmospheric phenomena since at least a wave must fit in the grid to describe it. There are a number of physical phenomena whose descriptions cannot explicitly be accommodated in the model equations. To include their effect, some statistical formulations are used to describe these sub-grid processes in resolvable scales. This process is called physical parameterization, and it introduces a lot of approximations.
(x) After the integration, computed values of meteorological variables are available at grid points which are to be interpreted at desired locations and also in terms of weather parameters by utilizing statistical techniques. This adds up to the approximations.

Based on the fundamentals of the prediction systems, several reputed organizations/institutes have developed forecast systems that are mainly physical-dynamical based and used to provide various scales of forecasts. Since these types of dynamical-physical based models need very high-performance computation systems, which are very costly, therefore, simulating of weather/climate events is performed in real-time by a few of the institutes/organizations. The beauty is that most of the organizations make available the hindcast (past) and real-time forecast products, and any user from any corner of the world can access those products through the internet. The details of those products with a status of monthly and seasonal forecasts along with the verification methods are provided in the next chapter.

Chapter 5
Status of Monthly and Seasonal Forecast

5.1 Introduction

The term 'weather' by definition means an instantaneous state of the atmospheric parameters, and to some extent, conditions of the atmospheric state averaged over a few hours. Thus, the weather forecast is helpful to daily life and urgent preparedness for sustainability against the short-time scale extreme weather events like flash floods, thunderstorms, tropical cyclones, heavy rainfall, etc. On the other hand, 'climate' usually connotes the mean behaviour of the atmospheric state averaged over typically from two weeks to a longer period such as a month, season, year, etc. The seasonal climate forecast which is beyond several weeks is one of the most challenging because of its dependency mainly on the slowly varying atmospheric processes (soil moisture, sea-ice, sea surface temperature, etc.) and complex interactions with the small-scale physical processes with the large-scale atmospheric flows. Seasonal climate forecast is essential since it is designed to investigate the slowly evolving processes of the atmospheric state and predict the deviation/departure from the mean behaviour of the atmosphere for the coming month, season, or so on. Thus, the seasonal forecast is inevitable for policy planning, management, and strategies to mitigate the short-term (monthly and seasonal) climate hazards that are vulnerable to different sectors (agriculture, husbandry, farming, fisheries, water management, hydro-power, etc.) of a sustainable society. In addition, the integration of monthly and seasonal climate forecasts with the weather forecast is structured as a seamless forecast from a few hours to a season that is significantly useful for the benefit of all mankind.

The need for weather and climate forecast for enhancing agricultural production and minimising loss has been realised more than thousands of years back. In the early date, traditional methods were based on divination and later by past experience on perceived patterns over a specified region, like a wet season with

U. C. Mohanty et al., *Climate Risk Management in Agriculture*,
https://doi.org/10.1007/978-3-031-51862-1_5

frequent rainfall occurring after a hot season over the Indian subcontinent. This kind of almanac calendar has been used for a long time. With the progress of civilisation and the enrichment of the knowledge about the weather and climate systems, notable improvement has taken place in the seasonal prediction method during the past one and half-century. In 1886, Sir Blandford for the first time provided an Indian summer monsoon rainfall forecast using a subjective method, and a few decades later, in the early nineteenth century, Sir Gilbert Walker used statistical relationship (linear regression technique) for the prediction of Indian summer monsoon rainfall. Initially, the development of the statistical forecasting methods is univariate (were predicting a particular atmospheric parameter say rainfall will be based on the past history of the same parameter) and multivariate (were predicting an atmospheric parameter will be a function of other parameters, for example, predicting rainfall using pressure and temperature). For long-term predictions, these methods are widely used; however, the skill of these methods is lower when compared to the present-day advanced dynamical models (numerical weather prediction models) skill at shorter times prediction.

With the advancement of technologies, and a better understanding of climate and weather systems, the past few decades have witnessed immense efforts to improve prediction skill worldwide, and as a result, a huge improvement in the seasonal climate forecast system is perceived. A few decades ago, seasonal climate forecasting was mainly purely academic research, while present day, several reputed national/international organizations/institutes are providing monthly and seasonal forecasts regularly throughout the year, which is tremendously in use by different stakeholders. These organizations are using statistical, dynamical, or hybrid (dynamical-statistical) methods to generate monthly and seasonal forecasts, and the methods of forecast are already presented in the previous chapter. A brief on the predictions of seasonal climate by several organizations/institutions and available forecasted parameters will be described below to facilitate the readers for making those products usable in agriculture and other allied sectors.

5.2 International Status on Monthly and Seasonal Forecast

The forecasts of atmospheric parameters at monthly and seasonal scales are in increasing demand because of their ability to predict with a satisfactory skill that is used for management and policy planning. In the context of weather and climate forecasts, the surface air temperature and precipitation are the two most important variables among all for their direct and indirect impact, which greatly influence livelihood. Many internationally reputed and recognized institutions/organisations are intimating seasonal climate forecasts in advance. These forecasts are mainly generated by using statistical or dynamical or dynamical-statistical approaches. Most of these organisations issue the seasonal forecast of precipitation (all forms of water that precipitate) and surface air temperature both by deterministic and probabilistic methods and disseminate the forecasted results in graphical representation. The user can easily understand the monthly/seasonal departure of precipitation/surface air temperature from the long-term mean for their respective region by observing the graphical charts. However, at the end of this chapter, a brief note will be provided to guide the user on how to transfer the graphical information to make it usable in different sectors. The probabilistic forecast makes the user confident to use the deterministic forecast for a specific region. Since these organizations provide the seasonal forecast at global/continental scale, the forecast is generated at low-resolution. These forecasts are mostly uniform for regional/local scales due to coarse resolution, however, this forecast provides a signal of deviation from the long-term mean, and one can make those forecasts usable; the way to make them useful is described in the next chapter. A glimpse of the available seasonal forecasts from different international organisations/institutions is delineated in the next.

5.2.1 World Meteorological Organizations (WMO) Regional Climate Centres

The World Meteorological Organization (WMO) is an intergovernmental organisation that involves around 193 countries/states/territories. The WMO is the prime organisation in the world to provide high-quality weather and climate-related services (https://public.wmo.int/en/programmes). Many of the leading institutes provide seasonal forecasts by using advanced atmosphere–ocean coupled models for global and specific regions. WMO has designed and is running a program Global Seasonal Climate Update (GSCU) in which global scale seasonal forecasts, including precipitation and surface temperature and also take into account ENSO as well as other climate drivers from thirteen Global Producing Centres that are collected into a grand multi-model ensemble prediction (https://public.wmo.int/en/our-mandate/climate/global-seasonal-climate-update). However, the GSCU of WMO does not provide any official forecasts and seasonal forecasts, however, the forecasts can be obtained from the relevant regional climate centre (RCC) for the interested domain. In the GSCU seasonal forecast/outlooks, one can get precipitation, surface air temperature at 2 m, and the other standard variables at a monthly/seasonal scale (at fixed cycles) with a lead time of 0–4 months. A brief on seasonal products of these thirteen leading centers under the WMO Global Producing Centres for Long-Range Forecasts (GPCLRFs) is given below.

5.2.1.1 China Meteorological Administration (CMA)/Bejing Climate Center (BCC), Beijing

The WMO proposed to establish Regional Climate Centers (RCCs) focussing on providing products of seasonal to Interannual climate prediction for the National Meteorological and Hydrological Services (NMHS) of its members. In this connection, the China Meteorological Administration (CMA) has long recognised the need to build the capacity of climate prediction. By using existing infrastructure and expertise, the CMA established the Beijing Climate Center (BCC) in 2003, in order to make tailored climate forecast products on a monthly and seasonal scale that can contribute to the regional sustainable development in Asia (http://bcc.ncc-cma.net/channel.php?channelId=22). BCC was designated as a WMO RCC in RA II (Asia) at EC-LXI in June 2009, and it has an obligation to deliver climate service for neighbouring and surrounding countries in Asia and the international climate community. Since 2006, BCC has incorporated functions as a WMO East Asian Monsoon Activity Centre (EAMAC), a WMO Global Producing Centre (GPC) for long-range forecasts, and a Centre for Extreme Events Monitoring in Asia (CEEMA, since 2010). BCC organises and coordinates research efforts on regional climate, operational predictions, and climate application and services, among other international organisations. It also provides climate services to China and other Asian countries for disaster prevention and mitigation and for socio-economic development.

The BCC forecast products are ensemble predictions and delivered by the ensemble mean and probabilistic using the terciles concept. Three types of weighted mean, such as equal weight, linear weight, and parabolic weight products, are available, and these outputs are also sent to the WMO Lead Centre for Long-Range Forecast Multi-Model Ensemble (LC-LRFMME) and Asia–Pacific Economic Cooperation (APEC) Climate Centre (APCC) for the operational forecasts every month. The digital seasonal forecasts and hindcast products can be obtained from the BCC website after successful registration in the web site mentioned above. The data are available from 1982 to the current period. The monthly forecasts are available in graphical and represented in a departure from the long-term average (for temperature, the departure is provided in actual magnitude, while for precipitation, the deviation is provided in % of departure from the long-term mean).

5.2.1.2 Center for Weather Forecasts and Climate Studies (CPTEC)/National Institute for Space Research (INPE), Brazil

The Centre for Weather Forecasts and Climate Studies (Centro de Previsão de Tempo e Estudos Climáticos, CPTEC) was planned and designed in the late ‘80s and implemented in the early ‘90s by the National Institute for Space Research (Instituto Nacional de Pesquisas Espaciais, INPE). This centre started producing operational global numerical weather forecasts using a spectral atmospheric general circulation model (CPTEC/COLA model) with a resolution of T62L28 in January 1995. CPTEC is the first and foremost operational centre in Latin America to have both NWP and seasonal numerical forecasting capabilities. A brief detail of the products of this centre is available at https://www.wmo.int/pages/prog/www/DPFS/ProgressReports/1999/Brazil-CPTEC.html. They provide seasonal forecasts (0–2 month's lead) using four different global models for the whole globe and also for the different regions. They have also provided the forecast using persistent method along with the NWP model predicted products. In the persisted products, one can find seasonal anomalies along with the probabilistic terciles (with upper and lower terciles), while in predicted products, the seasonal anomaly is provided. The persisted and predicted products are available in graphical representation (http://previsaonumerica2.cptec.inpe.br/bam).

5.2.1.3 European Centre for Medium-Range Weather Forecasts (ECMWF)

The European Centre for Medium-Range Weather Forecasts (ECMWF) is an intergovernmental organisation established on 1st November 1975 and is based at Shinfield Park, University of Reading, United Kingdom. The first real-time medium-range forecast of ECMWF was made in June 1979, and operational since 1st August 1979, and the ensemble prediction system is also started on 24th November 1992. Since then, ECMWF has continued to develop and improve its predictions skill. Progress over the years has included beginning to monitor the Earth system, installation of a new supercomputer, upgrades to the forecasting model, and contributions to climate studies.

This Centre has one of the largest supercomputer facilities and meteorological data archives in the world and produces global numerical weather predictions and other data for their Co-operating States and the broader community and also assists the WMO in implementing its programmes. ECMWF Integrated Forecasting System (IFS) produces forecasts for multiple time ranges to address different user requirements. These provide key aspects of forecast evolution and the associated uncertainty. Specific products are designed to highlight potential severe weather events. These predictions cover the medium range, up to two weeks ahead, extended range, up to a month ahead, and long range, up to a year ahead.

Each ECMWF forecast is made of one member (HRES) at 9 km horizontal resolution and the other 51 members at 18 km horizontal resolution. These are separate forecasts made by the same computer model, all activated from the same starting time. The starting conditions for each member of the ensemble are slightly different, and the physical parameter values used also differ slightly. The differences between these ensemble members tend to grow as the forecasts progress, which is as the forecast lead time increases, and this ensemble-based weather prediction describes the range of possible scenarios and their likelihood of occurrence, which can help in planning and decision making for policymakers and various risk management sectors. ECMWF routinely processes data from around 90 satellite data products as part of its operational daily data assimilation and monitoring activities. A total of 40 million observations are processed and used daily; the vast majority of these are satellite measurements, but ECMWF also benefits from all available observations from non-satellite sources, including surface-based and aircraft reports which can improve the initial conditions of the model and the increased observational network leads to increase the prediction skill.

The ECMWF provides monthly and seasonal scales forecast products for the whole globe and different regions, and these are freely available in a graphical mode in different types such as anomaly, deterministic, probabilistic, and the other score (https://www.ecmwf.int/en/forecasts/charts/catalogue/?facets=Type,Forecasts). The ECMWF has facilitated to obtain public data of reanalysis/model products at sub-seasonal to seasonal scale from their web portal (https://www.ecmwf.int/en/forecasts/datasets/reanalysis-datasets/era5). The operational data can also be archived after successful registration.

5.2.1.4 Met Office (United Kingdom), Exeter

The Met Office was established in 1854 as a small department within the Board of Trade under Vice Admiral Robert Fitz Roy as a service to mariners. The loss of the passenger's vessel, the Royal Charter, and 459 lives off the coast of Anglesey in a violent storm in October 1859 led to the first gale warning service. The first NWP forecast in the UK was completed by F. H. Bushby and M. K. Hinds in 1952 under the guidance of J. S. Sawyer FRS. These experimental forecasts were generated using a 12×8 grid with a grid spacing of 260 km, a one-hour time-step, and required four hours of computing time for a 24-h forecast on the EDSAC computer at Cambridge and the LEO computer of Lyons Co. In January 1959, a Ferranti Mercury computer, known as 'Meteor', was installed at the Met Office in Dunstable. This was a significant milestone due to it being the first computer dedicated to NWP research. Meteor was used to trial an operational forecasting suite in early 1960, using the improved Bushby–Whitelam three-level model covering Western Europe and the North Atlantic. It consisted of a six-hour forecast from 0000 UTC analysis, followed by a 0600 UTC reanalysis and a 24-h forecast. The forecasts took too long to run, and it was concluded that a more powerful and reliable computer was required. The first operational computer forecast was produced by the Met Office on 2nd November 1965, using the Comet computer. The media were invited to witness this landmark in the history of the Met Office and gave it unprecedented coverage both in the press and on TV.

Now, Met Office makes meteorological predictions across all timescales from weather forecasts to climate change using a unified model. The principal weather products for UK customers are 36-h forecasts from the operational 1.5 km resolution UKV model covering the UK and surroundings, 48-h forecasts from the 12 km resolution NAE model covering Europe and the North Atlantic, and 144-h forecasts from the 25 km resolution global model. The Met Office's Global Model forecast has consistently been in the top 3 for global weather forecast performance in independent verification to WMO standards.

The Met Office also makes seasonal and long-range forecasts using global models and distributes them to customers and users globally. The Met Office was the first climate and weather forecast provider to be recognized as a Global Producing Centre of long-range forecasts by the WMO and continues to provide forecasts to the WMO for dissemination to other national meteorological services worldwide. They disseminate probabilistic forecasts in graphical form for the whole globe/other regions and 3-months seasonal forecasts at 1–3 months lead time. The user can find on their web the maps for tercile categories, outer quintiles categories, and two categories and a brief description on how to use those three categories forecast (https://www.metoffice.gov.uk/research/climate/seasonal-to-decadal/gpc-outlooks/glob-seas-prob). If the readers/users have further interest, they can find the detailed documentation of the seasonal forecast methods and system, forecast diagram, and guide to skill assessment to evaluate the forecast for a specific region.

5.2.1.5 Bureau of Meteorology (BOM) (Australia), Melbourne

The Bureau of Meteorology (BOM) is an Executive Agency of the Australian Government responsible for providing weather services to Australia and surrounding areas. It was established in 1906 at Melbourne Docklands under the Meteorology Act, and it has grown into a world-class agency providing much-needed environmental intelligence to the community and industry. This is the main provider of weather forecasts, warnings, and observations to the Australian public. The BOM, Australia, provides different time scale forecasts (weekly to seasonal and longer-range climate outlooks) for the Australia region using the Australian Community Climate Earth-System Simulator–Seasonal (ACCESS–S). It is a state-of-the-art dynamical (physics-based) forecast modelling system which uses ocean, atmosphere, ice, and land observations to initiate outlooks for the season ahead. The ACCESS–S climate model is a collaboration between the Bureau of Meteorology and the UK Meteorological Office (UKMO). The atmosphere and land model components of ACCESS–S operate at an approximate resolution of 60 km in the Australian region. At this resolution, the model can represent the markedly different climates of the Great Dividing Range and the eastern seaboard in Australia's east. The ocean model component of ACCESS–S operates at an approximate resolution of 25 km in the Australian region. At this resolution, the model can resolve small-scale currents and eddies. ACCESS-S outlooks are based on a 99-member ensemble. This is a common climate forecasting technique where the model is run 99 times with slightly different initial conditions to capture a range of likely future scenarios. Being a dynamic model, ACCESS–S is undergoing continuous research and development. Advances in the science of seasonal prediction, improvements in the observations and how they are fed into the model, as well as increases in supercomputing power are just some of the ways the model's accuracy will increase over time. The Bureau uses the output from ACCESS–S for several products, including forecasts of the ENSO and the Indian Ocean Dipole (IOD), as well as weekly to seasonal rainfall and temperature outlooks and sea surface temperature (SST) anomaly forecasts. They disseminate these forecasts in graphical mode with the main features of the seasonal outlook in the documentation. For the user, they also provide a video of the seasonal outlook for a better understanding and time evolution of the seasonal forecast. The documented outlook and video are helpful to get an overview of the atmospheric state of the forthcoming seasons. They provide a one-month lead forecast for the next season of three months. They also provide the chance of above median, chances of the median, chances at least, and outlook scenarios for different sub-seasons of having time lengths 1-week, 2-week, 1-month, and 3-months (http://www.bom.gov.au/climate/outlooks/#/rainfall/median/seasonal/0).

5.2.1.6 Meteorological Service of Canada (MSC), Montreal

The Meteorological Service of Canada (MSC) is a division of Environment and Climate Change Canada, which primarily provides public meteorological information and weather forecasts and warnings of severe weather and other environmental hazards. The MSC has the specific objectives to predict the atmospheric state and associated trend of the meteorological parameters for the next seasons. Since September 1995, MSC has produced seasonal forecasts of surface air temperature and precipitation anomalies (1–3 month outlooks) for Canada based on dynamical atmospheric models. Since December 1, 2011 these forecasts and their extensions to longer lead times have been obtained from the Canadian Seasonal and Inter-annual Prediction System (CanSIPS), which consists of two coupled atmosphere–ocean-land physical climate models, CanCM3 and CanCM4. On July 3rd 2019 CanCM3 was replaced by GEM-NEMO. Since then, the seasonal forecasts have been made from an ensemble of 20 dynamical model runs, 10 from each of two atmosphere–ocean-land coupled models. One of the models, CanCM4, developed at the Canadian Centre for Climate Modelling and Analysis (CCCma), uses the atmospheric model CanAM4 with a horizontal resolution of 315 km (T63) and 35 vertical levels and the ocean model CanOM4 with an approximate horizontal resolution of 100 km and 40 vertical levels. In contrast, the second model GEM-NEMO, developed at the Recherche en Prévision Numérique (RPN), uses the atmospheric model GEM with a horizontal resolution of about 155 km and 79 vertical levels and the ocean model NEMO with a horizontal resolution of about 110 km and 50 vertical levels. The MSC provides both deterministic and probabilistic forecasts. The deterministic forecast is the anomaly from the long-term mean, while the probabilistic forecast is provided in 3 categories (below, near, and above normal). Both the deterministic and probabilistic forecasts are provided in the graphical mode (https://weather.gc.ca/saisons/prob_e.html).

5.2.1.7 Hydro-Meteorological Centre of Russia, Moscow

The North EurAsia Climate Centre (NEACC) is designed to provide weather and climate information and predictions of the atmospheric states for better climate services to meet the required climate information of the Commonwealth of Independent States (CIS-Armenia, Azerbaijan, Belarus, Georgia, Kazakhstan, Kyrgyzstan, Moldova, the Russian Federation, Tajikistan, Turkmenistan, Uzbekistan, and Ukraine). However, the forecasts are not publicly available.

5.2.1.8 Deutscher Wetterdienst (DWD), Offenbach

The German Climate Forecast System (GCFS), which provides routine seasonal forecasts for 3 months with 1–2 months lead time, is operated jointly by the Deutscher Wetterdienst (DWD), the Center for Earth System Research and Sustainability. This seasonal forecast system is based on the Earth system model of the Max Planck Institute for Meteorology (MPI-ESM). The current GCFS (GCFS2.0) uses the so-called nudging (Newtonian relaxation), where observation-based data are compared to the computed model state. This difference between observation-based data and model state is nudged into the model with time-weighting and forms the basis for the next model time step, which will then be compared to the observation-based data from this next time step. As such, this method produces a consistent adaptation of the whole model climate towards the observed climate. In GCFS2.0 the ensemble is generated by different methods in the atmosphere and ocean. In the atmosphere, a parameter in the uppermost atmospheric layer responsible for air diffusion is disturbed to create different solutions. In the ocean, so-called bred vectors lead to the growth of the most unstable mode of the temperature and salinity fields over the whole ocean depth, which is then applied as disturbances to the assimilated climate conditions. For the forecasts, GCFS2.0 starts with 30 ensemble members for each calendar month in the years 1990–2017. Since 2018, real-time forecasts have been performed with 50 ensemble members per forecast. They provide real-time deterministic as well as probabilistic forecasts in tercile for the whole globe and Europe in graphical representation. In the deterministic forecast, the ensemble mean anomaly is depicted. Along with the forecast, different skill metrics are also presented to understand the confidence level of the forecast (https://www.dwd.de/EN/ourservices/seasonals_forecasts/charts.html?nn=641552).

5.2.1.9 South African Weather Services (SAWS), Pretoria

The SAWS provides different types of the forecast, such as weather forecasts, seasonal forecast, long-range forecast, marine forecast, travellers forecast, and regional forecast (https://www.weathersa.co.za/). They use ECHAM4.5 global models (30 ensemble members) and SAWS operational prediction systems (40 ensemble members) to generate the seasonal forecast for the 3-months period with 1–3 months lead time. They produce a long-range forecast for the whole globe, and regional scale and the forecast is presented in graphical mode (https://www.weathersa.co.za/images/data/longrange/gfcsa/scw.pdf). They have also provided the forecast glimpse in the documentation.

5.2.1.10 Korea Meteorological Administration (KMA), Seoul

KMA started to run a numerical weather forecast model in 1999 with the introduction of 1st supercomputer (NEC/SX-5). The performance of the KMA model has been significantly improved since KMA introduced Unified Model (UM) from UK Met Office in 2010. In addition, KMA's next-generation global model, named KIM (Korean Integrated Model), was developed by KMA/KIAPS KIAPS: Korea Institute of Atmospheric Prediction System for 9 years since 2011 and operationally launched on April 2020 (https://web.kma.go.kr/eng/biz/forecast_02.jsp).

KMA is currently operating the global models, the local models, and the ensemble models as weather prediction models, the ocean wave/surge models, the Asian dust model, and the statistical models, as the application models. Along with the all-time sales forecasts, the KMA provides 1-month and 3-month outlooks (probabilistic) in graphical mode for Korea (except North Korea) and digital forecast information at a regional scale in South Korea. They also provide 1-week, 2-week, 3-week, and 4-week outlooks along with a monthly outlook for South Korea and its different regions (https://www.kma.go.kr/eng/weather/forecast/long-range3.jsp).

5.2.1.11 Japan Meteorological Agency (JMA)/Tokyo Climate Centre (TCC), Tokyo

The Central Meteorological Observatory (CMO) became an affiliate agency of the Ministry of Transport (MOT) under the name of the Japan Meteorological Agency (JMA) in 1956 and now serves as one of the most advanced and leading National Meteorological Services in the world. The JMA, as one of the most important WMO World Meteorological Centre (WMC) among the others, operates the ensemble prediction system of an atmospheric global circulation model (AGCM) for one-month prediction and atmosphere–ocean coupled global circulation model (CGCM) for three-month and warm/cold season prediction. The JMA generates the ensemble forecast for the monthly scale using the atmospheric general circulation model (GCM), while for three months forecast, they use coupled GCM. Both the monthly and three months forecasts are available for Asia (20°S–45°N; 30°E–180°E) and extended tropical regions (60°S–60°N; 180°E–180°W). The forecasts for the next 1st, 2nd, 3rd and 4th weeks, and an average of 28 days are available in advance when using the monthly forecast maps (anomaly) for different variables. In the three months forecast, with the mean forecast of the three months, the forecast for the individual months within that season is also available.

The ensemble mean, anomaly, and spread of the model's forecast are available. All the forecasts are available in graphical format (https://www.jma.go.jp/en/longfcst/000_0_10.html).

5.2.1.12 Météo-France, Toulouse

The model of Meteo-France is RA6 Regional Climate Centre node on Long Range Forecasts (RCC-LRF) of the WMO and provides a seasonal forecast by using Météo-France S7, CMCC3, DWD2, ECMWF SEAS5, NCEP, UKMO15 models (http://seasonal.meteo.fr/content/PS-previ-cartes?language=en). They provide 1-month (0–6 lead) and 3-months (0–4 lead) outlooks with both the deterministic and probabilistic forecasts graphical mode for Europe, Africa, Europe-North Africa, Tropical Atlantic and the entire globe. The deterministic forecast is the anomaly from the long-term mean, while the probabilistic forecast is provided in 3 categories (below, near, and above normal). Both the deterministic and probabilistic forecasts are provided in the graphical mode (http://seasonal.meteo.fr/content/PS-previ-cartes?language=en).

5.2.1.13 Climate Prediction Center (CPC)/National Oceanic and Atmospheric Administration (NOAA) (United States of America), Washington

The CPC, National Weather Service (NWS), NOAA provide different scales of forecasts ranging from extreme weather events to long-range forecasts, climate predictions, and space weather. The forecasts are mainly produced by using various numerical weather prediction models suitable for different spatial scales. The NOAA has used the Global Ensemble Forecast System (GEFS) for generating medium and extended-range forecasts and to issue watches and warnings during high-impact weather events, including hurricanes, blizzards, and extreme heat and cold waves across the world. The continuous efforts are undergoing significant upgrades to expand its capabilities and improve weather forecasting. GEFS now uses the Finite-Volume Cubed-Sphere (FV3) dynamical core with increased resolution approximately from 33 to 25 km, and the number of individual forecasts inputs into the ensemble has increased from 21 to 31, and these upgrades have better accuracy. There are also additional upgrades, including extending the forecast length from 16 to 35 days, along with improvements to the physics.

For the monthly and seasonal scale forecasts, the NOAA National Center for Environmental Prediction (NCEP) developed the climate forecast system (CFSv1) and made it operational in 2004. The CFSv1 has been upgraded to CFS version 2 (CFSv2) and made operational since 2011. The major components of CFSv1, such as the atmosphere, ocean, and land surface components, have been upgraded and operationalized as CFSv2. The GFDL Modular Ocean Model (MOM3) in CFSv1 has been replaced with a higher version of the GFDL Modular Ocean Model

(MOM4) in CFSv2. Further, the initial conditions for the atmospheric component of CFSv1 were derived from NCEP/DOE Atmospheric Model Inter-comparison Project (AMIP) II Reanalysis (R2) and for the ocean model from the NCEP Global Ocean Data Assimilation (GODAS). In CFSv2, the same is obtained from the NCEP CFSR (Saha et al. 2014). CFSv2 runs at a high resolution (T126, almost 100 km), while CFSv1 at a coarser resolution (T62, almost 210 km). Another important upgrade is that the CFSv2 is initialized four times a day, whereas CFSv1 is initialised once a day. The two-level Oregon State University (OSU) land surface model of CFSv1 was upgraded to a four-level NOAH model in CFSv2. A three-layer interactive sea–ice model has also been introduced in CFSv2 instead of climatological sea–ice in the older version.

The NOAA NCEP CPC forecasts are provided for the United States of America, and outlooks of 1st week, 2nd week, 3rd & 4th week, one month, and three months are available in graphical mode (https://www.cpc.ncep.noaa.gov/). The probabilistic forecast is available for different meteorological parameters/indices.

Besides the above thirteen independent regional climate centres, the WMO has designed to provide a long-range multi-model probabilistic forecast using all the 13 centres forecast (https://www.wmolc.org/seasonPmmeUI/plot_PMME) for global and different continents/regions. These forecasts are provided on a monthly and seasonal scale with one month lead time. The forecasts are available in graphical format.

5.2.2 International Research Institute for Climate and Society (IRI), (USA), New York

The IRI is one of the major centres that provide seasonal forecasts for the whole world and different continents/regions. The forecast is generated using a set an ensemble of predictions from several atmospheric/coupled GCMs obtained from the U.S. National Oceanographic and Atmospheric Administration (NOAA)'s North American Multi-Model Ensemble Project (NMME). The IRI first re-calibrated the GCMs before using those model products and then ensembling those products to generate a probabilistic forecast. The seasonal forecast (three months forecast) is disseminated for each month with a lead time from 1 to 4-months in advance for the upcoming four overlapping 3-months period. The IRI has also provided the forecast system and post-processing techniques for generating the seasonal forecast in detail. The probabilistic forecast is provided in three categories (below normal, normal, and above normal) in the percentage of the most likely category. The seasonal forecast is available in graphical format.

The beauty of the IRI is the data library, where about 300 datasets from different observational sources and model products associated with earth science and climate-related topics are available (https://iridl.ldeo.columbia.edu/index.html?Set-Language=en). Model forecasted data for the upcoming month/season, including observational and model data for historical periods is also available. The IRI has developed a map room, an extension of the data library, which is designed to facilitate the user to view past climatic information at different times and spatial scales. The map room also allows users to view the model forecast (such as anomaly) of the upcoming season, results of several statistical analyses for observed characteristics and verification of the model products, and many more. The user can also freely download raw model products from the IRI websites (if those are not restricted) and generate a skillful forecast for the domain of interest. The details of the methodologies on how to prepare a single forecast using multi-model and multi-methods are described in the next chapter.

5.2.3 *Asia–Pacific Economic Cooperation (APEC) Climate Center (APCC), Republic of Korea*

The APCC provides the seasonal (3 months) forecast of surface air temperature at 2 m and precipitation with 1 month and 4 months lead time (e.g. for January-March and April-June seasons of 2020, the forecast is disseminated in December 2019). They have also issued predictions for large-scale variables for 6-months period with one month lead time. They use several AGCMs/CGCMs and multi-model ensemble techniques to generate a probabilistic seasonal forecast with three terciles. The spatial scales of the seasonal forecasts are global and issued in a graphical format. In addition to the graphical forecast, a description of the outlooks, model, and methods is also provided.

The above-stated centres are the major organisations to provide monthly and seasonal scale forecasts globally and regionally. Besides these, many national/regional organizations have developed seasonal forecast systems using dynamical or statistical or dynamical-statistical approaches primarily focused to serve their own country or specified region by disseminating real-time/operational monthly and or seasonal forecasts. The skill of the forecast products available from any organisation/centres can be assessed for a particular region. The suitable and skillfull forecast products can be useful to different sectors for application and preparing advisory.

5.3 National (India) Status on Monthly and Seasonal Forecast

5.3.1 *IITM Seasonal Forecast*

Indian Institute of Tropical Meteorology (IITM), an autonomous institute of the Ministry of Earth Sciences, Govt. of India has a focal goal to carry out basic research on the Ocean–Atmosphere climate system and improvement of weather and climate forecasts. IITM located in Pune, India in collaboration with IMD has developed a global climate model based on Climate Forecast System (CFS) ocean–atmosphere coupled model for seasonal prediction, particularly for the Indian monsoon. Currently, the Indian CFSv2 is generating forecast products at higher resolutions (~36 km horizontal resolution), and simulated results show that the model CFSv2 has a potential skill in predicting the Indian summer monsoon. After the development of the dynamical seasonal forecast system through plenty of research work in different stages and achieving satisfactory skill, the model has been transferred to the India Meteorological Department (IMD), Govt. of India, for its operational use. The IMD is currently using customised CFSv2 developed at IITM Pune for dynamical seasonal predictions.

5.3.2 IMD Seasonal Prediction

The India Meteorological Department (IMD), established in 1875 after the series of the feminine in 1870 and beyond, is the pioneer organisation for providing seasonal forecasts in 1884 by Sir Blandford using the empirical relationship between the snow cover of the Himalayan region and the Indian monsoon. Sir Gilbert Walker used the linear regression method in 1910 for the prediction of the Indian summer monsoon (ISM). With a better understanding of the monsoon dynamics and its teleconnection with the large-scale fields as well as dependencies on the slowly varying meteorological variables, 8–10 potential predictors have been identified to use in the linear regression methods for ISM prediction. Later in 1988 and up to 2002, IMD used 16 predictors in the linear regression equation for the forecast of ISM. Since 2003, IMD has been using the power regression method for the prediction of the ISM. Besides the statistical approaches as stated above, IMD is also using a dynamical model developed at IITM for generating seasonal forecasts since 2015.

The Indian summer monsoon is the most important aspect since it contributes around 80% of the annual total rainfall in India, and a major part of the agricultural activities are inclined with the ISM rainfall. The IMD is issuing the summer monsoon (June–September) rainfall using statistical approaches in two stages, one in April and another in June. Currently, the statistical forecast of ISM rainfall is generated using 6-parameter Statistical Ensemble Forecasting System (SEFS).

On the other hand, the dynamical seasonal forecast is prepared using coupled Ocean–Atmosphere CFSv2 model every month around the year. The IMD is issuing a dynamical forecast for India, South Asia, and the globe, available in the form of anomaly and probabilistic way and both forecasts are available in the graphical mode for precipitation, surface air temperature at 2 m, and SST. The forecasts are available at a monthly scale with 1 to 4-months (for the next four months, each month separately) lead and the next two seasons having three consecutive months in a season with 1 to 2 months lead. The probabilistic forecasts are available in three terciles (below normal, above normal, and normal categories), and all deterministic and probabilistic forecasts are available in graphical mode (http://www.imdpune.gov.in/Clim_Pred_LRF_New/Models.html).

5.4 Making Usable of Long-Range Forecasts

The seasonal forecasts disseminated by several operational/research institutes/organisations are mainly deterministic or probabilistic. The user can collect these informations as different model forecasts and convert those forecasts into usable in different sectors, particularly in agriculture and allied sectors. Most of the stakeholders need the weather/climate forecast in actual magnitude, sometimes may be in deviation. However, in the deterministic forecasts, most of the institutes/organisations provide the information in 'anomaly' or 'percentage of departure' form which is a deviation from the long-term mean. It is quite simple to convert the deterministic forecast into a useful forecast for application. However, there is no simple way to directly convert the probabilistic forecast into a deterministic single value, but the probabilistic forecast has the strength to narrow down the uncertainty which in turn reduces the risk in the application of the probabilistic forecast than the deterministic forecast. We will next describe how to convert the available seasonal/monthly forecast to make it usable in different application sectors.

5.4.1 Convert Deterministic Forecast

As stated earlier, the deterministic forecasts are provided in the 'anomaly' or 'percentage of departure' form. Thus, the forecast of monthly/seasonal rainfall and temperatures with the percentage of departure and anomalies for a forecasted year are respectively needed to convert as a magnitude by using respective climatological mean. Users need to note down the anomaly or percentage of departure of precipitation/surface air temperature for a particular region/location from the forecasted graph provided by the operational institutes/organisations.

The following are conversion formulas from percentage departure and anomalies to mean/total (magnitude) of rainfall and temperature forecasts, respectively.

Rainfall
If the forecast is provided in the percentage of departure, then

$$R_f = \left(\frac{100 + RPD}{100}\right) \times \overline{Y} \tag{5.1}$$

Since,

$$RPD = \left(\frac{R_f - \overline{Y}}{\overline{Y}}\right) \times 100 \tag{5.2}$$

where, R_f is the predicted monthly/total seasonal rainfall of the forecasted year. *RPD* is the percentage of departure of monthly/seasonal rainfall for the forecasted year (noted by the user from the graphical forecast).

$\overline{Y}$ is the observed climatological mean of monthly/seasonal total rainfall computed using long-period data. A few institutions/organisations, such as 'maproom' of IRI, Columbia University now facilitate users to compute the climatological mean of the variable online. If the long-period observational/analysis data are available for the historical era (as per WMO classification, climatological mean can be computed using at least 30 years of data), one can compute the climatological mean of the rainfall for that region/location.

For example, the forecast for winter (December–February) seasonal precipitation at Jammu & Kashmir (J&K) region in India is provided in the **percentage of departure** and the value is -3% for the forecasted year 2017–18 with November 2017 initial conditions and the observed climatological mean of seasonal winter precipitation for the past period (1982–2017), referred as hindcast period, is 216 mm. The forecasted seasonal **total** winter precipitation for the year 2017–18 at J&K is:

$$\begin{aligned} \textit{Forecasted winter seasonal total rainfall at J\&K} &= \left(\frac{100+(-3)}{100}\right)\times 216 \\ &= 0.97\times 216 = 209.52\,\text{mm} \end{aligned}$$

If the forecast is provided in anomaly, then

$$R_f = \overline{Y} + R_a \quad (5.3)$$

where, R_a is the predicted monthly/seasonal rainfall anomaly of the forecasted year. $\overline{Y}$ and R_f are the climatological rainfall and forecasted total rainfall, respectively.

For example, consider the above case where the forecast is provided in 'anomaly' instead of 'percentage of departure' and the forecasted rainfall anomaly is –6.48 mm for the winter season of 2017–18 over the J&K region. Therefore,

$$\textit{Forecasted winter seasonal total rainfall at J\&K} = (216\text{–}6.48) = 209.52\ \text{mm}.$$

The user may kindly note that, sometimes, the forecasted anomaly is provided in mm/day, or the climatological value is provided in mm/day. In this situation, the users are advised to translate these anomaly/climatological values into monthly/seasonal total by multiplying the number of days of the corresponding month/season before using the above formula.

Temperature

$$T_f = \overline{T} + T_a \quad (5.4)$$

where,

T_f is the monthly/seasonal average temperature for the forecasted year.
T_a is the monthly/seasonal anomaly of average temperature for the forecasted year.
$\overline{T}$ is the observed climatological mean of monthly/seasonal average temperature for the hindcast (historical) period.

For example,

(i) The forecast for the anomaly of winter (December-February, DJF) seasonal average temperature at J&K is 0.55 °C for the forecasted year 2017–18 with November 2017 initial conditions and the observed climatological mean of winter (DJF) seasonal average temperature for the hindcast period (1982–2017) is 6 °C. The forecasted winter seasonal average temperature for the year 2017–18 at J&K is:

$$\textit{Forecasted winter seasonal average temperature at J\&K} = 0.55\,^{\circ}\mathrm{C} + 6\,^{\circ}\mathrm{C} = 6.55\,^{\circ}\mathrm{C}$$

(ii) The forecast for the anomaly of winter (DJF) seasonal minimum temperature at J&K is 0.69 °C for the forecasted year 2017–18 with November 2017 initial conditions and the observed climatological mean of winter (DJF) seasonal minimum temperature for the hindcast period (1982–2017) is 1 °C. The forecasted winter seasonal minimum temperature for the year 2017–18 at J&K is:

$$\textit{Forecasted winter seasonal minimum temperature at J\&K} = 0.69\,^{\circ}\mathrm{C} + 1\,^{\circ}\mathrm{C} = 1.69\,^{\circ}\mathrm{C}$$

(iii) The forecast for the anomaly of winter (DJF) seasonal maximum temperature at J&K is 0.64 °C for the forecasted year 2017–18 with November 2017 initial conditions and the observed climatological mean of winter (DJF) seasonal maximum temperature for the hindcast period (1982–2017) is 12 °C. The forecasted winter seasonal maximum temperature for the year 2017–18 at J&K is:

$$\textit{Forecasted winter seasonal maximum temperature at J\&K} = 0.64\,^{\circ}\mathrm{C} + 12\,^{\circ}\mathrm{C} = 12.64\,^{\circ}\mathrm{C}$$

5.4.2 Understanding Probabilistic Forecast

Many organisations/institutions provide a probabilistic forecast, mainly in three terciles as above normal, normal, and below normal. The normal precipitation is generally classified if the precipitation amount lies between the $\pm 1\sigma$ (σ is the standard deviation) of the long-period average. Thus, if above normal, normal, and below normal precipitation are indicated by R_A, R_N, and R_B, respectively, and the long-period precipitation average is indicated by $\overline{Y}$, then

$$\begin{aligned} \overline{Y} - \sigma \leq R_N \leq \overline{Y} + \sigma \\ R_A > \overline{Y} + \sigma \\ R_B < \overline{Y} + \sigma \end{aligned} \tag{5.5}$$

The probability forecast over a region is represented by these three terciles. If the probability forecast is 33.3% of each of the tercile (below normal, normal, and above normal) over a region, then all the three categories have an equal probability, and that forecast will not have any use. However, a forecast of a particular tercile with a higher probability can be found for a region, and that information can be useful for application. For example, let us assume that the above normal with a 70% probability, normal precipitation with a 20% probability, and 10% probability with the below normal category are predicted for precipitation occurrence over a region for next season. Therefore, the chances of occurrence of above-normal precipitation are higher than the other two categories, which gives confidence that the region most likely will experience above-normal precipitation. While the chances of occurrence of below-normal precipitation are negligible. The user may consider this kind of information and cross-check the 'anomaly' prediction from other forecasts and will have confidence in those forecasts which are in the same direction as the probabilistic forecast. However, the different operational centers may predict differently and that information altogether will provide a signal with the uncertainty range in the forecast.

5.5 Forecast Evaluation Procedure

Any weather and climate services mainly provide the public with warnings, forecasts, and other meteorological information in support of the safety of life and property and to make proper plans for climate risk management in various sectors such as agriculture, health, hydrology, hydro-power, and other water resource management sectors in a timely and reliable manner. Several methods, such as subjective or objective, or in a different way of telling that the forecast methods used by a great number of international/national/local agencies for precipitation or surface air temperature are mainly persistent or model based. The model may be physical (dynamical), statistical, or dynamical-statistical, and most of the forecasts in the world are model-based. However, these models are developed using mathematical equations based on hypotheses and/or established relationships. The forecasts are qualitative (description of the forecast outlook), deterministic (quantitative), or probabilistic. Since the forecast of the meteorological parameters, particularly precipitation and surface air temperature, are in demand by different stakeholders, thus, weather and climate forecast service programmes must include a system to evaluate whether this task is being fulfilled and to regularly assess the program's performance. The aim of the evaluation is twofold: first, to ensure that products such as warnings and forecasts are accurate and skillful from a technical viewpoint, and secondly, that they meet user requirements and that users have a positive perception of, and are satisfied with the products. The two evaluation components are essentially complementary since even a highly accurate and skillful forecast will not produce an effective public weather services programme if it does not respond to user needs. The main goal of the evaluation and verification process is to constantly improve the quality (skill and accuracy) of the forecasts and associated early warning services. On the other hand, it is necessary to evaluate the forecast to build up confidence in the methodologies and reliability when used in different application sectors.

The verification methods vary for various types of forecasts, such as deterministic, categorical, and probabilistic forecasts. We will describe some useful and important verification methodologies for evaluating the weather and climate forecast.

5.5.1 Statistical Skill Scores for Deterministic Forecast Verification

5.5.1.1 Root Mean Square Error (RMSE)

It is routinely used as the measure of the differences between model values and actually observed values, and it is generally called a residual or error of the model. The RMSE serves to aggregate the magnitudes of the errors in model for various times into a single measure of estimating power. The RMSE computed as

$$RMSE = \sqrt{\frac{\sum_{i=i}^{i=n} (X_i - O_i)^2}{n}} \tag{5.6}$$

where X_i is the ith time estimated value of the model.

O_i is the ith time value of actual observed.

5.5.1.2 Mean Bias (MB)

Mean Bias indicates that the method is overestimated or underestimated or exactly matched with actually observed climatology. It is computed as

$$MB = \overline{X} - \overline{O} \tag{5.7}$$

where $\overline{X}$ indicates the climatological mean value of the model and

$\overline{O}$ indicates the climatological mean value of reference data (observation).

5.5.1.3 Modified Mean Bias (MMB)

It provides a measure of the bias of the model with the observation that performs symmetrically with respect to under and over prediction and is bounded by the value −2 to +2. It is computed as

$$MMB = \frac{2}{n}\sum_{i=1}^{i=n}\left(\frac{X_i - O_i}{X_i + O_i}\right) \tag{5.8}$$

5.5.1.4 Fraction Gross Error (FGE)

The fractional gross error also gives a measure of the overall error of the model with observation, and it is bounded by the value 0 to 2. It is computed as

$$FGE = \frac{2}{n}\sum_{i=1}^{i=n}\left|\frac{X_i - O_i}{X_i + O_i}\right| \tag{5.9}$$

5.5.1.5 Index of Agreement (IOA)

It is a standardised measure of the degree of model error with observation, and it ranges from 0 to 1 (Willmott 1981). '1' indicates a perfect agreement between the estimated values of the model and actual observation, and '0' represents no agreement. The IOA represents the ratio between the mean square error and potential error. The potential error is the sum of the squared absolute values of the distances from the estimated values to the mean of actual observed values and distances from actual observed values to the mean actual observed value. The IOA detects additive and proportional differences in the observed and estimated means and variances. It computed as

$$IOA = 1 - \left(\frac{\sum_{i=1}^{i=n}(X_i - O_i)^2}{\sum_{i=1}^{i=n}\left(\left|X_i - \overline{O}\right| + \left|O_i - \overline{O}\right|\right)}\right) \tag{5.10}$$

5.5.1.6 Nash–Sutcliffe Efficiency Coefficient (NSEC)

Nash–Sutcliffe Efficiency Coefficient (NSEC) is generally used to assess the estimation power of the model (Nash and Sutcliffe 1970). In addition, it can be used to describe the accuracy of the model quantitatively. It is defined as

$$NSEC = 1\left(\frac{\sum_{i=1}^{i=n}(O_i - X_i)^2}{\sum_{i=1}^{i=n}\left(O_i - \overline{O}\right)^2}\right) \tag{5.11}$$

Nash–Sutcliffe efficiencies skill score range from $-\infty$ to 1. 1 indicates a perfect match between estimated values of the model and actual observations. Whereas an NSEC less than zero ($-\infty<\text{NSEC}<0$) occurs when the actually observed mean is a better predictor than the model.

5.5.1.7 Correlation Coefficient (CC)

The correlation coefficient is a commonly used skill score for a measure of the strength and direction of the linear relationship between two variables. It is defined as

$$CC = \frac{Cov(X, O)}{\sigma_x \sigma_0}$$

$$CC = \frac{\sum_{i=1}^{i=n} \left(X_i - \overline{X}\right) \cdot \left(O_i - \overline{O}\right)}{\sqrt{\sum_{i=n}^{i=n} \left(X_i - \overline{X}\right) \cdot \sum_{i=1}^{i=n} \left(O_i - \overline{O}\right)^2}} \tag{5.12}$$

where σ_x and σ_o are the the standard deviations of precipitation for the model and observation, respectively. The total number of points over space or time is n, and X_i and O_i are the model and observed precipitation for *ith* point/time, and $\overline{X}$ and $\overline{O}$ are the mean value averaged over space/time for model and observed precipitation.

5.5.1.8 Phase Synchronisation

The sign agreement between forecast and observed anomalies is quantified by using phase synchronisation (η). Phase synchronisation is defined as

$$\eta = \frac{(n - \acute{n})}{n} \tag{5.13}$$

where 'n' is the total number of events.

'$\acute{n}$' is the number of cases for which the two anomalies (forecast and observation) have opposite signs (out of phase).

Thus $\eta = 0$ if all the anomalies from the two sets are out of phase, while $\eta = 1$ when all of them are in phase.

5.5.2 Statistical Scores for Categorical Forecast Verification

This is one of the very simple ways of forecast in which an event is forecasted in a way whether the event will happen or not, like a 'yes' or 'no' forecast. Thus, four kinds of situations may occur when a prediction yes/no is made, such as:

(a) The event is forecasted to occur, and actually, it occurred, this is known as 'hit'.
(b) The event is forecasted not to occur, and actually, it occurred, this is known as 'misses'.
(c) The event is forecasted to occur, and actually, it did not occur, this is known as 'false alarm'.
(d) The event is forecasted not to occur, and actually, it did not occur, this is known as 'correct negative'.

These yes/no forecasts can be represented in a tabular form as shown in Table 5.1.

For a perfect prediction, the forecast will always fall either into 'hits' or 'correct negative' categories. If forecasts for several times and for over different regions are available, one can have different values for 'hits', 'false alarm', 'misses', and 'correct negative' (assuming that there is no such model which can give perfect forecast always for each region). Therefore based on the values of the contingency table, several statistics can be generated which will reflect the skill of the model in the forecast. To accesses the model's ability in predicting different precipitation categories (excess, deficit, and normal) and temperature categories (warm, cold, normal) with different aspects can be calculated by using various categorical statistical skill scores such as Accuracy (ACC), Bias Score (Frequency bias), Probability of detection (POD), False alarm Rate (FAR), Probability of false detection (POFD) or false alarm rate), Success Ratio (SR), Threat Score (TS), Equitable threat score (ETS) or Gilbert skill score, Hanssen and Kuipers discriminant (HK) or Peirce's skill score (PSS), Odds ratio skill score (ORSS), etc. by using the Contingency Table 5.1. Brief details of the above skill scores are given in Table 5.2, and the full description is available in Murphy and Winkler (1987), Murphy et al. (1989), and Ghajarnia et al. (2015) studies or at https://www.cawcr.gov.au/projects/verification/.

5.5.3 *Multi-categorical Skill Scores*

In addition to the above statistical skill score, some multi-categorical skill scores like Accuracy (ACC), Heidke Skill Score (HSS), and Hanseen Kuipers discriminant (HK) can be used to compute the combined skill for all the categories by using the below contingency table (Table 5.3). The description of the contingency table and the above three multi-categorical skill scores as studied by Murphy and Winkler (1987), Murphy et al. (1989) and Ghajarnia et al. (2015).

Table 5.1 A contingency table

	Observed		
Forecasted		*Yes*	*No*
	Yes	Hits	False alarm
	No	Misses	Correct negative

Table 5.2 Details of skill scores (single category)

Name of skill score	Formula	Range and perfect value
Accuracy (ACC)	$ACC = \frac{(\text{Hits} + \text{Correct Negatives})}{\text{Total}}$	Range: 0 to 1 Perfect score: 1
Bias score (Frequency bias)	$BIAS = \frac{(\text{Hits}+\text{Falsealarms})}{(\text{Hits}+\text{Misses})}$	Range: 0 to ∞ Perfect score: 1
Probability of detection (POD)	$POD = \frac{(\text{Hits})}{(\text{Hits}+\text{Misses})}$	Range: 0 to 1 Perfect score: 1
False alarm rate (FAR)	$FAR = \frac{\text{Falsealarms}}{(\text{Hits}+\text{Falsealarms})}$	Range: 0 to 1 Perfect score: 0
Probability of false detection (POFD) or false alarm rate)	$POFD = \frac{\text{Falsealarms}}{(\text{CorrectNegatives}+\text{Falsealarms})}$	Range: 0 to 1 Perfect score: 0
Success ratio (SR)	$SR = \frac{\text{Hits}}{(\text{Hits}+\text{Falsealarms})}$	Range: 0 to 1 Perfect score: 1
Threat score (TS)	$TS = \frac{\text{Hits}}{(\text{Hits}+\text{Misess}+\text{Falsealarms})}$	Range: 0 to 1 Perfect score: 1
Equitable threat Score (ETS) or Gilbert skill score	$ETS = \frac{\text{Hits}-\text{Hits}_{\text{random}}}{(\text{Hits}+\text{Misess}+\text{Falsealarms}-\text{Hits}_{\text{random}})}$ Where $Hits_{random} = \frac{(\text{Hits}+\text{Misess})(\text{Hits}+\text{FalseAlarms})}{\text{Total}}$	Range: −1/3 to 1 Perfect score: 1
Hanssen and Kuipers discriminant (HK) or Peirce's skill score (PSS)	$HK = \left(\frac{\text{Hits}}{\text{Hits} + \text{Misses}}\right) - \left(\frac{\text{Falsealarms}}{\text{Falsealarms} + \text{CorrectNegatives}}\right)$	Range: −1 to 1 Perfect score: 1
Odds ratio skill score (ORSS)	$ORSS = \frac{(\text{Hits}^*\text{Correct Negatives})-(\text{Misess}^*\text{Falsealarms})}{(\text{Hits}^*\text{Correct Negatives})+(\text{Misess}^*\text{Falsealarms})}$	Range: −1 to 1 Perfect score: 1

Table 5.3 Contingency table for multi-category

	Observed category					
Forecasted category	*i,j*	*1*	*2*	⋯	*K*	*Total*
	1	$n(F_1, O_2)$	$n(F_1, O_2)$	⋯	$n(F_1, O_K)$	$N(F_1)$
	2	$n(F_2, O_1)$	$n(F_2, O_2)$	⋯	$n(F_2, O_K)$	$N(F_2)$
	⋯	⋯	⋯	⋯	⋯	⋯
	K	$n(F_K, O_1)$	$n(F_K, O_2)$	⋯	$n(F_K, O_K)$	$N(F_k)$
	Total	$N(O_1)$	$N(O_2)$	⋯	$N(O_K)$	N

5.5.3.1 *Accuracy (ACC)*

ACC indicates that the overall fraction of the forecasts were in the correct category, this skill score range is 0 to 1. 1 indicates perfect to match all categories. ACC is computed as

$$ACC = \frac{1}{N}\sum_{i=1}^{i=k} n(F_i, O_i) \quad (5.14)$$

5.5.3.2 Heidke Skill Scores (HSS)

HSS compares the accuracy of the given forecast against the accuracy of a forecast of random chance (Heidke 1926). This skill score ranges from $-\infty$ to 1. 1 represents a perfect skill score. A negative HSS indicates that a forecast is worse than a random based/generated forecast. HSS is defined as

$$HSS = \frac{\frac{1}{N}\sum_{i=1}^{i=k} n(F_i, O_i) - \frac{1}{N^2}\sum_{i=1}^{i=k} N(F_i)N(O_i)}{1 - \frac{1}{N^2}\sum_{i=1}^{i=k} N(F_i)N(O_i)} \quad (5.15)$$

5.5.3.3 Hanssen and Kuipers Discriminant (HK)

The Hanssen and Kuipers discriminant score is nearly similar to the Heidke skill score, except that in the denominator, the fraction of correct forecasts due to random chance is for an unbiased Forecast (Wilks 2011). This skill score range from -1 to 1. '0' indicates no skill and '1' represents a perfect agreement. It is computed as

$$HK = \frac{\frac{1}{N}\sum_{i=1}^{i=k} n(F_i, O_i) - \frac{1}{N^2}\sum_{i=1}^{i=k} N(F_i)N(O_i)}{1 - \frac{1}{N^2}\sum_{i=1}^{i=k} (N(O_i))^2} \quad (5.16)$$

5.5.4 Statistical Skill Scores for Probabilistic Forecast Verification

In deterministic methods used above, the inherent uncertainty in the forecasts cannot be represented. Predictions are more useful if this uncertainty is also estimated, especially for climate risk management in agriculture. It is well known that the probabilistic forecast is the only way to represent the uncertainty in the forecast. A probabilistic forecast gives a probability of a particular category of rainfall or temperature occurring, with a value between 0 and 1 (or 0 and 100%). In general, it is difficult to verify a single probabilistic forecast. Instead, a set of probabilistic forecasts, p_k, is verified using observations that those events either occurred ($p_k = 1$) or did not occur ($p_k = 0$). Probabilistic forecasts have recently been demonstrated to be a key way to use monthly and seasonal scale predictions (Kulkarni et al. 2012). Therefore, it is better to convert the deterministic forecast into the probabilistic forecast to represent the uncertainties associated with it.

A parametric approach is used to build probabilistic predictions of rainfall/temperature under the assumption of normal distribution. These probabilistic predictions are generated for tercile categories: (i) below normal, (ii) near-normal, and (iii) above normal, which are based on the observed climatology. The final deterministic forecast is used as the mean of the forecast distribution, whereas the spread is calculated by the correlation method. Rank Probability Skill Score (RPSS) is one of the skill metrics for assessing probabilistic prediction. It represents whether a probabilistic prediction is better or worse than a climatological forecast of equal probabilities for each tercile category (0.33 of each category). RPSS measures the improvement of the multi-category probabilistic forecast relative to a reference forecast and strictly takes climatological frequency into account. The RPSS is computed as

$$RPSS = \frac{\overline{RPS}_{reference} - \overline{RPS}}{\overline{RPS}_{reference}} = 1 - \frac{\overline{RPS}}{\overline{RPS}_{reference}} \tag{5.17}$$

where Ranked Probability Score (RPS) is

$$(RPS) = \frac{1}{M-1}\sum\nolimits_{m=1}^{M}\left[\left(\sum\nolimits_{k=1}^{m} p_k\right) - \left(\sum\nolimits_{k=1}^{m} O_k\right)\right]^2$$

Here M is the number of forecast categories, p_k is the predicted probability in forecast category k, and O_k is an indicator (0 = no, 1 = yes) for the observation in category k. RPS measures the sum of squared differences in cumulative probability space for a multi-category probabilistic forecast. Penalises forecasts more severely when their probabilities are further from the actual outcome. RPS ranges from 0 to 1, and the perfect score is '0' while RPSS ranges from $-\infty$ to 1 and the perfect score is 1.

In this chapter, the availability of real-time monthly to seasonal scale forecasts along with the hindcast (past simulations) generated by the GCMs (atmospheric and coupled) by reputed organisations/institutes have been discussed. In addition, the conversion of anomalous/percentage departure of forecast (digital or graphical format) into expected deterministic/actual forecast has also been presented. Several standard methods are based on statistical techniques for the verification of the model products. However, as it is well established that GCM products are available in coarser resolutions and involved with both systematic and asymmetric biases, thus it would be very risky to use those products directly for application in different sectors, particularly at local/regional scales. Thus, it is indeed a necessity to estimate and remove the biases. It can be noted here that the systematic biases can be removed using statistical techniques if a user has a long period of historical data and that process is simple and easy compared to the removal of asymmetric biases that involves mainly the model's internal dynamics and interactions. On the other hand, the performances of GCMs vary from region to region and season to season and also for different weather/climate events. Besides these, the performance of a single GCM is not uniform over space and for events and varies model-to-model when intercomparison among the models is carried out. As indicated earlier, the GCMs products are coarser, and it is required to translate this coarser (low-resolution i.e., higher grid spacing) information into finer scale or high-resolution (smaller grid spacing) to make it usable. All these aspects, including methods for generating a single forecast using multi-GCMs and multi-statistical techniques, have been addressed in the next chapter.

Chapter 6
Generation of Regional Scale Monthly and Seasonal Forecast

6.1 General Circulation Models (GCMs) Products and Necessity of Assessment

General Circulation Models (GCMs) or Global Climate Models contain fundamental principles of the equation of motion, energy balance equation or thermodynamic equation, continuity equation or conservation of mass, gas law or equation of state, and water vapour equation which are well-proven as the governing equations of the atmospheric motion. In addition, since the atmosphere has a two-way interaction with land-air-sea, thus the GCMs involve the mathematical equations that represent exchanges of mass and energy between land-air-sea. The atmosphere, as we all are aware is a mixture of gases, various aerosols and dust particles having different size, shape, characteristics, water vapour, etc., that not only play a key role in determining the net balance of the atmospheric energy but many of them are involved in the chemical reactions and change the number density of the constituents. Besides these, the atmosphere is bounded only on the top and bottom and has no lateral boundary (an open boundary through the horizontal space), and has a complex interaction between the small-scale physical processes such as radiation, eddy, turbulence, etc. with the large-scale flow such as monsoonal flow, jet stream, etc. which are represented through mathematical formulations and parameterised concepts. The representation of the land-surface characteristics, such as elevation, soil type, soil texture, vegetation category, fractions, etc., are heterogeneous over space, and each of the land surface characteristics plays a different role in modulating the local and influencing global climate. The different parameters and/or mathematical equations are interlinked strongly or weakly with the other variables/equations in different ways; therefore, the GCMs are one of the most complex numerical models that involve an interaction between non-linear dynamics, small-scale physical processes with the large-scale flow, energy and momentum exchange among land-air-sea, chemical reactions and the other equations.

U. C. Mohanty et al., *Climate Risk Management in Agriculture*,
https://doi.org/10.1007/978-3-031-51862-1_6

The present-day GCMs, coupled model of Land-Atmosphere-Ocean (CGCMs) are the most important tools to simulate global circulations and associated events and reproduce the atmospheric variables in spatiotemporal scales. The GCMs are manifested in such a way that it reproduces the past climate that has been observed. Thus, if a GCM has the potential skill in generating the past climate, then one will have confidence in the products of that GCM for future prediction. Therefore, performance evaluation of a GCM is necessary in order to understand the credibility of past climate simulations and identify the skills in producing different climate variables. Although, continuous and immense efforts are being dedicatedly given for the improvement of the model physics and dynamics, there are a number of issues that are needed to be taken care of to enhance the prediction skill. Thus, the evaluation of model products in different aspects will be helpful to the scientists and model developers, and some pathways are mentioned for the evaluation of a GCM.

First and foremost, two types of model evaluations are widely performed, such as (i) evaluation of predicted products or operational/real-time results and (ii) deep diagnosis of atmospheric fields or processes that directly or indirectly affect the results. The scrutiny of operational results can be made through various ways, i.e., evaluation of prediction in different time scales from a few hours or a day to season or beyond. How the model performs in predicting extreme weather events? Or is there any systematic bias in forecasting a particular event over space, time, frequency, and intensity? Evaluation of operational forecast using statistical techniques/approaches not only gives the fidelity of the model products over specific regions/events but also estimates significant model errors. Diagnosis and analysis actually represent model internal behaviour in simulating the physical and dynamical processes. In other words, model diagnostic studies will indeed give insight into the reason for the errors in prediction or operational forecast, particularly if the model encompasses diversity in nature in the forecast compared to observations. Besides the above aspects, one may consider examining the model sensitivity on resolution (horizontal and vertical), physical parameterisation schemes, interaction and exchange coefficients, etc. A simple example of the performance evaluation is presented next.

6.2 Performance of GCMs Raw Products

A total of five GCMs products are used in this example, and a brief description of each model is presented in Table 6.1, and a full description of these models is also available in Mohanty et al. (2013, 2019b). The five GCMs are evaluated against the observations over North India for the period 1982–2009 during the winter months (December–January), and a brief analysis is provided in the next sub-section.

Table 6.1 A brief detail of the GCMs

Model	Resolution	AGCM	OGCM	Ensemble member	References
ECHAMCFS	(T42) $2.7^\circ \times 2.8^\circ$	ECHAM4p5	CFS-predicted SST	24	Roeckner et al. (1996)
ECHAM GML	(T42) $2.7^\circ \times 2.8^\circ$	ECHAM4p5	CFS-predicted SSTs prescribed over the tropical Pacific basin (semi-coupled)	12	Roeckner et al. (1996), Lee and De Witt (2009)
MOM 3 AC1	(T42) $2.7^\circ \times 2.8^\circ$	ECHAM4p5	MOMS (anomaly-coupled)	24	Roeckner et al. (1996), Pacanowski and Griffies (1998)
MOM3_DC2	(T42) $2.7^\circ \times 2.8^\circ$	ECHAM4p5	MOMS (direct-coupled)	12	Roeckner et al. (1996), Pacanowski and Griffies (1998)
NCEP_CFSv2	(T126) $0.9^\circ \times 0.9^\circ$	GFS (2009 version)	MOM4	24	Saha et al. (2006)

6.2.1 Analysis Procedure

The first step in diagnosing these GCMs is to examine the climatology and inter-annual variability (IAV) that provide a very basic step in evaluating how well these GCMs tend to capture the observed characteristics. Several skill metrics (RMSE, Correlation Coefficient, Bias, etc.) have also been used to evaluate the performance of the GCMs in predicting the seasonal mean precipitation over India. In the previous chapter, a number of methods have been described for evaluating a model's performance, and some basic and standard methods are used here in the example. The model data are verified with the India Meteorological Department (IMD) rainfall analysis at $1^\circ \times 1^\circ$ horizontal resolution for the same period, and the model products are interpolated to the observed grid points to avoid dependence on resolution. A comparison between the model-simulated climatology and IAV against the observed climatology and IAV is made. The basic means of analysing how the model behaves compared with observation is to determine the correlation coefficient between the two-time series, which gives a degree of association between them and is defined as

$$CC = \frac{Cov(X, O)}{\sigma_x \sigma_O}$$

$$CC = \frac{\frac{1}{(n-1)}\sum_{i=1}^{n}\left(\left(X_i - \overline{X}\right)\left(\overline{O}_i - \overline{O}\right)\right)}{\sqrt{\left(\frac{1}{(n-1)}\sum_{i=1}^{n}\left(X_i - \overline{X}\right)^2\right) \times \left(\frac{1}{(n-1)}\sum_{i=1}^{n}\left(O_i - \overline{O}\right)^2\right)}} \tag{6.1}$$

where σ_x and σ_o are the standard deviations of precipitation for model and observation. The total number of points over space is n, X_i and O_i are the model and observed precipitation for ith point and $\overline{X}$ and $\overline{O}$ are the mean value averaged over space for model and observed precipitation. The root mean square error (RMSE) is used as one of the skill measures of GCMs. RMSE is computed as

$$\text{RMSE} = \sqrt{\frac{\sum_{i=1}^{n}(X_i - O_i)}{n}} \tag{6.2}$$

Bias in the model predictions is defined as

$$\text{Bias (in \%)} = \frac{(\overline{X} - \overline{O})}{\overline{O}} \times 100 \tag{6.3}$$

A comparison of IAV between observation and model simulations is analysed using the standard deviation at each of the Indian grid points.

6.2.2 *Performance of GCMs for Predicting Winter (DJF) Seasonal Mean Precipitation*

The climatology of winter precipitation (seasonal total/accumulated) simulated by each state-of-the-art of five GCMs is compared against observation and is depicted in Fig. 6.1a–f. It is seen that only the NCEP_CFS2 (i.e. CFSv2) model is able to reproduce the high-rainfall zone to some extent, i.e., the region is analogous to what is observed over the Kashmir (head of the Indian) region (Fig. 6.1d). This model shows a zone of high precipitation over north India as well as over south eastern parts of peninsular India. The patch of high maxima zone seems to be extending southward in this model than observation. All the other models show a pocket of the high-rainfall zone over north India, but the magnitude over that region is found to be very less (less than 250 mm) compared to the observation. In Fig. 6.2, it is seen that the models generally produce very less rainfall in north India compared to observed rainfall. In 1994, observed rainfall during the season was more than 4 mm day^{-1}.

However, rainfall values from the models are within 1–2 mm day^{-1}, indicating a large bias in the model rainfall. Moreover, it is seen from the figure that IAV of the model-simulated rainfall does not agree with that of observed rainfall.

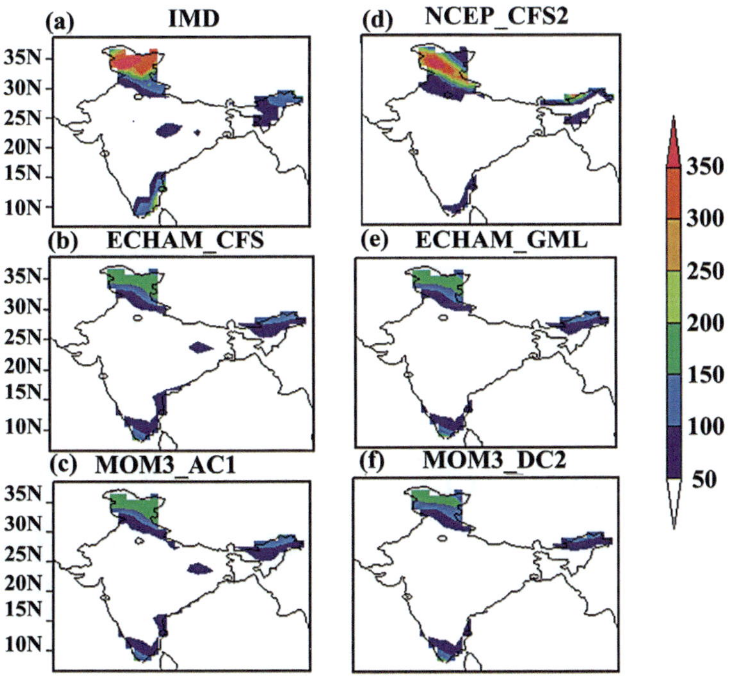

Fig. 6.1 Precipitation climatology (mm) from observation and models from 1982 to 2009 for December to February (DJF) (Tiwari et al. 2014)

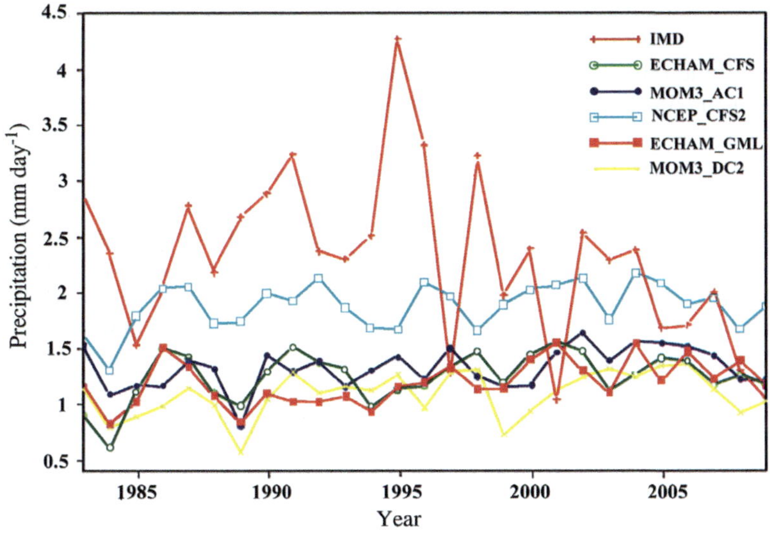

Fig. 6.2 Time series of observed and model predicted area-averaged precipitation (mm day^{-1}) over north India from 1982 to 2009 for December to February (Tiwari et al. 2014)

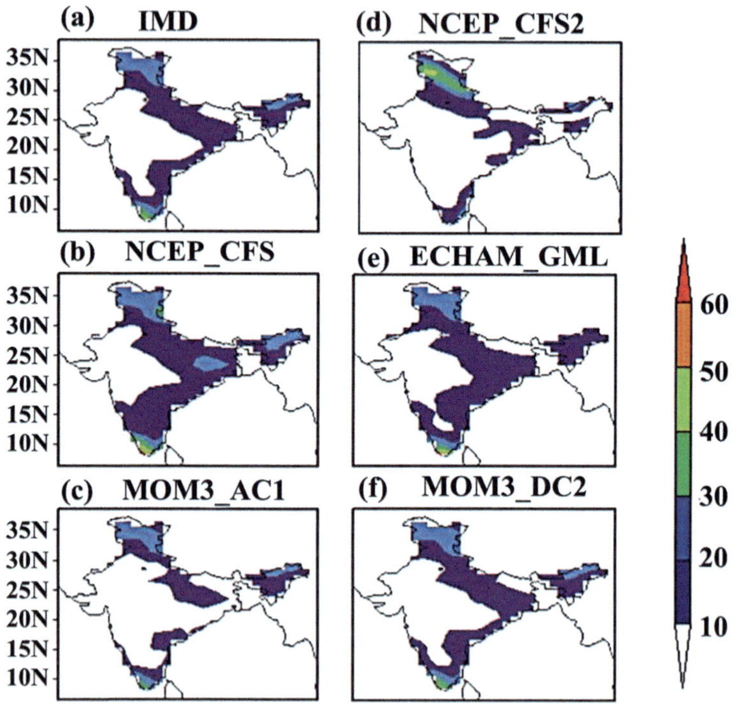

Fig. 6.3 Inter-annual variation (IAV) for precipitation between models and observation from 1982 to 2009 for December to February (Tiwari et al. 2014)

Figure 6.3b–f represents the standard deviation of precipitation from individual models. Almost all the models are able to show the total variance of precipitation as seen in the observation; however, MOM3_DC2 is somewhat able to show the region of maximum IAV, similar to the observation. In the case of NCEP_CFS2, the IAV is lesser than observations, especially over the eastern parts of India. Over northern India, the IAV is lesser than 30 mm in observations as well as in all the models, except it reaches up to 40 mm in NCEP_CFS2. Therefore, the figure suggests that most of the GCMs are able to predict the observed magnitude of IAV. Figure 6.4a–e represents the bias (in %) between the model and observed precipitation. It is seen that all the models (except NCEP_CFS2) show almost the same bias pattern, ranging from 50 to 200% over the northern parts (Jammu & Kashmir (J&K), Himachal Pradesh (HP), and Uttarakhand (UK) region) of India. On the other hand, the NCEP_CFS2 model has a stronger bias over the eastern part of J&K and the northeastern parts of India. As shown in Fig. 6.4c, this model has a higher bias, up to 500% of observed precipitation (Tiwari et al. 2014).

Figure 6.5a–e outlines the spatial correlation coefficient between the precipitation from the models and observation. It can be clearly seen from the diagram that the values of correlation coefficients for most models over the northern part of India are low except for the MOM3_DC2 and MOM3_AC1 models. The

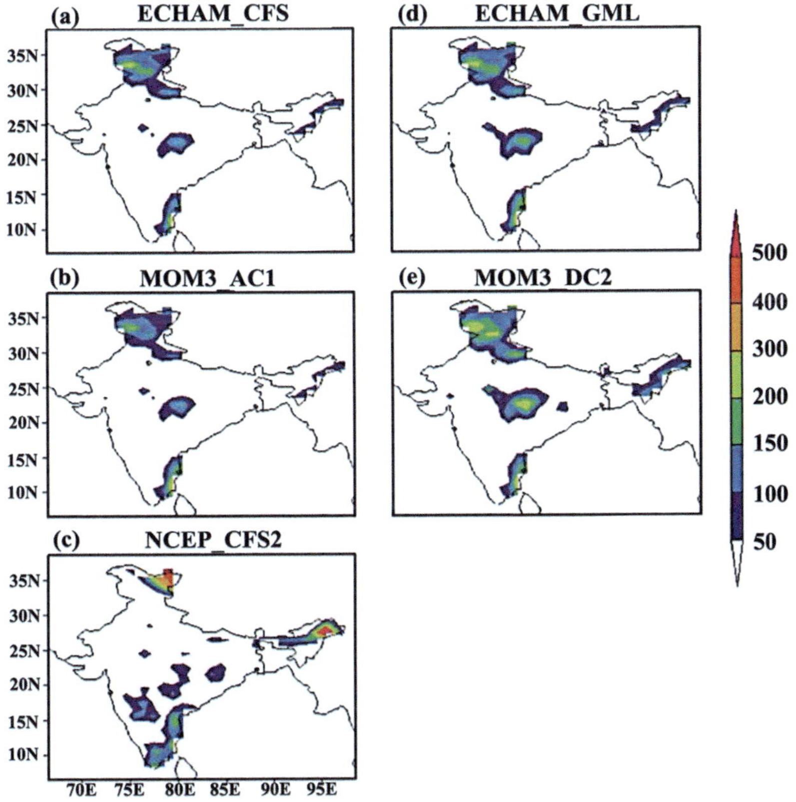

Fig. 6.4 Bias (in %) for precipitation between models and observation from 1982 to 2009 for December to February (Tiwari et al. 2014)

best among the models is MOM3_DC2 which has correlation coefficients of 0.4 and above over the region of interest. The other models have varying degrees of skill, and correlation coefficients range from 0.3 to as low as −0.2. It may be noted that for 27 years of data that are analysed, correlation values greater than 0.33 are statistically significant at a 95% confidence level. Therefore, most models do not have statistically significant correlation skills for precipitation over the north Indian region during the winter season. It may be noted that during the winter months, central India is a mostly dry region, and the model also shows dry months; thus, the correlation is high due to correct negative predictions. Figure 6.6 shows the model precipitation RMSE compared to observations. The RMSE values range from 0.5 to 3 mm day^{-1} over J&K, HP, and UK regions. The maximum RMSE is seen over northwest Kashmir in all the models (except NCEP_CFS2), while for the NCEP_CFS2 model, higher RMSE values are located over the eastern parts of J&K. In the Taylor diagram (Taylor 2001) the skill of individual GCM for prediction of rainfall during winter in terms of correlation, RMSE, and standard deviation are shown (Fig. 6.7). The figure clearly indicates that the significant

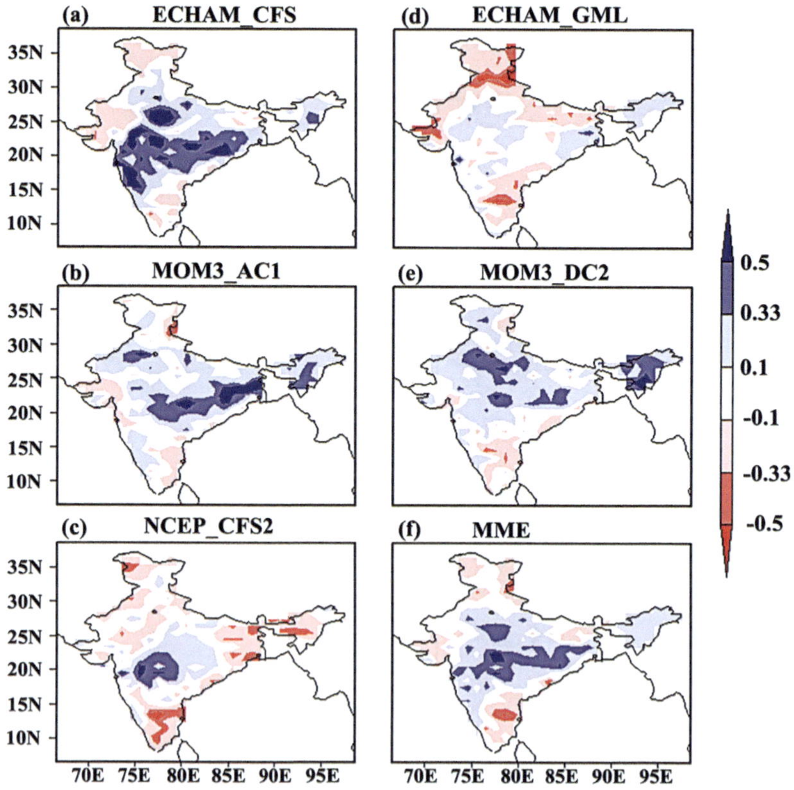

Fig. 6.5 Spatial correlation pattern of the GCMs. The correlations were calculated between observed precipitation and model simulations from the period 1982–2009 for December to February (Tiwari et al. 2014)

correlation skill of three coupled GCMs (MOM3_DC2, ECHAM_CFS, and MOM3_AC1) has a lesser RMSE. As shown in Fig. 6.7, a smaller, or even negative, correlation is observed in the case of ECHAM_GML and NCEP_CFSv2 with very large RMSE.

The evaluation of the GCMs is shown for seasonal scale precipitation prediction over northwest India during the winter (December–February) season. One can consider the other predicted variables, such as air temperature at 2 m height (T2M) for mean, maximum, or minimum instead of precipitation for their region of interest and specific time scale (daily, monthly, seasonal, annual, and so on). Besides the above basic analyses, several advanced statistical techniques can be employed for generating skill matrices that will delineate the robustness and reliability of a particular GCM for a specific event/parameter prediction over a region, along with the

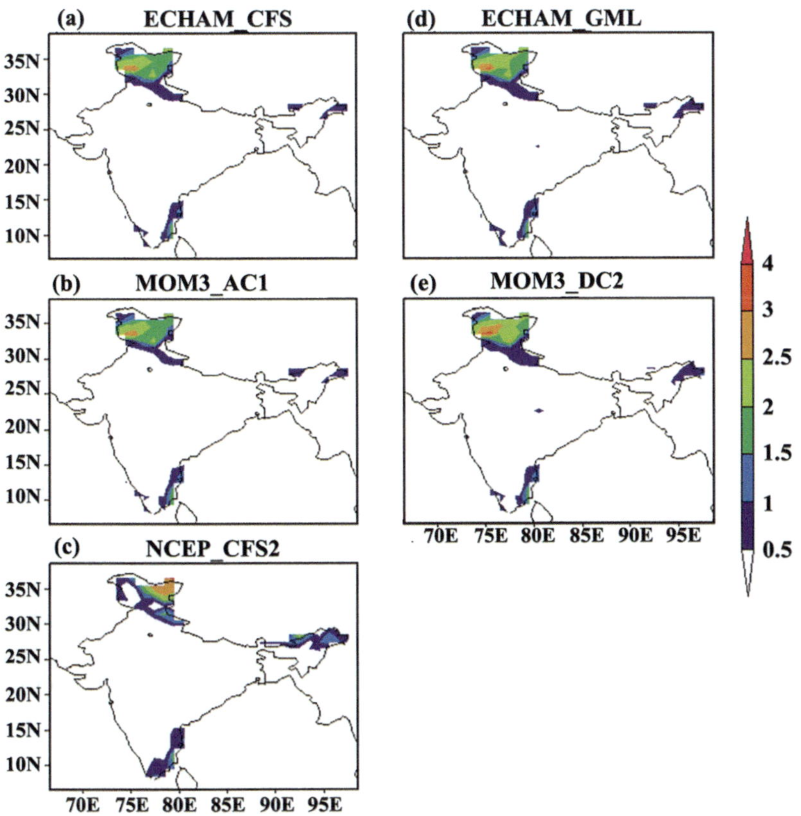

Fig. 6.6 RMSE (mm day^{-1}) of precipitation from the models against observations from 1982 to 2009 for December to February (Tiwari et al. 2014)

associated uncertainties involved in the prediction. It is found that none of the models is "perfect" in replicating the observation/truth, and the model has a systematic and asymmetric bias in predicting hydro-meteorological parameters, particularly rainfall.

Therefore, it is necessary to use appropriate statistical bias correction techniques to correct the forecasts (say rainfall) from the GCMs products. Statistical bias correction methodologies act on model outputs to remove the systematic bias. Consequently, the statistical properties of the corrected data match those observations. Past studies proposed different transfer functions, which remove bias from the GCM raw outputs. Kharin and Zwiers (2002) described the simple and regression-based bias corrections on 500 hPa geopotential height forecasts derived from the Atmospheric Model Intercomparison Project (AMIP). Hashino et al. (2006) evaluated three bias correction methods for ensemble stream flow volume forecasts for the Des Moines River basin. The results showed that all three bias-correction methods significantly improved the forecast quality by eliminating

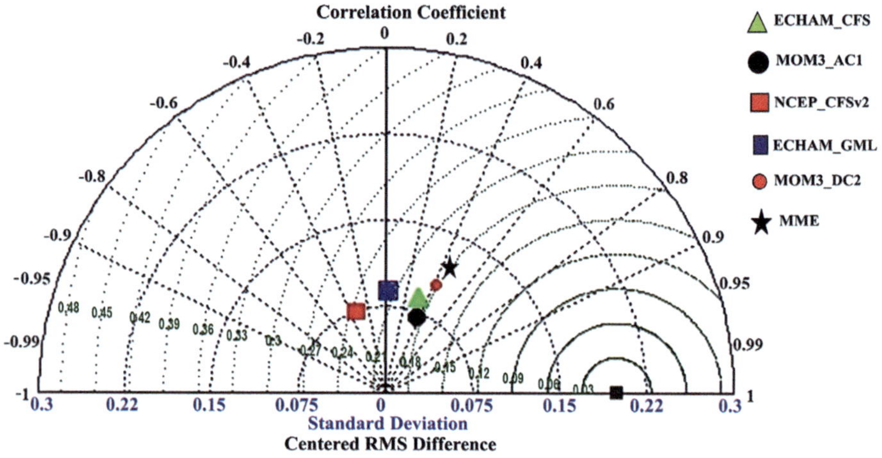

Fig. 6.7 Taylor diagram for the all-India average precipitation prediction skill of the GCMs (Tiwari et al. 2014)

unconditional biases and enhancing the forecast skill. Ines and Hansen (2006) showed the improvement of daily rainfall forecast by ECHAM4p5 GCM (developed at the Max-Plank Institute, Germany) after using a quantile-based bias correction. Li et al. (2010) also used the same technique for the Intergovernmental Panel on Climate Change AR4. Sajani et al. (2007) showed the skill of multiple linear regression-based bias correction implemented on the 12 ensemble members of the Meteorological Research Institute (MRI) model by the Japan Meteorological Agency for simulating the Indian Summer Monsoon Rainfall (ISMR). Feudale and Tompkins (2011) applied empirical orthogonal function-based bias correction techniques on the operational seasonal forecasts made by the European Centre for Medium-Range Weather Forecasts (ECMWF) for the West African monsoon region. The correction technique was found to improve the location of monthly and seasonal average precipitation anomalies. Piani et al. (2010) showed that an improvement in the forecast can be made by using statistical bias correction of the Danish Meteorological Institute (DMI) regional model over Europe. Haerter et al. (2011) proposed a cascade bias correction method on precipitation and temperature forecast by ECHAM5 GCM, forced by the IPCC B1 emission scenario 5 for future climate. Fan and van den Dool (2011) used a simple bias correction method to correct the daily operational ensemble week-1 and week-2 precipitation and 2 m surface air temperature (T2M) forecasts from the National Centre for Environmental Prediction (NCEP) Global Forecast System (GFS). Although it has been found that a large portion of these forecast errors is removable, the effectiveness is time and space-dependent. Nevertheless, these studies revealed the significant effect of bias correction on GCM products. However, these improvements are restricted to the specific time scale of the fluctuations that are considered. For example, the same statistical bias correction for daily values may not work on a monthly scale (Haerter et al. 2011).

Although the literature is discussed above in the context of bias correction of GCMs over the globe, however, it is highly necessary to apply these bias correction methods independently on a particular variable say precipitation, for a definite time scale (e.g., seasonal) over a fixed domain. This attempt to evaluate the skill of different statistical bias correction techniques implemented on the GCMs output will make confident to use a particular technique. Broadly, the evaluation of bias-corrected results is needed

- To find a robust bias correction method for a particular parameter (specific time scale and region).
- To examine whether the existing skill metrics such as climatology, root mean square error are reduced in the bias-corrected products.
- To examine how far the bias-corrected products such as rainfall can capture the extreme cases.

6.3 Bias-Correction Techniques

There are many techniques for bias corrections, which are divided into two groups. First is the transformation technique and the second one is without the transformation technique. While transformation method shows a statistical function that is constructed to remove the biases and in the transformation case, the biases are estimated explicitly and the model product is corrected by adjusting the biases. Suppose $F_{i,t}$ (i varies from 1 to n; n = number of members/events) is the rainfall predicted by the ith member of a model, and Y_t is observed rainfall at time t where the mean of ensemble members/events is referred to as $\overline{F_t}$.

6.3.1 Without Transformation Functions (WOTF)

In this process, bias is removed from the model's forecast without any transformation function. In WOTF, the following methods are used.

6.3.1.1 Mean Bias-Remove Technique

In this mean bias-remove technique (MBRT), the mean bias is adjusted every year. Mean bias is defined as the difference between observed climatology ($\overline{Y}$) and climatology of ensemble mean $\left(\overline{\overline{F}}\right)$:

$$b_t = \overline{Y} - \overline{\overline{F}} \tag{6.4}$$

For each year, this difference (b_t) is calculated in the leave-one-out cross-validation manner and adds this mean bias in the 'test' (t) year's model mean i.e.

$$F_t^c = F_t + b_t \tag{6.5}$$

where, F_t^c is the bias-corrected forecast for the tth year; b_t is the calculated bias.

If the data series is of moderate length, then one can use a leave-window method where the forecasted year is left in the calculation of the climatology.

For example, for each year, the difference (b_t) is calculated in the leave-one-out cross-validation manner and adds this mean bias in the 'test' (t) year's model mean, i.e., if a user has observed and model data for the period 1982–2008 and the bias correction for the year 2008 is calculated, then b_{2008} is calculated on the basis of the data from 1982 to 2007. This indicates that the model and observed climatology will be calculated for the 1982–2007 period (leaving the predicted year) to calculate b_t. Now, bias-corrected rainfall (F_{2008}^c) of the year 2008, the ensemble mean of that model (F_{2008}) for 2008 is added to the calculated b_{2008} (as defined in Eq. (6.5)). Kharin and Zwiers (2002) described this method as bias-removed individual forecast.

6.3.1.2 Multiplicative Shift Technique

Ines and Hansen (2006) applied this multiplicative shift technique (MST) which is a simple method. In this method, the ratio between observed climatology ($\overline{Y}$) and the climatology of the ensemble mean ($\overline{\overline{F}}$) is calculated and then applied to the forecasted value, that is,

$$m_t = \frac{\overline{Y}}{\overline{\overline{F}}} \tag{6.6}$$

The value m_t is calculated using observed and modeled climatology. Then it is multiplied by the model mean for obtaining the bias-corrected rainfall by the following equation:

$$F_t^c = \overline{F}_t \times m_t \tag{6.7}$$

The process where, F_t^c is the bias-corrected forecast for the tth year; if the data length is moderate, the user can consider the leave-one-out method as described above to calculate m_t.

6.3.1.3 Standardised Technique

Standardisation is the widely used bias-correction method to rectify the systematic error in the mean rainfall amount. The standardised technique (SZT) is also a simple bias correction technique introduced by Pan and van den Dool (1998). In their study, the standardisation process is implemented for both model and observation before making the probabilistic forecast. This technique adjusts the systematic error for the model. After standardization, the magnitude of the ensemble mean and observation lies in a certain range so that the bias is almost zero. However, it is not recommended to disseminate anomalies when giving a deterministic forecast to the end user. Therefore, after the standardisation of the ensemble mean, the observed climatology and standard deviation were reconstructed. In other words, it was assumed that the standardized anomaly of the ensemble mean is:

$$\overline{F'} = \left(\overline{F_t} - \overline{\overline{F}}\right)/\sigma_{\overline{F}} \tag{6.8}$$

where $\overline{F_t}$ is the ensemble mean of the tth year, $\overline{\overline{F}}$ is the climatology of model ensemble mean, and $\sigma_{\overline{F}}$ is the standard deviation of the model ensemble mean; In the example, the climatology and standard deviation are computed in the leave-one-out cross-validation mode by leaving the forecasted year (here tth year). The leave-one-out cross-validation method is the same as discussed above. In generating the forecast in a deterministic value, the forecast is reconstructed with the observed climatology ($\overline{Y}$) and observed standard deviation σ_Y by the equation below:

$$F_t^c = \overline{F_t'} \times \sigma_{\overline{Y}} + \overline{Y} \tag{6.9}$$

where F_t^c is the bias-corrected forecast for the tth year.

Therefore, in this method, the long-term climatology and standard deviation of the observation are directly projected in the ensemble mean by the cross-validation procedure.

6.3.2 *With Transformation Functions (WTF)*

In this process, bias is corrected by statistical transformation functions such as fitting or mapping between observation and model outputs. The following WTFs are considered in generating bias-corrected forecasts.

6.3.2.1 Regression Technique

In the regression technique (RT) method, a regression model is constructed between observation (Y) and the ensemble mean of the model ($\overline{F_t}$) by the following equation:

$$F_t^c = a_0 + a_t \times \overline{F_t} \tag{6.10}$$

where the coefficient of the regression equation is estimated by the least square estimates i.e.

$$a_t = Cov(Y, \overline{F})/Var(\overline{F}) \tag{6.11}$$

and

$$a_0 = \overline{Y} - a_t \times \overline{\overline{F}} \tag{6.12}$$

F_t^c is the bias-corrected forecast for the tth year. The regression coefficient a_t (slope) and regression constant a_0 (intercept) can be calculated by using the leave-one-out cross-validate manner for moderate data length, as discussed above. This regression coefficient rescales the forecast to correct systematic errors (Kharin and Zwiers 2002).

6.3.2.2 Quantile Mapping Method

A ‘quantile-based’ bias correction approach is useful to statistically transform rainfall simulated by a model to bias-corrected data and to make it applicable for use in the impact assessment model. This quantile mapping method (QMM) is also referred to as ‘histogram equalisation’ and/or ‘rank matching’ (Wood et al. 2004; Hamlet et al. 2002; Piani et al. 2010). The statistical bias correction method employed in this study is based on the initial assumption that both observed and simulated intensity distributions are well approximated by the gamma distribution.

In the quantile mapping method, empirical probability distributions of the observed and forecasted values are used. The bias-corrected outfput is the inverse of the cumulative distribution function (CDF) of observed values at the probability corresponding to the model output CDF at the particular value. In the quantile mapping method, the bias is not calculated explicitly. Suppose CDFs, F_y for observed data and $F_{\overline{F}}$ for ensemble mean of model forecast is known. For $\overline{F_t}$ the bias corrected value, Q will then be as follows:

$$Q = F_Y^{-1}\left(F_{\overline{F}}\left(\overline{F_t}\right)\right) \tag{6.13}$$

Here, F^{-1} is an inverse of CDF. Thus, the quantile mapping procedure is a transformation between two CDFs. In the example, the whole procedure is implemented in the leave-one-out cross-validation way. A schematic diagram to transfer the model information using bias correction method of quantile mapping is represented in Fig. 6.8. It may be noted that for the same CDF of observed and model the corresponding magnitudes are different.

6.3.2.3 Principal Component Regression

The principal component analysis (PCA) is found appropriate when a large number of observed variables exists, which may contain some redundancy and is necessary to develop a smaller number of new variables (called *principal components*) which are uncorrelated. Thus, PCA reduces a dataset containing a large number of variables to a dataset containing fewer new variables. These new variables are linear combinations of the original ones, and these linear combinations are chosen to represent the maximum possible fraction of the variability contained in the original data. Sajani et al. (2007) implemented multiple linear regressions on the member to find a bias-corrected weighted ensemble mean. They took 23 years to

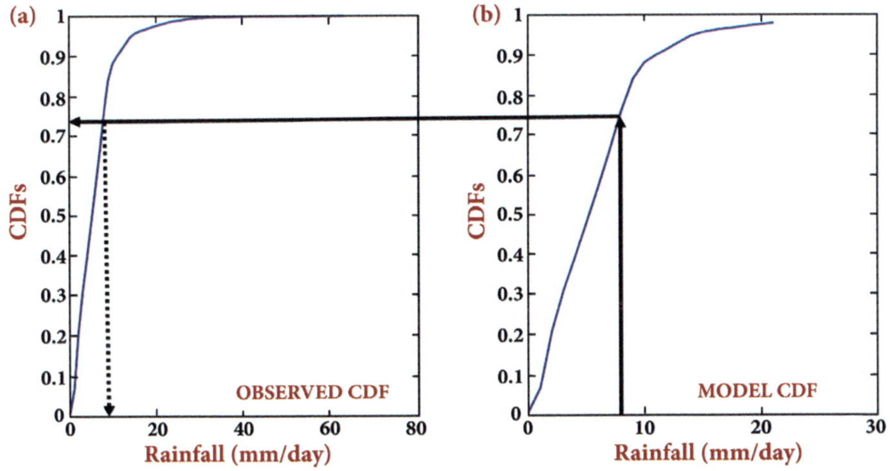

Fig. 6.8 Schematic diagram of observed (**a**) and model (**b**) CDF and its transformation

make the multiple linear regression models. Although the number of predictors is less than the total number of years, it is not sufficiently less. So, there is a high chance of over fitting and the presence of multicollinearity among the members may affect the estimated coefficient adversely (Fisher 1922). Therefore, instead of using all the members, one should screen them on the basis of principal component analysis. By use of principal component analysis, one can choose the predictors which contribute the maximum to the explained variance. The PCA based regression is introduced called principal component regression (PCR) for improvement over the work of Sajani et al. (2007). In this method, at first, the PCA is done on members and only such modes which explain a high amount of variance are retained for further simple multiple linear regression. It is general practice to consider a number of principal components (PCs) that can explain more than or equal to 80% of the variance. In the example, a minimum number of PCs is chosen to explain 90% of the variance for the regression. The PCs are orthogonal, and the multicollinearity is removed. This whole procedure is carried out in the same type of leave-one-out cross-validation mode as discussed earlier.

6.4 Standard Statistical Techniques for the Generation of a Single Forecast

The predictability of GCM has a strong dependence on the model's initial conditions. It is well established that if an error in initial conditions is present when model initial conditions are prescribed, that may amplify during the model integration and cause large uncertainties in the forecast. Besides this, uncertainty may also arise due to inappropriate representation of land surface characteristics and physical processes in the model. Therefore, the estimation of systematic bias is highly necessary to minimise the errors in the forecast. The degrees of accuracy in the prediction varies from model to model, region-to-region, and on different time scales (such as season-to-season, etc.). It is found that combining different GCMs (multi-model ensemble, MME) forecasts can greatly reduce the uncertainty in the prediction and the combined forecast significantly improve the forecast skill over the individual GCM. Thus, the forecast can be improved by improving the GCM physics and dynamics, and statistical corrections (or downscaling) of the GCMs output. The MME is one of the widely used techniques to improve forecasts at different space and time. There are several methods for combining multi GCM products to generate one single forecast. We will discuss here a few standard methods that are widely being used and one advanced multi-model multi-technique method for seasonal scale prediction.

6.4.1 Multi-model Ensemble (MME) Techniques

Two kinds of mathematical models, viz., statistical and dynamical, are being used for the prediction of rainfall. In the statistical approach, the first method is a time series analysis of historical rainfall observations. If the autocorrelation of year-to-year between the model and observation is significantly low, the model cannot work very well. Another approach is to build a regression model based on the relationship between rainfall and various atmospheric parameters like sea surface temperature, pressure tendency, etc. As this relationship also varies from year-to-year, this model also failed to forecast the rainfall correctly, especially during drought and flood years.

From a dynamic point of view, scientists are trying to give a forecast by formulating various physical processes occurring over the entire globe in the form of differential equations. These processes involve radiation budget, wind circulation, sea surface temperature, momentum transfer, soil moisture, vegetation cover, etc., and are responsible for the general circulation of the atmosphere. Therefore, these types of mathematical models are termed general circulation models. Though initial conditions for solving these equations are almost similar, some of the physical processes like momentum transfer, surface friction, cloud formation, etc., are parameterised in various ways in different models. Secondly, the initial conditions of a single model are also perturbed several times to get different outputs, which are known as members of the particular model. Since different models have different errors or biases, their output forecasts are not the same, and errors vary from region to region for a particular model.

With the availability of climate predictions produced by several dynamical models, multi-model ensemble (MME) forecasting has drawn some attention. Several approaches were attempted to combine MME forecasts into a single reliable forecast that carries higher skills when compared to the individual member models. These include the simple ensemble mean (Doblas-Reyes et al. 2000; Pavan and Doblas-Reyes 2000; Stephenson and Dolas-Reyes 2000; Peng et al. 2002; Palmer et al. 2004), the regression-improved ensemble mean (Peng et al. 2002; Kharin and Zwiers 2003), the bias-removed ensemble mean (Kharin and Zwiers 2002). The multi-model super ensemble technique showed a higher skill for short-range and seasonal forecasting compared with member model forecasts (Krishnamurti et al. 2000a, b; Yun et al. 2003, 2005). A few standards and widely used procedures of the Multi-model Ensemble (MME) technique are presented next.

Theoretical Foundation

Suppose F_1, F_2, F_3 ... F_m is m number of model simulated results with a hindcast run for t times (years) where each model simulation consists of several number of ensemble members. For example, let F_1 is the first model simulated outputs that have 15 ensemble members, F_2 is the second model simulated outputs that have 24 members, etc. Ensemble members are weighted equally to construct an ensemble mean as the prediction of an individual model. The main aim is to get a single forecast for $(t+1)$ year. There are different ways of a multi-model ensemble (MME) which are discussed in the following sections.

6.4.2 Arithmetic Mean Method

The simplest of all, the MME scheme in literature has been the method based on simple averaging of all the individual models. In other words, all the individual member models have been assigned the same weight while carrying out the ensemble average. Therefore, for $(t+1)$ year take the arithmetic mean of all model's individual forecasts. There is an assumption in this method that each model is relatively independent, and to some extent, it has the capability to forecast the regional climate well; therefore, one can expect a well model forecast by a simple composite of each model prediction from different models. Hagedorn et al. (2005) have described the rationale behind the success of such techniques for seasonal prediction. In addition, this simple scheme contains the common advantage and limitations of the model predictions; therefore, it could be a good benchmark for evaluating other MME schemes. This method is referred to as MME_{av}.

Before using the MME_{av} method, all models' forecasts are normalised with their own long-term mean (climatology) and long-term standard deviation (interannual variation). The simple mean of all normalised values was calculated. After that, these values were multiplied by observed interannual variation and added to the observed climatology for getting the final forecast.

The MME_{av} forecast was constructed with bias-corrected data and the mathematical formulation is given below:

$$S_t = \overline{O} + \left[\frac{1}{N}\sum_{i=1}^{N}\left(\frac{F_{i,t} - \overline{F}_i}{\sigma_{F_i}}\right)\right]\sigma_o \tag{6.14}$$

where S_t = MME prediction at time t; $F_{i,t}$ = ith model forecast at time t; $\overline{F}_i$ = Climatology of the ith model forecast; $\overline{O}$ = Climatology of observations for the same period; σ_{F_i} = Interannual variation of the ith model forecast; σ_o = Interannual variation of the observations; N = Total number of models.

6.4.3 Multiple Linear Regression Using Singular Value Decomposition

In this method for carrying out a weighted multi-model ensemble (MME) mean, the multiple linear regression method has been employed. Krishnamurti et al. (2000a, b) have introduced this approach to get betterment over the MME_{av} scheme and called "Superensemble". In this method, the regression coefficients are obtained using Gauss–Jordan elimination with pivoting. In general, the use of the Gauss–Jordan elimination method for solving the regression problem is not recommendable as it is not numerically robust. This type of problem appears when the covariance matrix of models is singular which means the determinant of the covariance matrix is zero. In other words, the covariance matrix is not full-ranked. Therefore, the calculation of the inverse of the covariance matrix is not possible in a classical way because of multicollinearity problems which means models are not independent. This problem can be solved with the use of principal component analysis, ridge regression, etc. by retaining only higher modes. The singular value decomposition (SVD) technique can be applied for computing inverse when the matrix is not full-ranked. Mandel (1982) shows the use of SVD in regression. SVD-based MME techniques were introduced by Yun et al. (2003) for finding robust weighted MME forecasts. Here onwards this technique is referred to as MLR_{svd}. A detailed mathematical description of this method is given below.

The MLR_{svd} forecast is constructed with bias-corrected data and can be mathematically expressed as

$$S_t = \overline{O} + \left[a_i\sum_{i=1}^{N}\left(\frac{F_{i,t} - \overline{F}_i}{\sigma_{F_i}}\right)\right]\sigma_O \tag{6.15}$$

where a_i = Regression coefficient obtained by a minimisation procedure during the training period between models forecasts F_i's and observation O and other terms have already been described in the previous section.

Estimation of a_i:

The regression equation between observation and model is given below

$$O = F \times \beta + \varepsilon \tag{6.16}$$

$$\text{where, } \underset{t\times 1}{O} = \begin{pmatrix} O_1 \\ O_2 \\ \cdot \\ \cdot \\ \cdot \\ O_t \end{pmatrix} \quad F = \begin{bmatrix} F_{11} & F_{21} \ldots F_{n1} \\ F_{12} & F_{22} \ldots F_{n2} \\ \ldots & \ldots \\ \ldots & \ldots \\ \ldots & \ldots \\ F_{1t} & F_{2t} \ldots F_{nt} \end{bmatrix} \quad \beta_{t\times 1} = \begin{pmatrix} a_1 \\ a_2 \\ \cdot \\ \cdot \\ \cdot \\ a_n \end{pmatrix} \quad \varepsilon_{t\times 1} = \begin{pmatrix} \varepsilon_1 \\ \varepsilon_2 \\ \cdot \\ \cdot \\ \cdot \\ \varepsilon_n \end{pmatrix}$$

Now, the minimisation of $E(\varepsilon^2)$ is computed by the least square method and one can get regression coefficients. These coefficients are obtained up to time t (training phase) and use this regression coefficient for year $(t+1)$ (forecast phase). The following flow diagram shows how to calculate and use these regression coefficients.

For obtaining the weights (a_i), the Covariance matrix of models is built with the seasonal cycle-removed anomaly (F')

$$C_{i,j} = \sum_{t=0}^{Train} F'_{i,t} F'_{j,t} \tag{6.17}$$

where, Train = Training period (1, 2, …, t); $i \rightarrow i$th model forecast; $j \rightarrow j$th model forecast.

The goal of regression is to express a set of data as a linear function of input data. For this, we construct a set of linear equation. By the least-square method, we get,

$$\begin{aligned} C_{F\cdot F} \cdot \beta &= C_{F\cdot O} \\ \beta &= C^{1}_{F\cdot F} \cdot C_{F\cdot O} \end{aligned} \tag{6.18}$$

where,

$$C_{F\cdot F} = \begin{bmatrix} C_{11} & C_{12} \ldots C_{1n} \\ C_{21} & C_{22} \ldots C_{2n} \\ \ldots & \ldots \\ \ldots & \ldots \\ \ldots & \ldots \\ C_{n1} & C_{n2} \ldots C_{nn} \end{bmatrix} \qquad C_{F\cdot O} = F'^{T} O' O' = \left(O_t - \overline{O}\right) F' = \left(F_{j,t} - \overline{F}_j\right)$$

When C_{FF} is singular, calculating $C^{-1}_{F\cdot F}$ is not possible by the conventional method. Therefore, singular value decomposition is applied for calculating $C^{-1}_{F\cdot F}$ and hence β. A representational diagram of multi-linear regression using

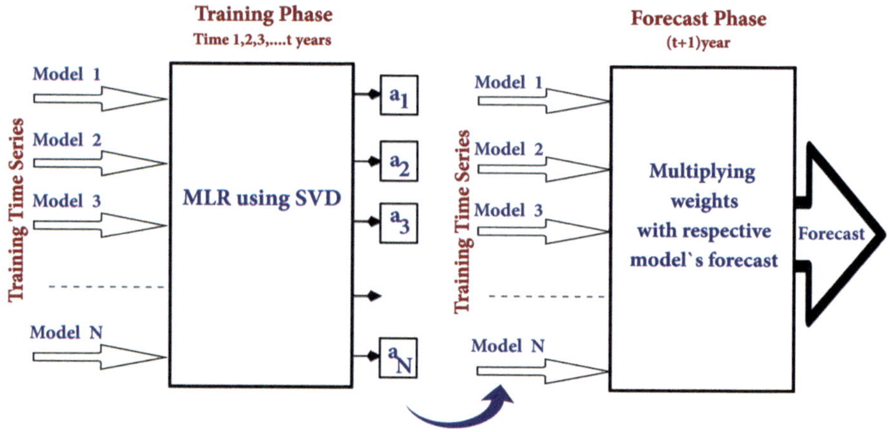

Fig. 6.9 Schematic diagram for computing statistical weights obtained in the training phase and the same passed on to the forecast phase

singular value decomposition is shown in Fig. 6.9. In this figure, the two phases i.e. Training phase and Forecast phase illustrate that the coefficients found in the Training phase will be applied in the Forecast phase in which the inputs will be multiplied by the respective weights calculated in the Training phase.

Use of Singular Value Decomposition (SVD)

Singular value decomposition (SVD) is a very robust method to solve the singularity problem. Singular value decomposition is a particular type of factorisation of a matrix into a product of three matrices, of which the second is a diagonal matrix that has, as the entry, entries on its diagonal the singular values of the original matrix. The decomposition of a matrix is useful when the matrix is not of full rank. That is, the rows or columns of the matrix are linearly dependent.

Let A $(n \times p)$ be a rectangular matrix, then the SVD theorem states:

$$A_{n\times p} = U_{n\times n} \cdot S_{n\times p} \cdot V^{T}_{p\times p} \tag{6.19}$$

where, $U^TU = I_{n\times n}$, $V^TV = I_{p\times p}$ and $I =$ Identity matrix. (i.e. U and V are orthogonal).

Calculating the SVD consists of finding the eigen values and eigen vectors of AA^T and A^TA. The eigen vectors of A^TA make up the column of V, and the eigen vectors of AA^T make up the column of U. In addition, the singular values in S are square roots of eigen values from AA^T or A^TA, which are arranged in descending order. If A is symmetric, then the left and right singular vectors of U and V are equal. If matrix A is a real matrix, then U and V are also real.

For $n = p$, matrices **U** and **V** are square and the inverse matrix of $\boldsymbol{A}$ is

$$A^{-1} = \left(U\,S\,V^T\right)^{-1} = \left(V^T\right)^{-1} S^{-1} U^{-1} = V\,S^{-1} U^T \tag{6.20}$$

where the diagonal elements of $\mathbf{S}^{-1}$ are $1/S_i$. The matrices **U** and **V** are orthogonal and their inverse matrices are equal to their transposes i.e.

$$U^{-1} = U^T \text{ and } V^{-1} = V^T \tag{6.21}$$

SVD is applied to the computation of the regression coefficients for a set of different model forecasts. The SVD of the covariance matrix $C_{F \cdot F}$ is its decomposition in to a product of three matrices.

$$C_{F \cdot F} = U \cdot S \cdot V^T \tag{6.22}$$

Because the covariance matrix $C_{F \cdot F}$ is a square symmetric matrix, $C_{F \cdot F}^T = VSU^T = C_{F \cdot F}$. This proves that the left and right singular vector U and V are equal.

Therefore,

$$C_{F \cdot F}^{-1} = \mathbf{V} \cdot \mathbf{S}^{-1} \cdot \mathbf{U}^{\mathrm{T}} \tag{6.23}$$

Hence,

$$\beta = C_{F \cdot F}^{-1} \cdot C_{F \cdot O}$$
$$\beta = \left(V \cdot S^{-1} \cdot U^T\right) \cdot C_{F \cdot O}$$

where, $S_i \geq \varepsilon$, else 0, where ε is the so-called *singularity threshold*. In other words, if S_i is zero or close to zero (smaller than ε), one must replace $1/S_i$ with 0. The value of ε depends on the precision of the hardware. This method can be used to solve linear equations systems even if the matrices are singular or close to singular. Detailed is described by Eubank and Webster (1985).

6.4.4 Principal Component Regression Technique

The use of multiple regression models often depends on the estimates of the individual regression coefficients. Some examples of inferences that are frequently made include

1. Identifying the relative effects of the predictor variables,
2. Prediction and/or estimation, and
3. Selection of an appropriate set of variables for the model.

If there is no linear relationship between the predictors, they are said to be orthogonal. When the predictors are orthogonal, the inferences such as those illustrated above can be made relatively easily. Unfortunately, in most applications of regression, the predictors are not orthogonal. However, in some situations, the predictors are nearly perfectly linearly related, and in such cases, the inferences based on the regression model can be misleading or erroneous. When there are near-linear dependencies among the predictors, the problem of multicollinearity is said to exist. Due to multicollinearity, there can be no clear interpretation of the regression coefficients as measures of original effects. Moreover, the coefficients will be unstable in the sense that small data revisions could have a disproportionate effect on the calculated coefficient and thus prediction. A high degree of multicollinearity can also cause computer software packages to be unable to perform the matrix inversion that is required for computing the regression coefficients. To alleviate the problem of multicollinearity, Principal Component Analysis (PCA) of predictors is recommended. This methodology is found appropriate when we have a large number of observed variables, which may contain some redundancy and wish to develop a smaller number of artificial variables (called principal components) which are uncorrelated. Thus, PCA reduces a dataset containing a large number of variables to a dataset containing fewer new variables. These new variables are linear combinations of the original ones, and these linear combinations are chosen to represent the maximum possible fraction of the variability contained in the original data. Geometrically, these linear combinations represent the selection of a new coordinate system obtained by rotating the original system so the new axes represent the directions with maximum variability and provide a simple more parsimonious description of the covariance structure. When these principal components are used as predictors, the regression is termed Principal Component Regression. Rajeevan et al. (2004) used this methodology to forecast the southwest monsoon rainfall in 2003 over India. Cannon and McKendry (1999) used the principal component of 200 hPa Geopotential height set in May and found it to yield forecasts that were better than climatology. Yu et al. (1997) employed this methodology to forecast rainfall in 10 USAPI stations using SSTs in the Pacific Ocean as predictors and found good skill. In light of the above discussions, it is clear that Principal Component Regression can be used as a tool for prediction.

Theoretical and Numerical Aspects

Given multiple observations of a $(K \times 1)$ data vector $\boldsymbol{x}$, PCA finds $(M \times 1)$ $(M << K)$ vectors $\boldsymbol{u}$ whose elements are linear combinations of the elements of the $\boldsymbol{x}$s, which contain most of the information in the original collection of $\boldsymbol{x}$s. This transformation ranks the new orthogonal variables in order of their importance, and the procedure then involves eliminating some of the principal components to effect a reduction in variance. After the elimination of the least important principal components, a multiple regression analysis of the response variable against the reduced set of principal components is performed using ordinary least squares estimation (OLS). Because the principal components are orthogonal, they are pairwise independent, and hence OLS is appropriate. Once the regression coefficients for the reduced set of orthogonal variables have been calculated, they are mathematically transformed into a new set of coefficients that correspond to the original or initial correlated set of variables. These new coefficients are principal component estimators (Gunst and Mason 1980).

Centering and Scaling

Suppose we have n years and k variables $x_1, x_2, \ldots x_k$ which are independent. Let $y_1, y_2, \ldots y_k$ be the dependent variable. Then the regression model will take the following form

$$y_i = \beta_0 + \beta_1 x_{1i} + \beta_2 x_{2i} + \cdots + \beta_k x_{ki} \quad (i = 1, 2, \ldots, n) \tag{6.24}$$

where $\beta_0, \beta_1, \beta_2, \ldots \beta_k$ are the regression coefficients. The above equation in matrix form can be written as

$$Y = X \cdot \beta + \in \tag{6.25}$$

Suppose that the independent variables (each column of X) are centered and scaled, i.e. x_{ji}, the ith year measurement on the jth climatic variable (x_j) in the natural units, is transformed into x_{ji}^* as follows:

$$x_{ji}^* = \frac{x_{ji} - \overline{x}_j}{S_j} \tag{6.26}$$

where, $S_j = \frac{1}{n-1}\sqrt{\sum_{i=1}^{n}\left(x_{ij} - \bar{x}_j\right)^2}$. The process of centering and scaling allows for an alternative formulation of Eq. (6.24) as follows:

$$y_i = \beta_0^* + \beta_1^*\left[\frac{x_{1i} - \overline{x}_1}{S_1}\right] + \beta_2^*\left[\frac{x_{2i} - \overline{x}_2}{S_2}\right] + \ldots + \beta_k^*\left[\frac{x_{ki} - \overline{x}_k}{S_k}\right] + \in_i \tag{6.27}$$

Consider the model formulation in Eq. (6.27). Separating the first column of ones (1) from the X matrix results in the model form

$$\boldsymbol{y} = \beta_0^* 1 + \mathbf{X}^* \beta^* + \in \tag{6.28}$$

where, in this form, $\beta^* = \left(\beta_1^* \beta_2^* \ldots \beta_k^*\right)'$ is the vector of coefficients, apart from the intercept, and X^* is then $n \times k$ matrix of centered and scaled independent variables. The notation 1 is used to denote an n-vector of ones.

Computational Technique

Let X^* be the centered and scaled $n \times k$ data matrix as given in Eq. (6.5). The $k \times k$ correlation matrix of the climatic variables is then $C = \boldsymbol{X}^{*\prime}\boldsymbol{X}^*$. Let $\lambda_1, \lambda_2, \ldots \lambda_k$ be the eigenvalues of the correlation matrix, and $V = [v_1, v_2 \ldots v_k]$ be the $k \times k$ matrix consisting of the normalized eigenvectors associated with each eigenvalue. The eigenvalues are the solution of the determinant equation $\left|X^{*\prime}X^* - \lambda I\right| = 0$ (Draper and Smith 1981), and associated with each eigenvalue, λ_j, is a vector, $\boldsymbol{v}_j$, that satisfies the set of homogeneous equations $\left(\mathbf{X}^{*\prime}\mathbf{X}^* - \lambda_j \boldsymbol{I}\right) = 0$. The vectors, $\boldsymbol{v}_j = \left(\boldsymbol{v}_{1j}, \boldsymbol{v}_{2j} \ldots \boldsymbol{v}_k\right)'$, are the normalized solutions such that $v_j' v_j = 1$ and $v_j' v_i = 0$

for $\boldsymbol{i} \neq \boldsymbol{j}$. That is, the eigenvectors have unit lengths and are orthogonal to one another. Hence the eigenvector matrix $\boldsymbol{V}$ is orthonormal, i.e., $\boldsymbol{V}=\boldsymbol{I}$.

Now consider the model formulation given in Eq. (6.28). That is $y = \beta_0^* 1 + X^*\beta^* + \in$. Since $\boldsymbol{V}=\boldsymbol{I}$ one can write the original regression model (Eq. 6.28) in the form

$$y = \beta_0^* 1 + \boldsymbol{X}^*\boldsymbol{V}\boldsymbol{V}'\beta^* + \in \tag{6.29}$$

$$y = \beta_0^* 1 + \boldsymbol{Z}\alpha + \in \tag{6.30}$$

where $\boldsymbol{Z} = \boldsymbol{X}^*\boldsymbol{V}$ and $\alpha = \boldsymbol{V}'\beta^*$. $\boldsymbol{Z}$ is an $n \times k$ matrix of principal components and $\alpha = (\alpha_1\alpha_2 \ldots \alpha_k)$ is a $k \times 1$ vector of new coefficients. The model formulation in Eq. (6.30) can be expanded as $y = \beta_0^* + \alpha_1 z_1 + \alpha_2 z_2 + \cdots + \alpha_k z_k + \in$, where $z_1, z_2 \ldots z_k$ are the k new variables called principal components of the correlation matrix. Hence, the model formulation in Eq. (6.30) is nothing more than the regression of the response variable on the principal components, and the transformed data matrix $\boldsymbol{Z}$ consists of the k principal components.

For the model in Eq. (6.30), the principal components are computed using:

$$\boldsymbol{Z} = X^* V \tag{6.31}$$

where, $\boldsymbol{X}^*$ is the $n \times k$ matrix of centered and scaled climatic variables without a column of ones, and $\boldsymbol{V}$ is the $k \times k$ orthonormal matrix of eigenvectors. The principal components are orthogonal to each other, that is:

$$\boldsymbol{Z}'\boldsymbol{Z} = \left(\boldsymbol{X}^*\boldsymbol{V}\right)'\left(\boldsymbol{X}^*\boldsymbol{V}\right) = \boldsymbol{V}'\boldsymbol{X}^*\boldsymbol{X}^*\boldsymbol{V} = \boldsymbol{V}'\boldsymbol{C}\boldsymbol{V} = diag(\lambda_1, \lambda_2, \ldots \lambda_k) \tag{6.32}$$

Equation (6.32) shows that $z_j'z_j = v_j'Cv_j = \lambda_j$ and $z_j'z_j = 0,\ i \neq j$. From Eq. (6.31), one can see that the principal components are simply linear functions of the centered and scaled climatic variables, and the coefficients of this linear combination are the eigenvectors. For example, the elements of the jth principal components, z_j, are computed as follows:

$$z_j = v_{1j}x_1^* + v_{2j}x_2^* + \ldots + v_{kj}x_k^* \tag{6.33}$$

where $\boldsymbol{v}_{1j}, \boldsymbol{v}_{2j}, \ldots \boldsymbol{v}_{\boldsymbol{k}}$ are elements of the eigenvector associated with λ_j, and x_j^*'s are the centered and scaled climatic variables obtained using Eq. (6.26). Note that $\sum_{i=1}^{n} z_{ji} = 0$ the sum of squares of the elements of $z_j\left(\sum_{i=1}^{n} z_{ji}^2\right)$ is λ_j. Since $\sum_{j=1}^{k} \lambda_j = k$ then the total sum of squares, $\sum_{j=1}^{k}\left(\sum_{i=1}^{n} z_{ji}^2\right) = kz_j$ accounts for λ_j the total variance.

If the dependent variable (y) is regressed against k principal components using the model in Eq. (6.30), then the least squares estimator for the regression coefficients in vector α is the vector $\hat{\alpha} = \left(\boldsymbol{Z}'\boldsymbol{Z}\right)^{-1}\boldsymbol{Z}'y$, and the variance–covariance matrix of the estimated coefficients in vector is given by

$$\boldsymbol{Var}(\hat{\alpha}) = \hat{\sigma}^2\left(Z'Z\right)^{-1} = \hat{\sigma}^2\boldsymbol{diag}\left(\lambda_1^{-1}, \lambda_2^{-1}, \ldots \lambda_k^{-1}\right) \tag{6.34}$$

If all of the k principal components are retained in the regression model of Eq. (6.30), then all that has been accomplished by the transformation is a rotation of the k original climatic variables.

Elimination of Principal Components

Even though the new variables are orthogonal, the same magnitude of variance is retained. But if multicollinearity is severe, there will be at least one small eigenvalue. Eliminating one or more principal components associated with the smallest eigenvalues will reduce the total variance in the n model and thus produce an appreciably improved diagnostic or prediction model (Draper and Smith 1981; Myers and Myers 1990). The principal component matrix **Z** contains the same information as the original centered and scaled climatic dataset ($\mathbf{X}^*$), except that the data are arranged into a set of new variables which are completely uncorrelated with one another and which can be ordered or ranked with respect to the magnitude of their eigenvalues (Draper and Smith 1981; Myers and Myers 1990). Note that z_j corresponding to the largest λ_j accounts for the largest portion of the variation in the original data. Furthermore, z_j's explain smaller and smaller proportions until all the variation is explained; that is, $\sum_{j=1}^{k} \lambda_j = k$. Thus, the z_j are indexed so that $\lambda_1 > \lambda_2 > \ldots > \lambda_k > 0$. In the regression model of Eq. (6.30), one does not use all the z's but follows some selection rule. The property that makes PCR unique and more complex is that there is no universally agreed upon procedure in selecting the z_j's to be included in the reduced model of Eq. (6.30) (Draper and Smith 1981). Methods used to determine which and how many principal components should be removed to gain a substantial reduction in variance include:

a. The strategy of elimination of principal components should begin by discarding the component associated with the smallest eigenvalue. The rationale is that the principal component with the smallest eigenvalue is the least informative. Using the procedure, principal components are eliminated until the remaining components explain some pre-selected percentage of the total variance (for example, 80% or more). That is, one selects the set of largest contributors (principal components) in the total variance, that should be given priority in the selection to reach pre-selected percentage of variance, e.g. if it is 85%, then

$$\frac{\sum_{j=1}^{r} \lambda}{k} > 0.85$$

b. Some researchers use the rule that only principal components associated with eigenvalues greater than 1.00 are of interest (Draper and Smith 1981). This method is often referred to as the "Kaiser-Gutman Rule" (Loehlin 1998).
c. Others use the selection rule that keeps the first principal components whose combined eigenvalue product is greater than 1.00 (Guiot et al. 1982).

d. A more objective statistical strategy is to treat the principal component reduction as if it were a standard variable screening problem. Since the principal components are orthogonal predictor variables, a reasonable criterion to control the order of reduction are the *t*-statistics given by

$$t = \frac{\hat{\alpha}_j}{S_{\hat{\alpha}_j}} = \frac{\hat{\alpha}_j \sqrt{\lambda_j}}{S} \tag{6.35}$$

where, $s_{\hat{\alpha}_j}$ is the standard error (s.e.) of $\hat{\alpha}_j$ (Myers and Myers 1990). From Eq. (6.34) $Var(\hat{\alpha}_j) = S^2 \lambda_j^{-1}$ where $S^2 = \hat{\sigma}^2$ and hence, $s.e.(\hat{\alpha}_j) = S(\sqrt{\lambda_j})^{-1}$. In this procedure, *t*-values should be rank ordered and components should be considered for elimination beginning with the smallest *t*-value.

Suppose that some such selection rule results in the elimination of *r* principal components, that is, the model in Eq. (6.30) will now use only *k–r* components. Let us denote the reduced **Z** matrix of Eq. (6.30) by $\boldsymbol{Z_{k-r}}$ ($n \times (k-r)$ matrix). Let the reduced vector of coefficients (α) be $\alpha_{k-r} = (\alpha_1 \alpha_2 \ldots \alpha_{k-r})$. The reduced model, after the elimination of *r* principal components, can be written as

$$y = \beta_0^* 1 + Z_{k-r}\alpha_{k-r} + \epsilon^\circ \tag{6.36}$$

The superscript ∘ symbol on ∈ in Eq. (6.30), since they are not the same. But the predicted values and residuals of the model in Eqs. (5.29) or (5.30) are the same as those in Eqs. (6.24) or (6.25), (6.27) or (6.28). Note that

$$\boldsymbol{Z_{k-r}} = \boldsymbol{X}^* \boldsymbol{V}_{k-r} \tag{6.37}$$

where, $\boldsymbol{V_{k-r}} = [\boldsymbol{v}_1 \boldsymbol{v}_2 \ \ldots \ \boldsymbol{v}_{k-r}]$ is a $k \times (k-r)$ matrix of eigenvectors associated with the retained eigen values or principal components. The least squares procedure is then used to obtain a diagnostic or prediction equation for the response **y** as a function of the selected **z**'s; that is fitting the model in Eq. (6.36) using ordinary least squares. Once the fitted equation is obtained in terms of the selected **z**'s, it can be transformed back into a function of the original **x's** as described in the following section.

The Transformation Back to the Original Climatic Variables

Suppose with *k* variables and hence *k* principal components, $r<k$ components are eliminated. From Eq. (6.30) with the retention of all components, $\alpha = \boldsymbol{V}'\beta^*$ and the coefficients for the centered and scaled climatic variables are obtained as:

$$\beta = \boldsymbol{V}\alpha \tag{6.38}$$

If one eliminates r components and fits the model given in Eq. (6.36), the principal component estimators of the regression coefficients, in terms of the centered and scaled climatic variables for all k parameters of the model in Eq. (6.28) are given by (Gunst and Mason 1980)

$$b_{pc}^{*} = V_{k-r}\hat{\alpha}_{k-r}$$

where, $\boldsymbol{V}_{k-r}$ is defined as in Eq. (6.37), $\hat{\alpha}_{k-r}$ is the vector of estimated coefficients in the model of Eq. (6.36), and b_{pc}^{*} is a vector of estimated coefficients of the parameters in vector β^{*} of Eq. (6.28). Note that the elements of b_{pc}^{*} are principal component estimators of the coefficients of the centered and scaled climatic variables, and subscript pc is simply used to denote that the estimators are principal component estimators rather than ordinary least squares estimators. Since the $\mathbf{x}$'s are centered and scaled, the estimate of the constant term (β_0^{*}) in the model of Eqs. (6.28) and (6.36) is $\bar{y}$, that is, $\hat{\beta}_0^{*} = \bar{y}$.

Development of PCR Model

The above-mentioned PCR methodology is used to predict rainfall in different regions and seasons. The predictor set for the PCR model is developed using outputs from different Global Circulation Models (GCMs). The GCMs provide real-time predictions and have long series of hindcast runs. For example, consider rainfall data from a few GCMs such as CCM3v6, CFS, ECHAM4p5CASST (CASST), ECHAM4p5CFSSST (CFSSST), ECHAM4p5GML (GML), ECHAM-MOM3AC1 (AC1), ECHAM-MOM3DC2 (DC2), and JAMSTEC (JAM). The members of these GCMs are treated as predictors. These GCMs outputs can be downloaded from IRI data library (http://portal.iri.columbia.edu/portal/server.pt). A brief description of these GCMs is presented in Table 6.2.

In order to develop the model, one needs to train the data using a common period and for the above example, trained the data from 1982 as the GCMs have different hindcast periods. In the example, the mean of the ensembles (except the CFS model) is considered, so in all 22 variables are present, which will be treated as predictors (input variable). The observation obtained from India Meteorological Department (IMD) is treated as predictand (output variable).

Table 6.2 A brief description of different GCMs with their hindcast data period

	CCM3	CFS	CASST	CFSSST	GML	AC1	DC2	JAM
Category	2-Tier	Fully coupled	2-Tier	2-Tier	Semi coupled	Anomaly coupled	Direct coupled	Coupled
Resolution	T42 19 levels	T62 64 sigma layers	(T42, L18)	(T42, L18)	(T42, L18)	T42 19 levels	(T42, L18)	T106 19 levels
Members	24	15	24	24	12	12	12	9
Hindcast period	1957–2008	1981–2008	1957–2008	1982–2008	1982–2008	1982–2008	1982–2008	1982–2008

Steps in Developing PCR Model

1. As mentioned earlier, the members from different models are treated as predictors (X) and the data from IMD is treated as predictand (Y) since the resolution of these members is different so they are first regridded/interpolated onto the same resolution to Indian grid (6.5–38.5, 66.5–100.5).
2. The interpolated data is then normalized to remove the bias.
3. A subdivisional average is taken for each of the members.
4. These members are then screened according to their correlations with the observation. A threshold is fixed at 0.2. The predictors with a correlation greater than 0.2 will be selected first.
5. Then principal components are calculated by first finding the correlation matrix and
6. Then calculating the eigenvectors of the correlation matrix.

$$Z = X * V; \quad Y = Z * \beta$$

   Here X is a vector of independent variables (predictors), Y is a vector of dependent variable (predictand), V is eigenvector matrix obtained from the correlation matrix of X, Z the resulting Principal components and β the regression coefficients.
7. Again correlation coefficient is found between these principal components (PCs) and the observation and the first three PCs with the highest correlation are selected.
8. Now these selected PCs will enter model (2) stepwise and the model with least RMSE is selected.
9. The above steps (steps 4–7) are repeated for seven different thresholds (0.2–0.5 in steps of 0.05) and a simple mean of all these models will be considered as a forecast.

The developed model based on the above method is defined as the supervised PCR (SPCR) method. It is found that the prediction using the SPCR method is better than the simple mean of all the GCMs in predicting the seasonal rainfall over the Indian region.

6.4.5 Canonical Correlation Analysis Technique

Canonical correlation analysis (CCA) is a multivariate statistical technique that transforms the original variable into a new set of canonical variables having a maximum correlation between them. In other words, a pair of patterns for the multivariate predictor and multivariate predictand is interrelated. This multivariate analysis was introduced by Hotelling (1936) and applied to the meteorological data first by Glahn (1968) where he compared it with the discriminant analysis and multiple regression. Furthermore, CCA was used by Barnett and Preisendorfer (1987) to evaluate the monthly and seasonal forecast skill for the surface air temperature over the United States. For that, he used some atmospheric variables like sea level pressure (SLP) and sea surface temperature (SST) for different selected domains as predictor variables. In 1994, Barnston first used CCA to evaluate the climate predictive skill for precipitation over the Northern Hemisphere. For the prediction of US or European seasonal average surface temperature, the previous temperature, SST and 700 mb height are used as predictors. The preceding SST and 700 mb height are used as predictors when the predictand is precipitation. Much of the higher skill was found over some pockets of the Pacific and North American regions in midwinter through mid-spring while the predictive skill for precipitation was also high but was lower than that for the temperature. These studies gave some positive indications that this statistical analysis has some potential and can be used for the prediction of atmospheric variables like precipitation and temperature. Several studies used this analysis, like Shabbar and Barnston (1996) on Canada, He and Barnston (1996) on the Tropical Pacific Islands, Hwang et al. (2001) over far Eastern Asia. Yu et al. (1997) used this statistical model to evaluate the predictive skill of seasonal rainfall variations over the U.S. Affiliated Pacific Islands and found significantly high skill, especially for the winter (December–February; DJF) and spring (March–May; MAM) seasons. They used the sea surface temperature in the Pacific Ocean as the predictor. Over the Indian domain, this analysis has been used in the study by Prasad and Singh (1996) where they estimated the monsoon rainfall over 29 meteorological subdivisions using some global variables like the 500 hPa ridge axis position of April, the Darwin surface pressure tendency, the sea-surface temperature of the central and eastern equatorial Pacific for the five successive months preceding the monsoon. They found some high and significant skill scores over the subdivisions of the west-central part of India. Sinha et al. (2013b) have found seven potential predictor fields over three domains and found the usability of CCA method which improved prediction over many places in India and forecast skill also higher than the GCM.

In view of the above, it is clear that the Canonical correlation analysis has some potential to predict seasonal rainfall and temperature; therefore, it is being used very broadly in several studies all over the globe.

Canonical Variables, Canonical Vectors and the Canonical Correlations

The canonical correlation analysis (CCA) is a statistical technique introduced by Hotelling (1936) that identifies a sequence of pairs of patterns in two multivariate data sets and constructs sets of transformed variables by projecting the original data onto these patterns. The approach thus bears some similarity to PCA, which searches for patterns within a single multivariate data set that represent maximum amounts of variation in the data. In CCA, the patterns are chosen such that the new variables defined by the projection of the two data sets onto them exhibit maximum correlation while being uncorrelated with the projections of the data onto any of the other identified patterns. That is, CCA identifies new variables that maximize the interrelationships between two data sets, in contrast to the patterns describing the internal variability within a single data set identified in PCA. It is in this sense that CCA has been referred to as a double-barreled PCA.

In ordinary, multiple regressions, one finds a weighted average, or pattern, of the vector of predictor variables x such that the correlation between the dot product $b'x$ and the scalar predictand y is maximised. Here the elements of the vector are the ordinary least-squares regression coefficients and $b'x$ is a new variable called the predicted value of y, or $\hat{y}$. Canonical correlation analysis identifies pairs of sets of weights analogous to the regression coefficients, such that the correlations between the new variables defined by the respective dot products with x and (the vector) y are maximised.

CCA extracts relationships between a linear combination of vectors of predictor x and predictand y that is,

$$v_m = a_m^T x' = \sum_{i=1}^{r} a_{m,i} x_i', \ m = 1, \ldots \min(r, s) \tag{6.39}$$

$$w_m = b_m^T y' = \sum_{j=1}^{s} b_{m,j} y_j', \ m = 1, 2, \ldots, \min(r, s) \tag{6.40}$$

where, x' and y' are the standardised values for the predictor and predictand, respectively. Here the new transformed variables which are called the canonical variables are v_m and w_m are the canonical variables. These vectors of weights a_m^T and b_m^T are canonical vectors. These canonical vectors are the unique choices that result in the canonical variables such that

$$Corr[v_1, w_1] = r_{C_1} \geq Corr[v_2, w_2] = r_{C_2} \geq \ldots Corr[v_m, w_m] = rC_m \tag{6.41}$$

and, (6.42)

$$Var(v_m) = Var(w_m) = 1 \tag{6.42}$$

For forecasting purpose, these canonical variables are regressed to obtain the regression coefficients. That is,

$$w_m = \hat{\beta}_{0,m} + \hat{\beta}_{1,m} v_m m = 1, 2 \ldots, \min(r, s) \tag{6.43}$$

In addition, the regression coefficients are obtained using the least square method which leads

$$\hat{\beta}_{1,m} = \frac{n\text{cov}(v_m, w_m)}{n\text{var}(v_m)} = \frac{n s_v s_w r_{v,w}}{n s_v^2} = r_{v,w} = r_{C_m} \tag{6.44}$$

Here r_{C_m} is the canonical correlation which is the correlation between the newly transformed variables (canonical variables). That is because the canonical variates are scaled to have unit variance the regression slopes are simply equal to the corresponding canonical correlations, and the predictor and predictand are the standardised values, which means that the mean is zero, which leads to the intercept term being zero. Therefore, after the evaluation of the canonical variables the estimated value of predictand can be obtained as

$$\hat{w} = R_C v \tag{6.45}$$

R_C is the matrix of canonical correlation matrix of order $M \times M$ that is

$$R_C = \begin{bmatrix} r_{C_1} & 0 & & 0 \\ 0 & r_{C_2} & & \\ & & \cdot & \\ & & & \cdot \\ & & & \quad\cdot \\ 0 & & & r_{C_M} \end{bmatrix} \tag{6.46}$$

These canonical variables are transformed back to original variables to obtain the estimated value for the predictand.

Development of the CCA Forecast Model

Let us assume that X is a predictor matrix of order $s \times t$ and Y the predictand matrix of order $r \times t$ where s and r are the number of points of predictor and predictand matrix in space and t is the number of observations. Before the canonical correlation analysis, the standardization and the orthogonalisation of the predictor and predictand matrix are highly advised. The orthogonalization is done using the concept of empirical orthogonal function (EOF). There are several reasons for doing this. First of all, the large number of spatial points can cause difficulty in inverting the matrices and in the eigenvalue problem. The pre-processing procedure transforms the large spatial dimension into a smaller number of retained EOF modes, and this value is quite modest even for large fields. The computation of canonical modes is thus simplified after pre-processing. In addition, the small-scale noise is filtered out after EOF analysis since the use of EOFs allows the analysis to focus on the dominant modes of variability within each input dataset. Therefore, performing the EOF analysis on X and Y leads to

$$X_{s,t} \quad = E_{s,s} T_{s,t} \tag{6.47}$$

$$Y_{r,t} = E_{r,r} T_{r,t} \tag{6.48}$$

Here $E_{s \times s}$ and $E_{r \times r}$ are the EOF spatial modes of predictor and predictand and $T_{s,t}$ and $T_{r,t}$ are their time components. Assume that i and j are the retained EOF modes for the predictor and predictand enter in the canonical correlation analysis. Now we have to obtain the canonical variables Z and W, which are the linear combination of the retained EOF time components. That is,

$$Z = u' T_{i,t} \tag{6.49}$$

u and v are the canonical vectors, further, the following canonical vector matrices

$$W = v' T_{j,t} \tag{6.50}$$

are defined as

$$U' = \begin{bmatrix} u'_1 & u'_2 & \dots & u'_m \end{bmatrix}' \tag{6.51}$$

$$V' = \begin{bmatrix} v'_1 & v'_2 & \dots & v'_m \end{bmatrix}' \tag{6.52}$$

where, $m = \min(i, j)$.

Now as defined above the regression coefficients between the canonical variables are obtained using Eq. (6.44). Here, the estimated value for the predictand canonical variable W at time $t+l$, where l is the lead l is given as

$$W' = R_C Z \tag{6.53}$$

Here R_C is the matrix of canonical correlation of order $m \times m$ defined as above, m is minimum (i, j). Using the values of Z and W from Eqs. (6.49) and (6.50) in (6.53).

Using the orthogonal property of empirical orthogonal function and the Eqs. (6.47) and (6.48) we have,

$$T_{i,t} = E'_{s,t} X_{s,t} \tag{6.54}$$

$$Y_{r,t} = E_{r,j} T_{j,t} \tag{6.55}$$

The original rainfall predicted values at time $t+l$ could be obtained (Yu et al. 1997) after the back transformations using Eqs. (6.54) and (6.55)

$$\hat{Y}_{r,t+l} = E_{r,j} \left(V'\right)^{-1} R_C U' \left(E_{s,i}\right)' X_{s,t+l} \tag{6.56}$$

This Eq. (6.56) gives the forecast value at time $t+l$ at all r spatial points.

Calculation of Canonical Vectors

As discussed above, the canonical vectors are the unique choices; hence, in the methodology, these vectors are obtained using the concept of Singular Value Decomposition (SVD). SVD is based on a theorem from linear algebra, which says that a rectangular matrix A can be broken down into the product of three matrices—an orthogonal matrix U, a diagonal matrix S, and the transpose of an orthogonal matrix V. The theorem is usually presented something like this:

$$A = USV^T \tag{6.57}$$

where, $UU^T = I$, $VV^T = I$ the columns of U are orthonormal eigenvectors of AA^T, the columns of V are orthonormal eigenvectors of A^TA, and S is a diagonal matrix containing the square roots of eigenvalues from U or V in descending order.

Therefore, this property of SVD is used here in order to decompose the cross-covariance (or the correlation) matrix of predictor and predictand in these three components in which U gives the canonical vector for the predictor variable whereas, V gives the canonical predictand. The matrix S gives the matrix of canonical correlation (as discussed before the matrix R_C of canonical correlation) between the canonical predictor and predictand. These values are used in Eq. (6.55).

Use of CCA for Prediction

In view of the scope of CCA for the seasonal prediction of atmospheric variables, this analysis is used to improve the prediction skill of the General Circulation Models (GCMs). These GCMs may have systematic biases, which result in non-satisfactory skill over a regional domain, for example, if the ENSO teleconnection is displaced in the model. The CCA is applied over all the GCMs separately in order to improve their individual forecast skill over the domain and is particularly suited to correcting for systematic conditional biases in patterns, such as in climatic teleconnections. At the outset, simple arithmetic mean over ensemble members is evaluated to obtain a single ensemble-mean forecast for the interested atmospheric variable (precipitation and surface air temperature). An example is provided below.

The observation and the GCM outputs (Table 6.2) for rainfall are extracted for the domain that is a small extension of the Indian domain (approximately) which is from 0° to 40° N in latitudinal and 60° E to 110° E in longitudinal directions. The observation data set for rainfall is taken from IMD which is at the resolution of $0.25° \times 0.25°$ (Pai et al. 2014 in which 4964 points over the Indian landmass are present) and considered as predictand variables. The predictor (precipitation or surface air temperature corresponding to the similar predictand variable) spatial points depend on the resolution of the GCM outputs. These GCM outputs are treated as predictors, and the observations are separately pre-orthogonalised and standardised. Furthermore, these retained EOFs are entered into the CCA forecast model for the Indian domain. The number of hindcast years is 27; therefore, the leave-one-out cross-validation technique is used for the verification of the forecast model at each grid point as well as over the homogenous regions. Skill is evaluated after the canonical correlation analysis for each model as well as after combining all the forecast values (by taking their simple mean (CCA mean)). CCA shows a noticeable improvement in the skill of the country as a whole, which is a positive sign that this analysis may improve all India's performance for prediction as compared to the simple mean (the arithmetic mean of interpolated GCMs for rainfall). The point which may be highlighted is, this analysis is improving the skill of some of the atmospheric models. However, the improvement of prediction skills using CCA varies region-to-region, season-to-season, and predictor variables (Sinha et al. 2013b). Therefore, it is strongly suggested to find potential predictor fields before using CCA.

6.5 Generation of Single Forecast from Multi-technique Predictions

6.5.1 Deterministic Forecast

In this method, the outputs obtained from the above three techniques (MLR_{svd}, CCA, and SPCR) are further combined based on their hindcast skill in order to obtain a final consensus product (FCP). For generating one single forecast, the following method has been employed in FCP that is described in the next. Precipitation/temperature (T_{mean}, T_{max}, and T_{min}) can be predicted using all three different techniques (MLR_{svd}, CCA, and SPCR). Along with this, the simple mean of all bias-corrected GCMs (MME_{BC}) is also computed for a particular season and region. The FCP is compared with MME_{BC} and the other individual GCM products for understanding the strength of the FCP method.

In the below example, the prediction is made in the leave-one-out cross-validation mode, and the skill of the methods is examined based on the anomaly correlation coefficient computed between IMD observation and forecast for the Indian region. Only those predictions, which show a significant correlation coefficient at a 90% confidence level, are considered further. In some regions, during the dry season for precipitation prediction (like northwest India during March–May, MAM), it is noticed that the anomaly correlation coefficient is not significant using any of the methods. In that case, those predictions, which have a negative correlation, are rejected. After the selection of the methods for a particular season and the region, equal weight is given to precipitation obtained from the selected methods for that region and the season of the forecasted year. The statistical details with hindcast performance of each individual method with FCP product are well documented in the study by Mohanty et al. (2013, 2019b). The details results are provided next.

Skill of the FCP Deterministic Forecast

To know the skill of FCP forecast products, the prediction of FCP for precipitation for various seasons in the hindcast period (1982–2008) is evaluated and also compared with the skill of each GCM, a simple mean of the multi-model ensemble of GCM raw outputs (MME_{Raw}) and bias-corrected GCM outputs (MME_{BC}). In order to get details of the capability of GCMs, MME_{Raw}, MME_{BC}, and FCP in simulating observed characteristics, the climatological mean and standard deviation are evaluated and presented in Table 6.3, along with the observed values. In this example, rainfall data from a few GCMs such as CCM3v6, ECHAM4p5CASST (CASST), ECHAM4p5CFSSST (CFSSST), CFSv2, COLA, GFDL, GFDLA06, GFDLB01, SFM, ECMWF are considered. These GCMs, except SFM and ECMWF outputs, can be downloaded from the IRI data library (http://portal.iri.columbia.edu/portal/server.pt). The IMD-SFM and ECMWF model products are available through IMD and European Centre for Medium-Range Forecast, respectively.

Table 6.3 The climatology and standard deviation of GCMs, MME_BC and FCP along with IMD observation for all seasonal precipitation (mm/day) for hindcast period of 1982–2008 (Clim: Climatology, STD: Standard Deviation) (Mohanty et al. 2019b)

GCMS	Dec–Feb (DJF)		Mar–May (MAM)		Jun–Sep (JJAS)		Oct–Dec (OND)	
	Clim	STD	Clim	STD	Clim	STD	Clim	STD
Observation	0.69	0.22	1.59	0.28	7.47	0.65	1.42	0.32
CCM3v6	0.97	0.09	3.01	0.19	6.36	0.30	1.88	0.14
CASST	0.61	0.08	2.32	0.10	7.48	0.30	1.31	0.10
CFSSST	0.57	0.10	2.28	0.12	7.36	0.52	1.02	0.15
CFSv2	0.62	0.08	1.95	0.13	5.36	0.39	1.65	0.15
COLA	1.08	0.15	3.09	0.18	6.58	0.25	1.68	0.28
GFDL	0.46	0.08	1.29	0.19	7.32	0.53	1.95	0.18
GFDLA06	0.73	0.07	1.16	0.08	6.03	0.53	1.93	0.15
GFDLB01	0.73	0.08	1.17	0.07	5.86	0.47	1.85	0.34
IMD-SFM	3.35	0.44	1.29	0.26	3.00	0.63	1.65	0.27
ECMWF	1.01	0.09	1.95	0.06	6.15	0.29	1.66	0.08
MME_{RAW}	0.94	0.086	1.87	0.05	6.25	0.27	1.58	0.09
MME_{BC}	0.69	0.14	1.59	0.15	7.46	0.48	1.42	0.17
FCP	0.68	0.12	1.60	0.18	7.49	0.56	1.40	0.14

It is found that the climatological winter seasonal precipitation (mm/day) of FCP and MME_{BC} are reasonably agreed with observation than the MME_{Raw}. Among the individual GCMs, large errors are found with a maximum in the IMD-SFM model (3.35 mm/day) followed by COLA and ECMWF models. In the pre-monsoon season (MAM), the observed mean precipitation is 1.59 mm/day with a standard deviation of 0.28 mm/day. The climatology of the pre-monsoon precipitation is underestimated in all the versions of the GFDL and IMD-SFM model, while it is overestimated in the other models. On the other hand, the FCP and MME_{BC} are approximately equal to the observed climatological values while a large overestimation is noticed in MME_{Raw}. The results are similar for standard deviation, except IMD-SFM, which portrays the correct observed precipitation deviation; however, there is a larger underestimation in MME_{Raw}. During the summer monsoon season, the seasonal climatological mean precipitation of FCP, and MME_{BC} are very close to the observation followed by CASST and CFSSST. Large systematic errors are found in IMD-SFM (−4.47 mm/day), CFSv2 (−2.11 mm/day), GFDLB01 (−1.61 mm/day), GFDLB01 (−1.44 mm/day), ECMWF (−1.32) and it is also clearly reflected in MME_{Raw} (−1.22 mm/day). The reason for this difference may be due to warm SST bias in coupled ocean-atmosphere forecast systems over both the southeastern Pacific and Atlantic Oceans (Wang et al. 2005; Preethi et al. 2010; Rajeevan et al. 2012; Singh et al. 2012; Nageswararao et al. 2016). The observed precipitation variability during the JJAS season is much more than in the DJF and MAM seasons. Most of the GCMs show almost half the values of the observed variability during the JJAS season except GFDL and CFSSST.

However, the FCP is able to show a variability of about 0.56 mm/day, which is much closer to the observed variability than any individual GCM, MME_{Raw} and MME_{BC}. During the post-monsoon (OND) season, the GCMs and MME_{Raw} show the climatological error of the order of 0.1–0.5 mm/day, whereas the MME_{BC} and the FCP represent the observed counterpart with fidelity. The standard deviation in individual GCMs, MME_{Raw}, MME_{BC} and FCP is less than the observation during the OND season except for GFDLB01. Along with the climatological mean and the standard deviation, some primary statistical methods like correlation, root mean square error (RMSE), as well as inter-annual variation is considered to examine the skills of GCMs and their statistical corrections for a higher degree of confidence that can reveal the nobility of the GCMs, MME_{Raw}, MME_{BC}, and the FCP.

The skills of atmospheric GCMs are not satisfactory in simulating the seasonal precipitation for all seasons except DJF. As depicted by correlation and RMSE, a few models like CFSSST, GFDL, GFDLB01, and ECMWF are found to be the best in capturing the winter seasonal precipitation (Fig. 6.10). Tiwari et al. (2014) and Nageswararao et al. (2016) have studied the skill of precipitation predictions during DJF seasons over northwest India from various GCMs, and they found a large error in the GCMs seasonal precipitation predictions over the region. During the pre-monsoon season, the coupled dynamical model COLA shows significant positive correlation values followed by CFSv2 and CCM3v6, while other models show negative correlation values. For southwest monsoon season (JJAS), three coupled GCMs are significantly (at 90% confidence level) correlated with fairly high values, and the highest correlation is found in CFSv2, followed by the GFDL and GFDLB01. In the post-monsoon season, GFDLA06 is the only model that shows a significant correlation (at 90% confidence level) to the other models. The low skill in most of the models during the post-monsoon season is maybe because of the low signal-to-noise ratio of the GCMs and also because of the inefficacy of the GCMs in simulating the remote responses (Nair et al. 2015). The skill of multi-model ensemble mean of these raw GCMs (MME_{Raw}) and bias-corrected GCMs (MME_{BC}) have also been examined since this is now a common standard practice in seasonal scale forecasting. It is interesting to note that the MME_{Raw} shows a significant (at 90% confidence level) skill only in DJF, while in other seasons, it is either negative or insignificant. In the case of MME_{BC}, there is not much improvement in terms of correlation, while in RMSE, there is a slight depreciation indicating MME_{BC} is better than MME_{Raw}. The bias-correction method adopted here is the standardized reconstruction method (Acharya et al. 2013), which usually adjusts the mean that is reflected in the figure with a reduction in RMSE. Since, one individual GCM and the mean of ensemble members of that particular GCM, are not able to properly simulate the seasonal patterns, the multi-model ensemble mean through robust statistical technique is adopted in the FCP. Figure 6.10 clearly indicates that the FCP shows high (low) correlation (RMSE) value in most of the seasons and relatively higher (lower) than the individual GCMs, MME_{Raw}, and MME_{BC}. It can be noticed that the skill of the FCP is higher in representing the seasonal mean precipitation with the maximum skill in JJAS followed by

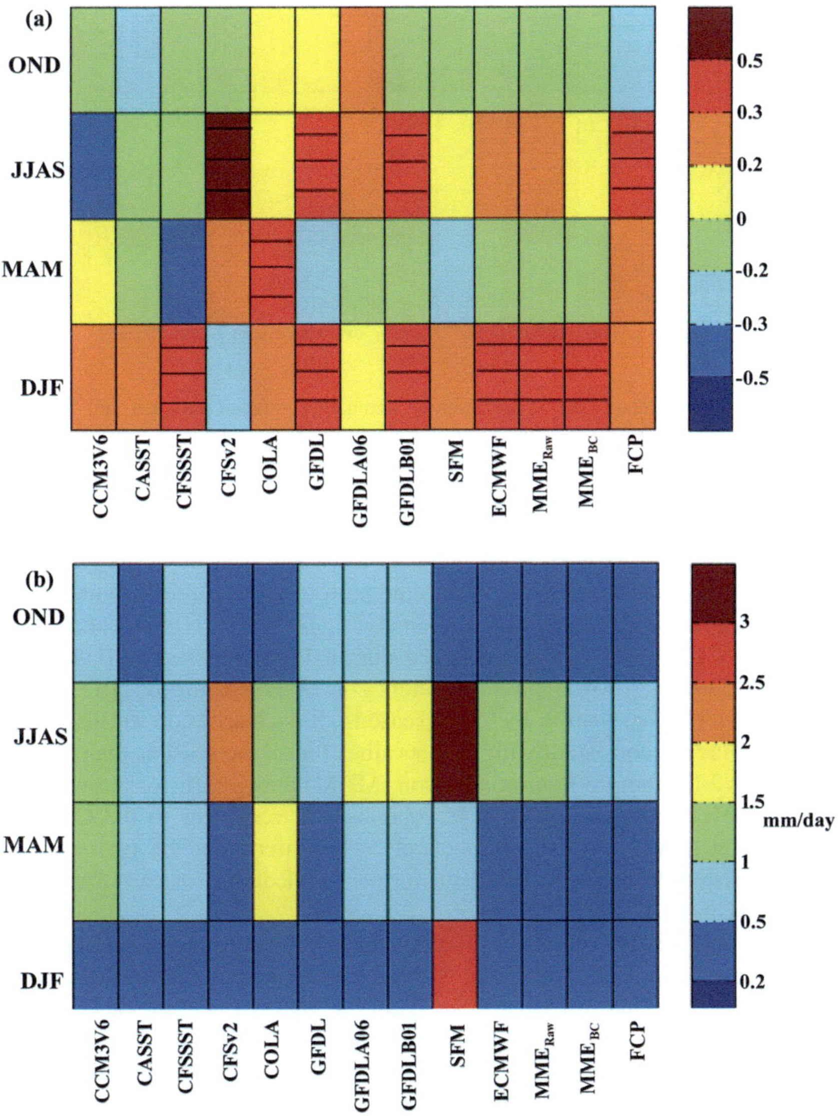

Fig. 6.10 **a** Correlation coefficient and **b** root mean square error of GCMs, MME, MME_BC and FCP for all the seasons in hindcast period 1982–2008 against IMD observation. The dashed lines in the Fig. 6a indicate significant positive correlations (Mohanty et al. 2019b)

DJF and MAM (Fig. 6.10). The only exception in OND when the skill of FCP is relatively less than the other seasons because of the insignificant hindcast skill of GCMs. In the case of RMSE, the error is relatively less in the FCP as compared to other individual GCMs, MME_{Raw}, and MME_{BC}.

In view of the above correlation and RMSE analysis, phase coherency (Sinha et al. 2015) of each individual GCMs and FCP for hindcast mode (1982–2008)

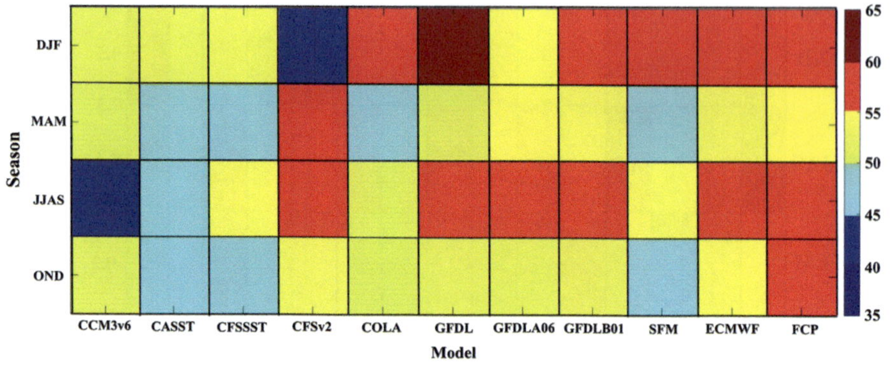

Fig. 6.11 Phase coherency (%) of precipitation anomaly computed between individual GCM and FCP with observation for all the seasons for the hindcast period (1982–2008). The calculation is made by pooling all the met-subdivisions together (Mohanty et al. 2019b)

with IMD observation are examined. The phase coherency is computed by pooling together all the metsubdivisions of India for all seasons, as depicted in Fig. 6.11. This skill score gives a clear insight into the performance of the individual GCMs and FCP in predicting precipitation anomaly signs. The maximum phase coherence in the GCMs and FCP is observed during DJF, followed by JJAS, OND, and MAM. However, lower phase coherency is observed during DJF in CFSv2 and JJAS in CCM3v6 than in the other seasons. It is clearly noticed that the FCP phase coherency is consistent with higher values for all the seasons than all GCMs except CFSv2 for the pre-monsoon season (MAM) and GFDL for the winter season (DJF). Moreover, the FCP phase coherency is very near to the CFSv2 and GFDL during MAM and DJF, respectively. This infers that the performance of FCP is consistent and can be useful for further applications instead of using individual GCMs.

However, these deterministic forecasts have some limitations: they do not measure uncertainty in it. Therefore, wrong deterministic predictions can mislead, and the use of this kind of forecast can lead to a huge risk. On the other hand, the predictions become more useful if the uncertainties in them are conveyed to the user along with the deterministic value. This can be done by using parametric and non-parametric probabilistic prediction approaches.

6.5.2 *Probabilistic Forecast*

In the parametric and non-parametric probabilistic approaches, the categories for the predictions are made on the basis of observed climatology. Often, terciles categories viz. Below-Normal, Near-Normal, and Above-Normal (Tippett et al. 2007; Kar et al. 2012) are used to generate the probabilistic predictions. These categories are equally likely, mutually exclusive, and collectively exhaustive. One of the simplest examples of a non-parametric approach is to count the number of ensemble members falling in each category (Rajagopalan et al. 2002; Robertson et al. 2004) and then divide that number by the total number of ensemble members. In the parametric approach, one has to assume a distributional form for the observations and forecasts and then estimate parameters such as mean and variance. Once the distributions are known, the probability of occurrence of a particular event can be computed. In this approach, forecast uncertainty is captured through the spread of the forecast distribution. Methods to compute the forecast distribution spread include ensemble variance (Kharin and Zwiers 2002, 2003), error residuals (Johnson and Bowler 2009), and potential predictability (Tippett et al. 2005, 2007). These methods essentially allow deterministic predictions to be converted to probabilistic ones. A nice illustration and skill comparison of these three techniques is given in Kulkarni et al. (2012).

Skill of the FCP Probabilistic Forecast

In the FCP, probability values are generated for three precipitation categories, namely below normal, near normal, and above normal. The skill of probabilistic forecast needs to be evaluated along with deterministic values so as to get confidence in the probabilistic forecast skills of the FCP. For this purpose, we have evaluated the Likelihood Skill Score (LSS) and Heidke Skill Score (HSS) for each of the subdivisions for all four seasons.

The Likelihood Skill Score (LSS) measures whether the higher probability associated with a certain rainfall category is further observed. In addition, a comparison is made with a reference forecast that is taken as a climatological forecast (0.33). The likelihood indicates the extent to which the forecasts assign high probabilities to the later observed outcomes. Such high probabilities are possible when these two conditions are satisfied: (1) the forecasts are able to distinguish successfully among the observed outcomes, and (2) the forecasts do not have substantial unconditional or conditional biases in their probabilities (Brocker and Smith 2007). Figure 6.12 shows the spatial extent of the above-described skill score. For DJF seasonal rainfall, positive LSS skill is observed in most of the subdivisions, in particular the northwest part of the country, where winter rainfall plays a major role. Some subdivisions in Central and North East also show high LSS values. In the pre-monsoon season, the LSS is of the order of 10 over Northeast and south peninsular India, which indicates that the FCP performed better in predicting the pre-monsoon rainfall. In the JJAS season, more than 70% area has high LSS values with the majority of subdivisions lying in MCZ. The Western Ghats and the subdivisions in North East also show high LSS values. Bihar, West Bengal-Sikkim and Punjab-Chandigarh-Delhi show an LSS of more than 10, indicating efficacy in predicting summer monsoon rainfall. In the post-monsoon season, the central part of India depicts high LSS values ranging from 4 to 10. However, over south peninsular India and North East India, there is insignificant LSS.

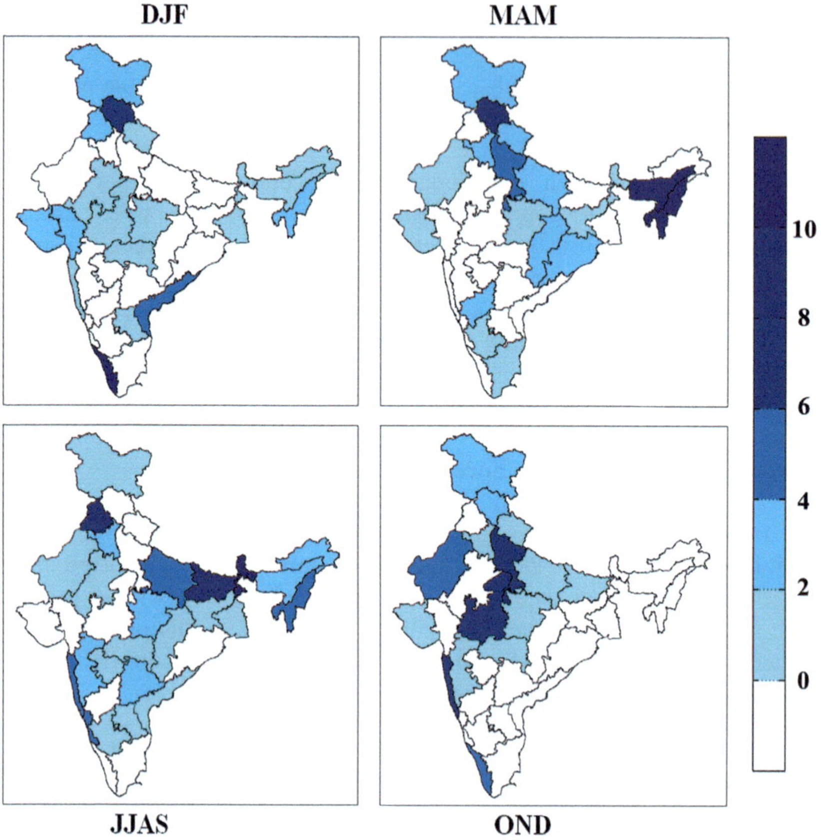

Fig. 6.12 Spatial distribution of Likelihood Skill Score for FCP probability forecast for the period (2009–2014) (Mohanty et al. 2019b)

With the intention of verifying real-time probabilistic forecast, Heidke Skill Score (HSS) is further evaluated (Fig. 6.13). It is seen that LSS measures the probability of the event as further observed and compares it with the reference forecast. Therefore, LSS takes into account the probability corresponding to observed rainfall events, but HSS corresponds to the hit proportion.

A hit is considered if the highest probability for a particular category is realised. For example, let us assume that the forecast probability for the below normal category is 55% (highest in comparison to the forecast probability for other categories). In case the precipitation is realized in the below normal category, it will be termed a hit. Higher HSS will signify higher accuracy in probability forecast to discriminate between categories. It may be seen from Fig. 6.13 that for DJF, most of the subdivisions show high HSS values (greater than 0.33) over most of the subdivisions, specifically over central India. Northeast and north India also exhibit higher HSS values. In the pre-monsoon season, the northwest and northeast parts of India exhibit a significant value of HSS, whereas, over the southern part, the

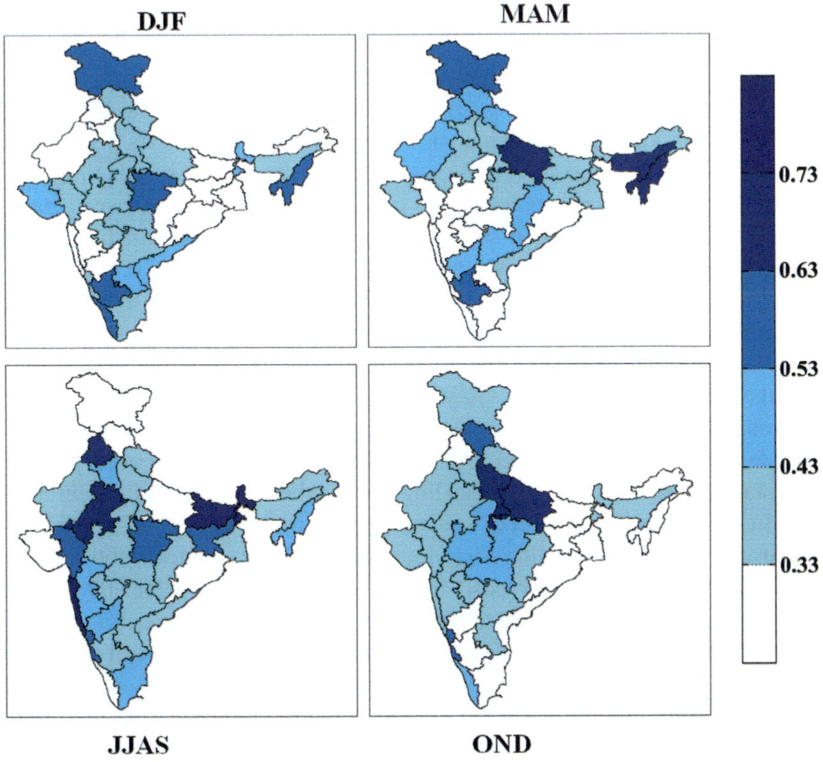

Fig. 6.13 Spatial distribution of Heidke Skill Score for the FCP probability forecast for the period (2009–2014) (Mohanty et al. 2019b)

HSS is not found significant. In the summer monsoon season, 90% of the subdivisions exhibit significant HSS values, except for some subdivisions where summer monsoon rainfall is sporadic. High HSS is found over Bihar, West Bengal-Sikkim, West Madhya Pradesh, Punjab-Haryana and Chandigarh and parts of Western Ghats. In the post-monsoon season, the subdivisions also exhibit high HSS values, similar to the summer monsoon. Over East and West Uttar Pradesh, the HSS skill is found to be significant. More than 80% of subdivisions exhibit high HSS skills.

Thus, the results clearly indicate the efficacy of the FCP in predicting rainfall over all of India and over its meteorological subdivisions in all seasons by means of various measures of performance.

6.6 Dynamical Downscaling

The general circulation models or global climate models (GCMs) are the most important tools in predicting atmospheric states at a monthly and seasonal scale or at a scale more than a season. The present-day GCMs are highly complex due to mathematical representations of all possible interactions among land, air, and seas as well as interactions between small and large-scale processes which have already been briefed early in this chapter. The GCMs produce predicted data at coarse resolution, and the average horizontal grid resolution is 1.25° (which is equivalent to 138 km near the equator). In other words, the GCMs provide single information of a particular meteorological variable for a grid box having an area of ~19,000 km^2, which is inadequate to represent the prediction over a targeted or required region that is much smaller than the GCM grid box. Therefore, it is necessary to reproduce the GCMs predicted information to the required region. There are several methods adopted for converting the GCMs output to the targeted region, and many of those methods are advanced with time. One of the simplest methods is to predict over a point of the representative required region by predicting the nearest grid point of GCM. This method is often unreliable owing to coarse resolutions GCMs, and the targeted point of the required region is often situated far away from the model grid. In order to improve the nearest point forecast, general calibration and downscaling approaches emerged as potential techniques. In the calibration method, statistical corrections of the GCM outputs are widely used, while downscaling technique scientifically connects the large-scale spatial or temporal (climate forecast) forecasts to the smaller-scale spatial or temporal forecasts (weather forecast). In other words, the downscaling technique translates coarse-resolution GCMs outputs in a smaller resolution of the required region. Thus, downscaling technique links the state variables of a larger space with the state variables of a smaller space. Statistical downscaling and dynamical downscaling are the two major methods, and the various methods of statistical downscaling have already been discussed in the previous sections. In the dynamical downscaling, the downscaled products/outputs will be consistent with the large-scale atmospheric state of that GCM that has been used as a driving force for downscaling. In the dynamical downscaling, high-resolution regional climate models (RCMs) are used to simulate the finer-scale physical processes consistent with the driving GCM.

The RCM can be used in "one way" or "two-way" nesting with the GCM. In the "one way" nesting technique, the initial and lateral boundary conditions of the RCM are provided from a GCM, but no feedback from the high-resolution simulations of RCM to the driving coarse resolution GCM is considered. Hence, a regional increase in resolution can be attained through the use of nested RCMs to account for sub-GCM grid-scale forcing. The most relevant advance in nested regional climate modelling activities was the production of continuous RCM multi-year climate simulations. Previous regional climate change scenarios were mostly created using samples of month-long simulations. The primary improvement represented by continuous long-term simulations consists of the equilibration of model climate with surface hydrology and simulation of the full seasonal cycle for use in impact models. In addition, the capability of producing long-term runs facilitates the coupling of RCMs to other regional process models, such as lake models, dynamical sea ice models, and possibly regional ocean (or coastal) and ecosystem models.

RCMs are a complementary research method, allowing more detailed process studies and simulation of regional and even local conditions. This is due to the complexity of the processes that determine regional climate change, which span a wide range of spatial and temporal scales, and to the difficulty in extracting fine-scale regional information from coarse resolution coupled Atmosphere-Ocean General Circulation Models (AOGCMs). An additional consideration is that in order to run an RCM experiment, high frequency (e.g. 6 h) time-dependent GCM fields are needed. The main theoretical limitations of the dynamical downscaling technique are the effects of systematic errors in the large-scale driving fields provided by GCMs output and the lack of two-way interactions between regional and global climate. In addition, for each application, careful consideration needs to be given to some aspects of the model configuration, such as physics parameterisations, model domain size and resolution, and the technique for the assimilation of large-scale meteorological forcing. A description of a widely used regional climate model is provided next.

The idea that limited area models (LAMs) could be used for regional studies was originally proposed by the Earth System Physics section of the Abdus Salam International Centre for Theoretical Physics, Abdus Salam International Centre for Theoretical Physics (ICTP) and they have developed Regional Climate Model RegCM that are being in use for dynamic downscaling of GCMs product. This idea was based on the concept of one-way nesting. The first generation NCAR RegCM was built upon the National Center for Atmospheric Research (NCAR)-Pennsylvania State University (PSU) Mesoscale Model version 4 (MM4) in the late 1980s. The dynamical component of the model originated from the MM4, which is a compressible, finite difference model with hydrostatic balance and vertical-coordinates. The use of a split-explicit time integration scheme was added along with an algorithm for reducing horizontal diffusion in the presence of steep topographical gradients. The physics of RegCM was based on that of the NCAR Community Climate Model version 3 (CCM3), and the mesoscale model MM5. In particular, the CCM3 radiative transfer package was used for radiation calculations, the non-local boundary layer scheme replaced the older local scheme, the mass flux cumulus cloud scheme was added as an option, and the latest version of the Biosphere–Atmosphere Transfer Scheme (BATS) was included in the model. In the CCM3 package, the effects of H_2O, O_3, O_2, CO_2, additional greenhouse gases (NO_2; CH_4; CFCs), atmospheric aerosols, clouds, and cloud ice were accounted for by the model. The solar radiative transfer was treated with a d-Eddington approach, and cloud radiation depended on three cloud parameters, the cloud fractional cover, the cloud liquid water content, and the cloud effective droplet radius. Scattering and absorption of solar radiation by aerosols are also included based on the aerosol optical properties (Absorption Coefficient and Single Scattering Albedo). A simplified explicit moisture scheme is included, where only a prognostic equation for cloud water is used, which accounts for cloud water formation, advection and mixing by turbulence, re-evaporation in sub-saturated conditions, and conversion into rain via a bulk autoconversion term. Prognoses cloud water variable is directly used in the cloud radiation calculations, and not diagnosed in terms of the local relative humidity, adding a very important and far-reaching element of interaction between the simulated hydrologic cycle and energy budget calculations. The solar spectrum optical properties are based on the cloud liquid water path, which is in turn based on the cloud liquid water amount prognostically calculated by the model, cloud fractional cover, which is calculated diagnostically as a function of relative humidity, and effective cloud droplet radius, which is parameterised as a function of temperature and land-sea mask for liquid water and as a function of height for ice phase.

The total cloud cover at a model grid box is given by a value intermediate between that obtained using the random overlap assumption (which maximises cloud cover) and that given by the largest cloud cover found in any single layer of the column overlying the grid box (which implies a full overlap and it is thus a minimum estimate of total cloud cover). A large-scale cloud and precipitation scheme which accounts for the subgrid-scale variability of clouds, parameterizations for ocean surface fluxes, and multiple cumulus convection schemes are the same as in RegCM, but a new "mixed scheme" is introduced: it allows the user to select one of the two schemes in the function of the ocean-land mask. In RegCM, the aerosol radiative forcing was based on three-dimensional fields produced by the aerosol model and included only scattering and absorption in the shortwave spectrum and infrared spectrum. This is especially important for relatively large dust and sea salt particles and it is calculated by introducing an aerosol infrared emissivity calculated as a function of the aerosol path and absorption cross section estimated from aerosol size distribution and long wave refractive indices. The RCM modelling system has four components: Terrain, ICBC, RegCM, and Postprocessor. Terrain and ICBC are the two components of the RegCM pre-processor shown in Fig. 6.14. Terrestrial variables (including elevation, land use, and sea surface temperature) and three-dimensional isobaric meteorological data are horizontally interpolated from a latitude–longitude mesh to a high-resolution domain on either a Rotated (or Normal) Mercator, Lambert Conformal, or Polar Stereographic projection. Vertical interpolation from pressure levels to the σ coordinate system of RegCM has also performed surfaces near the ground closely follow the terrain, and the higher-level σ surfaces tend to approximate isobaric surfaces. Since the vertical and horizontal resolution and domain size can vary, the modelling package programs employ parameterized dimensions requiring a variable amount of core memory, and the requisite hard-disk storage amount is varied accordingly.

6.6.1 Some Aspects of Dynamical Downscaling Products

The dynamical downscaling is performed using an RCM that usually treats the physical processes through a number of parameterisation schemes. An RCM obtained from the developer group is not generally suitable for any region. Thus, sensitivity tests through different parameterization schemes for a particular

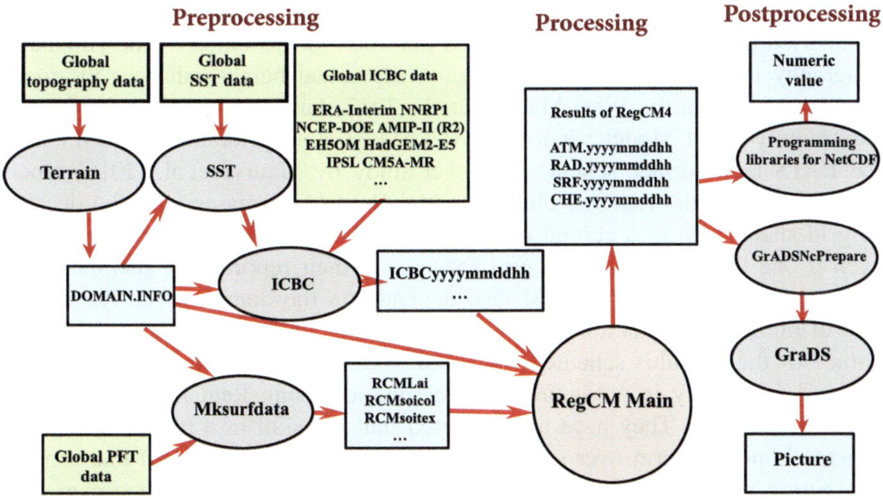

Fig. 6.14 Schematic representation of regional climate model (RegCM) [adopted from RegCM user guide provided in the RegCM website]

atmospheric event and for a specific domain are required. Extensive diagnostic studies of the RCM products and estimation of the model error are needed to understand the efficiency of the model. A few standard processes are described next for customisation of the model.

It is the first step to choosing a specific event for the customisation of the model. For example, one can consider simulating the summer monsoon rainfall or winter season mean/minimum/maximum temperature using an RCM. The user needs to examine the response of the model simulations by varying the domain size and resolution (one at a time, keeping the rest of the model configuration the same). After getting the best choice on the domain size and resolutions, one needs to perform several sensitivity experiments through physical parameterisation schemes such as cumulus scheme, land surface scheme, radiation scheme, planetary boundary scheme, etc. It is a general practice to keep the other model configurations the same and vary only one parameterisation scheme. After each set of experiments, an extensive diagnostic study is required to identify the suitable choice for the customisation of the model.

It would be of interest to explain the customisation of the regional climate model with examples. Let us consider the Indian summer monsoon (ISM), and one needs to predict the summer monsoon circulation and associated rainfall. Several sensitivity experiments (e.g., orography, cumulus parameterisation, land surface model, horizontal resolution, domain size, moisture flux adjustment, auto conversion coefficients, etc.) have been conducted to customize the regional climate model RegCM for the simulation of ISM (Sinha et al. 2013a, 2014, 2019; Maurya et al. 2017, 2018, 2020; Mohanty et al. 2019a). Sinha et al. (2014) have reported that the RegCM model underestimates the height of the orography, and a change in the Himalayan orography will lead to changes in the seasonal mean rainfall and its distribution. They have also reported that the representation of the Himalayan orography in the RegCM model closer to the actual height enhances the model skill in simulating the ISM. Maurya et al. (2017) have suggested that the use of Community Land Model version 4.5 (CLM4.5) is superior to the CLM3.5 and the BATS land surface schemes. Another study by Maurya et al. (2018) reports that the skill of the regional climate model RegCM increases with the decrease of grid spacing up to a certain resolution, and beyond that, the skill deteriorates even if one further reduces the grid spacing. In their report, they also mentioned that the performance of RegCM also depends on the domain size. The simulation of rainfall/precipitation in a model is very crucial and highly depends on the choice of the cumulus scheme. Sinha et al. (2013a, 2019) demonstrated that the degree of accuracy in predicting the precipitation using RegCM varies with the cumulus schemes. They have also reported that the skill of a particular cumulus scheme is not uniform over space. In their study, it was illustrated that a mixed convection scheme (one scheme over the ocean and another scheme over land) performs better than one cumulus scheme for the entire domain of the model. It is indeed helpful to tune the microphysics further for an improved forecast. For example, Mohanty et al. (2019a) examined the model's sensitivity to precipitation simulation by changing the autoconversion coefficient that determines how much available moisture will be condensed to precipitation. Their study infers that autoconversion adjustment greatly improves precipitation simulation, not only in magnitude but also in spatial distribution. Figure 6.15 represents how the skill of the RegCM model has been progressively enhanced as one will customize the model through sensitivity experiments.

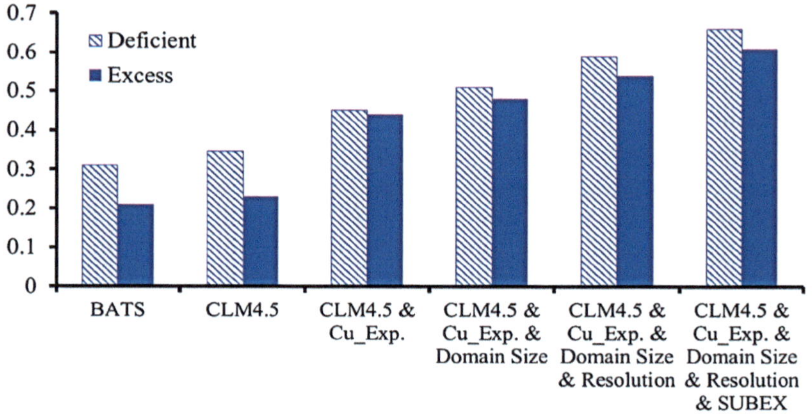

Fig. 6.15 Correlation coefficients of three composite excess (1983, 1988 and 1994) and deficit monsoon seasons (1982, 1986 and 1987) with default (BATS), addition of CLM4.5, CLM4.5 with customised Cumulus scheme, with domain customized, with resolution customized and with autoconversion coefficient customised, respectively (Mohanty et al. 2019b)

It may be worth noting that statistical downscaling can be applied to the GCMs outputs, bias corrections can be made to GCM raw products, dynamical downscaling approaches are useful for improving the GCMs forecast, and statistical downscaling can be applied to the dynamically downscaled products. A number of forecasts will then be available, and one can suitably combine using all the forecasts to prepare one consensus prediction.

This chapter presents the need for evaluation of the model products and different bias correction methods. It describes how to estimate the efficiency of the bias correction methods and how one can efficiently use the bias correction method even if the data length is not sufficiently large. The methods of some standard statistical downscaling techniques and their application on the GCMs products have been analysed. It is already indicated in Chap. 5 that a number of GCMs products are available, and many standard bias correction methods are described here. In addition to statistical downscaling approaches, dynamical downscaling techniques are also described in this chapter. If a user uses multiple GCM products, applies several bias correction methods on each GCM, and employs different statistical and/or dynamical and/or hybrid (dynamical–statistical) techniques, then the user will have a large number of raw, bias-corrected, and downscaled forecast for a particular region or point for a particular time. In this chapter, one robust and reliable method has been described to generate one single forecast using a grand number of GCMs, bias-corrected, and downscaled products. Although these products have satisfactory skill in the forecast, however, those are at a monthly or seasonal scale, which needs to be temporally downscaled to a daily sequence for use in crop models. The next chapter covers the aspects of disaggregation processes of monthly scale data into daily sequence and its linkage with the crop models.

Chapter 7
Disaggregation of Monthly and Seasonal Forecasts

7.1 Introduction

The key impact of weather and climate on a short time scale is extremely high on socio-economic structure, particularly in the agriculture and water management sectors. With the advancements in computer resources and more faithful representations of large-scale atmospheric processes, seasonal climate predictions have become more reliable (Balan Sarojini et al. 2012). Consequently, there are many efforts to use seasonal climate forecasts for agriculture and food security risk management. Since many farm management decisions (such as which crop to grow in how much area and under what management conditions) are to be taken without knowing how the weather will likely be during the crop growth period, it is hypothesized that advanced information about the possible seasonal conditions in the form of seasonal climate forecasts will help farmers in making more informed decisions and thereby make better use of the season by adopting practices that match with the seasons potential with lower risk. Applying seasonal climate forecasts to improve decision-making for particular crops requires first translating forecasts into production and economic outcomes. Locally adapted and tested crop simulation models quickly explore the various management alternatives production outcomes under different weather conditions (Meinke and Hammer 1995; Jones et al. 2000; Royce et al. 2001). However, the spatial and temporal scales of dynamic seasonal climate prediction and crop simulation models complicate the task. Operational seasonal climate forecasts are typically expressed as climatic anomalies or tercile probability shifts averaged in space at the horizontal scale of general circulation model (GCM) grid cells and averaged in time over months or seasons. Generally, forecasts are probabilistic in nature and hence are unfamiliar to the user. It is difficult for farmers to interpret and understand the implications of the forecast information presented by meteorological organisations/departments/scientists. One of the main constraints in using

U. C. Mohanty et al., *Climate Risk Management in Agriculture*,
https://doi.org/10.1007/978-3-031-51862-1_7

seasonal/sub-seasonal/monthly/extended climate information by farmers is to have timely access to and easily understand the format of weather information by the farmers.

Advance intimation of rainfall distribution within a season will be more appropriate than on the season averages forecast, which is typically provided for management decisions in high-value crop production. The time of growing season onset, probability of water deficit during critical periods for yield determination, and conditions during ripening, harvest, and drying are essential determinants of crop production. This climate information can be successfully used in crop models for simulating the yield well in advance, and climate risk management options can be generated.

Arguably, GCMs generally provide skillful data at the monthly scale for grid boxes of hundreds of kilometers. At the same time, agricultural system models require daily weather data and operate at a field or farm scale. Both dynamical and statistical (empirical) approaches are used to bridge the spatiotemporal gap. Dynamical downscaling is used to achieve a higher spatiotemporal resolution by nesting an RCM within GCM output fields. RCMs provide daily values at a spatial resolution of tens of kilometers. Though the spatio-temporal resolution of RCMs is greatly improved over GCMs, the direct use of RCM output in impact models should be avoided, as suggested in the IPCC guidance for using RCM output (Mearns et al. 2003). This is first because daily precipitation statistics at a station or point are different from those of the spatial averaging that occurs within grid cells distorts the temporal variability of daily weather sequences (Mearns et al. 1995; Guo and Senior 2006). Second, it is well known that RCMs have considerable systematic errors in predicting daily precipitation. As RCMs consistently over forecast the number of wet days and low-intensity daily precipitation events, yet underestimate rainfall amounts of high-intensity, daily precipitation can seriously distort crop model prediction (Semenov and Porter 1995; Mearns et al. 1996; Riha et al. 1996; Guo and Senior 2006; Semenov 2007; Maraun et al. 2010; van Roosmalen et al. 2010; Jakob Themeßl et al. 2011).

In essence, bias correction of GCM/RCM predictions is an integral part of the model output statistics (MOS) methods. Jakob Themeßl et al. (2011) evaluated seven different MOS methods to correct raw RCM output errors and provide relevant daily climate data at particular locations for impact studies. They reported that quantile mapping of daily precipitation between the native RCM grids and target locations best bridges the gap between RCM output and end-user needs. Those MOS methods can correct the RCM's precipitation intensity bias (overestimating light events while underestimating heavy precipitation). It has also adjusted precipitation frequency (number of wet days) but is unable to improve temporal structure (representation of the lengths of dry and wet spells) (Maraun et al. 2010). The proper representation of precipitation sequence is crucial for correctly simulating plant growth and surface hydrology.

This distortion can result in either under-prediction of crop yields due to increased evaporative loss from the soil surface or over-prediction due to reduced dry spell duration (DeWitt 2005; Mearns et al. 1996; Riha et al. 1996). The Spatio-temporal scale mismatch between dynamic climate models and crop models poses a challenge in using crop yield prediction to explore crop response to variations in the predicted climate. Still, applying intra-seasonal climate variation for crop model applications is challenging.

The monthly to seasonal forecasts that various operational agencies generate are reliable and have potential skill. These forecasts are deterministic and probabilistic in nature and given in a single value for each month or season at least one month in advance. Hence, to use these monthly or seasonal scale forecasts in the crop models, it is first needed to downscale those forecasts temporally from monthly/seasonal into daily scale. Several approaches are available for the temporal downscaling from monthly/seasonal into daily scale (Mavromatis and Hansen 2001).

Stochastic weather generators are statistical models used to simulate synthetic weather sequences, which are expected to be statistically similar to the observed counterparts. Weather generators (WGs) are numerical tools designed to simulate synthetic time series of various meteorological variables of theoretically infinite length for a given climate and location based only on historical observations. Many WGs exist, with different methods to compute the meteorological variables (i.e., stochastic-statistical or physical-dynamical approaches) for different spatial and temporal resolutions (Wilks 1999a; Fowler et al. 2007). They are usually used with hydrological and environmental models for water resource and ecological management (Semenov and Porter 1995; Mavromatis and Hansen 2001; Wheater et al. 2005). More often, weather generators have been used as downscaling tools to produce high-resolution climate change projections by linking their parameters to climate model outputs (Wilks 1992; Pruski and Nearing 2002; Zhang and Garbrecht 2003; Zhang and Liu 2005; Zhang 2005; Chen et al. 2012a, b; Chen and Brissette 2014). Compared to other statistical downscaling methods, such as regression-based approaches and weather typing schemes, the weather generator-based approach has the advantage of producing an ensemble of climate change projections for analysing risk-based environmental impacts. Two types of models are usually used to generate precipitation occurrence: the Markov chain-based model and the alternative renewal process. The most commonly used first-order Markov chain has been found to be adequate for simulating wet spells for various climates and dry spells for temperate climates (Katz 1977; Richardson 1981; Wilks 1989, 1992, 1999a; Wilks and Wilby 1999; Nicks et al. 1995; Zhang and Garbrecht 2003; Chen et al. 2009). However, higher-order (second and third-order, for example) or hybrid-order Markov chain-based models (for instance, the first-order Markov dependence for wet spells and higher-order dependence for dry sequences) are usually necessary for adequately representing long dry spells, especially for dry climates (Wilks and Wilby 1999; Chen et al. 2012a, b). Instead of simulating precipitation occurrence day by day, the alternating renewal process considers precipitation as a sequence of alternating wet and dry spells of varying lengths. Various distributions (e.g., logarithmic series, truncated negative binomial distribution, truncated geometric distribution, and semi-empirical distribution) are then used to fit the wet and dry spells (Buishand 1978; Roldán and Woolhiser 1982; Semenov et al. 2002).

7.2 Theoretical Background of Stochastic Weather Generator

The mean and other statistical properties of generated daily sequences through temporal disaggregation using stochastic weather generators match predicted monthly or seasonal means and other statistical properties. This stochastic weather generator's precipitation occurrence is modeled by a two-state, second-order hybrid Markov chain. The precipitation occurrence is modeled with a first-order chain if the previous day was wet. If the previous day was dry, the precipitation occurrence is modeled with a second-order Markov chain (Stern and Coe 1984; Wilks 1999a). As a cause, the number of parameters to be estimated becomes less than higher-order Markov chains. A three-parameter mixed exponential distribution, also known as a hyper-exponential distribution, is used to sample the precipitation amount on a wet day (Roldán and Woolhiser 1982).

Temperature data (minimum: T_{min} and maximum: T_{max}) are sampled from the Gaussian distributions conditioned on precipitation occurrence. The mean and standard deviation for the Gaussian distributions used for T_{min} and T_{max} are estimated separately for wet and dry days.

The stochastic weather generator samples a logit-transformation of daily clearness, K from the Gaussian distributions conditioned on precipitation occurrence for modelling solar irradiance. K is defined as H/H_0, where H is daily solar irradiance at the surface and H_0 is extraterrestrial solar irradiance. K is rescaled between upper and lower limits. Solar irradiance is then derived by applying the inverse logit-transformation to obtain K, then multiplying by H.

Auto- and cross-correlation among generated daily T_{max}, T_{min}, and H_0 are imposed by sampling standard Gaussian deviates from a trivariate, first-order autoregressive process, with correlation coefficients that are held constant across location and season. Autocorrelation, also called serial correlation, is the correlation of a series with itself. It's the similarity between observations as a function of the time lag between them. Cross-correlation is the correlation of a series with another.

An autoregressive process (AR) is a stochastic process in which future values are estimated based on a weighted sum of past values. It assumes that past values affect current values. The first-order AR process means that the current value is based on the immediately preceding value only. After the stochastic generation of daily rainfall, temperature, and solar irradiance, they are subjected to a stochastic disaggregation process that imposes monthly targets on the mean values. So, the variances of distributions used to generate T_{max}, T_{min}, and H are centered on the mean of each month in the historical time series rather than on the long-term mean for the given calendar month.

7.3 Mathematical Steps of Stochastic Weather Generator

A second-order Markov chain can substantially improve the modelled distribution of dry period durations relative to the more commonly used first-order precipitation occurrence model for some locations and seasons. But, the first-order model usually captures the distribution of wet period durations. Therefore, Stern and Coe (1984) proposed using a Markov chain of hybrid order that would simulate precipitation occurrence with a first-order chain if the previous day was wet and a second-order chain if the previous day was dry. In this hybrid-order Markov chain, the Markovian 'memory' extends backward in time only to the most recent wet day if that time lag is smaller than the order of the process (Wilks 1999a). WM2 uses this type of two-state second-order hybrid Markov chain (Hansen and Mavromatis 2001).

7.3.1 Markovian Occurrence Model Used in WM2

First order (day t − 1 was wet)
P_{11} = Pr {precipitation (precip) on day t | precip on day t − 1}

Second order (day t − 1 was dry)
P_{001} = Pr{precip on day t | no precip on day t − 1 | no precip on day t − 2}
p_{101} = Pr{precip on day t | no precip on day t − 1 | precip on day t − 2}
It is a two-state (dry or wet) model with three transition probabilities (p_{11}; p_{001}; p_{101}).

Precipitation Amount

Probability Distribution

A probability mixture of two exponential distributions usually fits observed distributions of daily precipitation amounts better than more commonly used alternatives, such as the single exponential and gamma. Furthermore, the interannual variability of monthly rainfall is strongly influenced by rare extreme events. By better representing these extreme events, a three-parameter mixed exponential (also known as a hyper-exponential) distribution improves the inter-annual variability of monthly precipitation total relative to the more frequently used two-parameter gamma distribution.

7.3.2 *Probability Distribution Used in WM2*

The three-parameter, the mixed exponential distribution used by WM2 for sampling precipitation amount on a wet day is given by

$$f(x) = \alpha \frac{e^{\left(\frac{-x}{\beta_1}\right)}}{\beta_1} + (1-\alpha)\frac{e^{\left(\frac{-x}{\beta_2}\right)}}{\beta_2} \tag{7.1}$$

where
Mean $\mu = \alpha\beta_1 + (1-\alpha)\beta_2$
Variance $\sigma^2 = \alpha\beta_1^2 + (1+\alpha)\beta_1 + (1-\alpha)(\beta_1 - \beta_2)$
It uses three parameters α, β_1 and β_2.

A (unitless): Mixing probability of a hyper exponential distribution, which determines the weights given to the two Exponential distributions with scale parameters β_1, and β_2.

$\boldsymbol{\beta_i}$ **(mm/day)**: Mean of the *i*th component of a hyper-exponential rainfall intensity distribution.

Temperature and Solar Irradiance

Temperature data (T_{min} and T_{max}) are sampled from Gaussian distributions conditioned on the occurrence of precipitation. The mean and standard deviation for the Gaussian distributions used for T_{min} and T_{max} are estimated separately for wet and dry days.

For modelling solar irradiance, the stochastic weather generator samples a logit transformation, $y' = \ln(y/(1-y))$, of daily clearness, K from the Gaussian distributions conditioned on precipitation occurrence. K is defined as H/H_0.

Auto- and cross-correlation among generated daily T_{max}, T_{min}, and H is imposed by sampling standard Gaussian deviates from a trivariate, first-order autoregressive process, with correlation coefficients that are held constant across location and season.

After the stochastic generation of daily rainfall, temperature, and solar irradiance, they are subjected to a stochastic disaggregation process that imposes monthly targets on the mean values. So, the variances of distributions used to generate T_{max}, T_{min}, and H are centered on the mean of each month in the historical time series rather than on the long-term mean for the given calendar month.

Disaggregation of Monthly Climate Targets

Precipitation

Conditioning stochastic model input parameters:

$$R_M = n\pi\mu_{ri} \tag{7.2}$$

where

R_M Rainfall in a given month
π Frequency
μ_{ri} Mean intensity
n Number of days

Let R'_M be the target total rainfall in a given month. The stochastic model will have a central tendency of R'_M, if we either adjust rainfall frequency while holding intensity parameters to their climatological values:

$$\pi' = \frac{R'_M}{\mu_i n} \tag{7.3}$$

or adjust mean rainfall intensity while holding occurrence parameters to their climatological

$$\mu'_i = \frac{R'_M}{\pi n} \tag{7.4}$$

or adjust the frequency and mean intensity in combination, subject to:

$$R'_M = n\pi\mu'_i \tag{7.5}$$

Adjusting transition probabilities to match a target rainfall frequency is relatively simple in the case of a first-order Markov chain. The transition probabilities that define a first-order, two-state Markov chain relate directly to rainfall occurrence probability:

$$\pi = \frac{p_{01}}{1 + p_{01} - p_{11}} \tag{7.6}$$

and persistence of dry days:

$$p_1 = p_{11} - p_{01} \tag{7.7}$$

If we assume that persistence remains constant when occurrence probability changes, the adjusted first-order transition probabilities are:

$$p_{01} = \pi(1 - p_1) \tag{7.8}$$

$$p_{0111} = p_1 + p_{01} \tag{7.9}$$

The transition probabilities for a given p_0 can be adjusted in the same manner for the hybrid second-order Markov chain rainfall occurrence model, holding both first-order (7.7) and second-order persistence of dry days:

$$P_{0111} = p_{101} + p_{001} \tag{7.10}$$

In simplified form, they are

$$P'_{11} = \pi'(1 - p_1) + p_1 \tag{7.11}$$

$$P'_{001} = \pi \frac{(1 - p_1) + (1 - \pi') + p_2(1 - p'_{11})}{1 - \pi'} \tag{7.12}$$

$$P'_{101} = P'_{001} + p_2 \tag{7.13}$$

The hyper-exponential distribution used to sample rainfall intensity has mean:

$$\mu_1 = \alpha \frac{e^{\left(\frac{-x}{\beta_1}\right)}}{\beta_1} + (1 - \alpha) \frac{e^{\left(\frac{-x}{\beta_2}\right)}}{\beta_2} \tag{7.14}$$

The parameters of the hyper exponential distribution are adjusted to match a given target μ'_1 by adjusting the mixing probability α:

$$\alpha = \frac{\mu'_1 - \beta_2}{\beta_1 - \beta_2} \tag{7.15}$$

(if μ'_1 is between the means β_1 and β_2 of the two component exponential distributions). Otherwise, the model samples from an exponential distribution of mean μ'_i:

$$f(x) = \frac{e^{\left(\frac{-x}{\mu_i}\right)}}{\mu'_1} \tag{7.16}$$

Constraining Output to Target Values

In this case, rainfall for a given month is generated repeatedly until the generated monthly total R_M is within 5% of the target R^1_M. The daily series generated for the month is then rescaled by multiplying the rain generated for each day by R'_M/R_M to match the target rainfall total. This model's performance is better than the previous method of conditioning input parameters in terms of the number of realisations required to achieve a given correlation or RMSE.

Temperatures and Solar Irradiance

To rescale generated sequences of T_{max}, T_{min}, and H to match target monthly means, simple additive shifts are applied.

$$y'_t = y_t + D(t) \tag{7.17}$$

Such that

$$\sum_{i=1}^{n} D(t) = n\left(Y' - y_t\right) \tag{7.18}$$

where

Y' Target monthly mean of a given climatic variable
y_t Value generated on day t

To avoid discontinuities at month boundaries, the monthly shift is interpolated to daily values. To make $D(t)$ continuous and smooth, a mean preserving, continuous, segmented linear interpolation function is used for $D(t)$ that only requires Y' for the preceding, current and subsequent months.

7.4 Types of Weather Generators

Stochastic weather generators (WGs) are statistical models that aim at quickly simulating realistic random sequences of atmospheric variables such as maximum and minimum daily surface air temperatures (T_{max} and T_{min}), solar radiation (R), and precipitation occurrence (P_o) and its amount (P_a) and wind (Johnson et al. 1996; Wilks and Wilby 1999). Ideally, spatio-temporal dynamics and correlation structures among the variables of interest and weather persistence and natural variability have to be reproduced accurately in a distributional sense by WGs.

A stochastic weather generator is a computer algorithm that can generate a long series of synthetic daily weather data using existing meteorological records. The generated data characteristics are expected to be similar to those of the actual data for a specified site. The weather generator produces a complete record for any desired period, thus enhancing continuous hydrologic models, while some historical weather records may have been missing the data (McKague et al. 2005). Moreover, it can generate daily weather data for ungauged areas through spatial interpolation of model parameters from adjacent gauged sites (Baffault et al. 1996). The advantage of weather generators are mainly serving as tools to generate the daily sequence of multiple-year climate change scenarios that are used to assess the impact of future climate change, and these tools are computationally inexpensive (Wilks 1992, 1999b; Pruski and Nearing 2002; Zhang 2005; Zhang and Garbrecht 2003; Zhang and Liu 2005; Minville et al. 2008). Model parameters of the weather generator can be readily manipulated to simulate arbitrary changes in mean and variance quantities for sensitivity analysis or be deliberately modified to mimic changes in mean and variance as predicted by global climate models (GCMs) for impact assessment. Over the years, several weather generators such as the Weather Generator (WGEN) (Richardson 1981; Richardson and Wright 1984), the CLImate GENerator (CLIGEN) (Nicks et al. 1995), Climate Generator (ClimGen) (Stockle et al. 1999), Long Ashton Research Station-Weather

Generator (LARS-WG) (Semenov et al. 2002), the Weather generator of École de Technologie supérieure (WeaGETS) (Chen et al. 2012b), Multi-site stochastic weather generator (MulGETS), etc. have been developed and widely used for climate generation and climate change impact studies. Several studies (Buishand 1978; Johnson et al. 1996; Wilks 1989, 1999a; Katz and Parlange 1993, 1998; Hansen and Indeje 2004; Zhang and Garbrecht 2003; Chen et al. 2009) found that weather generators have a good ability in producing long-term time series. It can be helpful to study the impact of rare occurrences of meteorological events. But, there is an underestimation of monthly and inter-annual variances by daily weather generators. It is mainly due to the weather generator not taking into account the low-frequency component of climate variability.

Most weather generators are single-site based and can only generate meteorological data at a single point or independent time series at several points. However, for hydrological studies, weather data are usually required to be known at several locations within a watershed. This is particularly true for precipitation, which often displays a large variability in time and space. Hence, proper representation of spatial correlation in meteorological variables is particularly important for generating multi-site climate data within a large watershed. Several approaches have been developed to simultaneously generate multi-site precipitation (Wilks 1998; Kottegoda et al. 2003; Wilby et al. 2003; Clark et al. 2004a, b; Fowler et al. 2005; Mehrotra and Sharma 2007; Eum et al. 2010). The brief details of various types of weather generators are described in the following subsection.

7.4.1 Auto-regressive Weather Generator Model

The autoregressive weather generator model has been used extensively worldwide, and was introduced by Richardson (1981) and Richardson and Wright (1984). This model can work based on the multivariate autoregressive process described by Matalas (1967). In many studies (Schoof and Robeson 2003), the original approach has been altered through changes in how the parameters are estimated. In this method additional variables inclusion (WXGEN, Nicks et al. 1995; GEM; Hanson 1998) where normality constraints relaxation for the added variables (Parlange and Katz 2000), however the basic structure of the model has remained unchanged. Schoof and Robeson (2003) found that the improvements to the model parameterizations in the autoregressive weather generators still occasionally produce fundamental simulation errors, such as negative diurnal temperature range (DTR). Additionally, monthly T_{max} and T_{min} probability distributions are generally skewed, and generating temperatures with the normal distribution can lead to physically unlikely values (Harmel et al. 2002).

7.4.2 Multi-site Stochastic Weather Generator (MulGETS)

The multi-site stochastic weather generator (MulGETS) is a new Matlab-based stochastic weather generator, and it is an extension of a single-site weather generator (WeaGETS) (http://www.mathworks.com/matlabcentral/fileexchange/29136-stochastic-weather-generator-weagets). MulGETS is used to generate multi-site precipitation and temperature. MulGETS generates the precipitation occurrence using a first-order two-state Markov chain. On the contrary, the temperature is generated by MulGETS using the first-order autoregressive model like an unmodified single-site weather generator. Instead of a single gamma distribution, MulGETS uses a multi-gamma distribution to handle the spatial correlation of precipitation amounts. By adjusting its parameters to supported climate model outputs, MulGETS can easily be used as a downscaling tool for global climate change impact studies.

7.4.3 CLIGEN (CLImate GENerator)

CLIGEN mainly generates climate data to simulate erosion processes and sediment delivery at the hillslope profile and small-scale watersheds (Nicks et al. 1995). CLIGEN can simulate the storm duration, peak rain intensity, and time to peak, along with the intensity patterns within a storm. CLIGEN can produce essential climate variables (such as precipitation, temperatures, solar radiation, and wind velocity) on a daily scale using observed monthly statistics (means, standard deviations, skew coefficients, etc.). These products are beneficial for any water resource studies. In CLIGEN, RAND is a pseudo-random number generator, and it produces a sequence of random numbers independent of each other statistically. A standard parameter (*.par) file containing minimum raw weather data such as precipitation and maximum and minimum temperatures is mainly required for CLIGEN to generate synthetic data of unlimited length. If other parameter data are not available at the location of interest, they may need to be interpolated from nearby stations.

Usually, the data can be obtained from NOAA with a *.dly format file. The *.dly file information is translated to a *.DAT file. By using python scripting (DAT2PAR.exe program), the *.DAT file can often be transformed into a *.par. This PAR file contains basic information including mean, variance, and skewness for rainfall and wet day, probability of occurrence of the wet day provided the previous day is wet or dry day; average and variance for max and minimum daily air temperatures; mean daily radiation and its standard deviation; temperature; and time to peak rainfall intensity. CLIGEN can handle missing data to some extent. But, if the input dataset has quite two months of continuous missing data, then the PAR file would not be generated. Temperatures are assumed to be normally distributed in CLIGEN. For simplifying the mathematics behind the simulation process, the CLIGEN produces the utmost, minimum, and temperature temperatures with the assumption that they're independent of each other. However, there are checks in situ to make sure that the values obtained are acceptable. For example, there are checks in situ within CLIGEN to make sure that the simulated daily minimum temperature does not exceed the simulated daily maximum temperature that may happen if the underlying hypothesis of minimum and maximum temperatures being normally distributed for each month is violated (Harmel et al. 2002). Precipitation is calculated by employing a probability distribution. Fourier series interpolation/disaggregation is considered to offer more consistent results than the linear interpolation method performed to preserve monthly averages for daily simulation from monthly data. The full details of CLIGEN are available at https://www.ars.usda.gov/ARSUserFiles/50201000/WEPP/cligen/CLIGENDescription.pdf.

7.4.4 LARS-WG (Long-Ashton Research Station Weather Generator)

A license-protected weather generator called LARS-WG uses 21 different parameters to signify interval bounds and calculate the number of events within each interval based on a semi-empirical distribution. The amounts and occurrence of precipitation are quantified (Racsko et al. 1991). Three distinct steps are generated in Weather data. The foremost is model calibration, where the observed dataset is analysed statistically. In the second stage, the data (synthetic and observed data) is checked for any statistical differences. Finally, a parameter file is created using monthly statistics as determined from observed data. For LARS-WG, even though a single year of data is adequate to generate the synthetic climate data, 20–30 years of data are suggested for improved representation of the climate for the site (Semenov and LARS-WG-5, 2017). For extreme events, a long-term observed data record may help capture less frequent climate events. Parameters like semi-empirical distributions for minimum and maximum temperatures, radiation, and the length of dry and wet series precipitation are calculated for dry and wet days. Also, correlation and autocorrelation coefficients are found in the LARS-WG file. Long-term climate series can still be generated even though missing data alter statistics from the observed data set.

The output results of the statistical tests are formed by comparing generated data with the observation. The random seed generator can be set to any number which makes up LARS-WG stochastic a defaulting value set to 541. The precipitation threshold to define if a day is dry or wet is 0.0 mm, differing from CLIGEN and WeaGETS, where the threshold is 0.1 mm. In a way to generate daily maximum and minimum air temperatures daily, means and standard deviations are accustomed based on whether a day is dry or wet. Residuals are calculated following a normal distribution, whereas a finite Fourier series of order 3 is used for generating the daily minimum and maximum air temperatures. The pre-set autocorrelation is about 0.6 for minimum and maximum temperatures, indicating at least 60%-time autocorrelation for minimum and maximum air temperatures dissected from the noted residuals by eliminating the mean value from the observed dataset. Analogous to CLIGEN, a verification is done during the simulation process to preclude a situation where simulated minimum temperatures are higher than the maximum.

7.4.5 Weather Generator (WeaGETS)

Weather Generator (WeaGETS) is an adaptable and stochastic, MATLAB-based weather generator that simulates daily precipitation, and minimum and maximum air temperatures in a sequence of the climate of unlimited length. The exclusivity of WeaGETS is that it can use Markov chain models of all three orders to generate precipitation occurrence and four distributions (exponential, gamma, skewed normal and mixed exponential) for simulating precipitation quantity. A total of 12 sub-models are used to simulate weather data within WeaGETS. In addition, an alternative is available for smoothing precipitation parameters and low-frequency correction. For the precipitation parameters smoothing, Fourier harmonics can be utilized (Richardson 1981). To simulate the minimum and maximum air temperatures on a daily scale, conditional and unconditional schemes are accessible. The spectral correction approach can be used to rectify the underestimated variability at monthly and interannual time scales (Chen et al. 2010). Biweekly the transitional probabilities are computed to account for the occurrence of precipitation. Four different orders of Fourier harmonics can be used within the smoothing process for smoothening of variation in a 2-week time period, which can account for true yearly distributions of the transitional probabilities associated with precipitation occurrence. The input data mandatory for weather generation in WeaGETS consist of daily precipitation and minimum and maximum air temperatures. The date of 29 February is not taken into account by the weather generator, and any significant precipitation event occurring on 29 February is redistributed equally between 28 February and 1 March. The leap year (29 February) temperatures are removed (Chen et al. 2012a), and WeaGETS handles the missing values in the observed data set as there is not much information on random seed generation within WeaGETS, and although it is presumed that it utilises the control random number generation function within MATLAB for generating random numbers that may be repeatable or different.

7.5 Performance Evaluation of Stochastic Generator

Examining the performance of daily rainfall sequences generated by stochastic weather generators is undoubtedly a remarkable fact. For an, e.g., Ghosh et al. (2015) disaggregated the observed monthly as well as total seasonal rainfall of Bhubaneswar station to daily weather sequences during the summer monsoon (June through September; JJAS) for the period 1971–2007 using the method suggested by Mavromatis and Hansen (2001). In general, in a stochastic weather generator, the monthly rainfall total is considered a product of the frequency and intensity of rainfall (Ines and Hansen 2006). Stochastic disaggregation targets the monthly/seasonal rainfall total by adjusting the rainfall frequency and intensity. Even if, in theory, one can generate an infinite number of realisations, yet, in their study, one thousand realisations are generated for disaggregated series by taking into account 5% deviations in parameters that are considered to achieve maximum forecasting skill for rainfall frequency and intensity in a particular month and season.

For a whole summer monsoon (JJAS) and also for individual months with possible combinations for intensity and frequency of rainfall, correlations are calculated between observed and disaggregated realisations average of different sample sizes (e.g., sample size 10, the average correlation of 1–10, 2–11, 3–12,..., 991–1000). The rainfall intensity and frequency's correlation value gets increased with the increase of sample size of realisations up to 200 and then stabilise. The correlation coefficient for July rainfall intensity was maximum, i.e., 0.90 (Fig. 7.1), and had a significant correlation of 0.71, 0.88, and 0.88 for June, August, and September, respectively. Likewise, the frequency of the disaggregated rainfall sequences was also evaluated, and similar patterns were found. The maximum

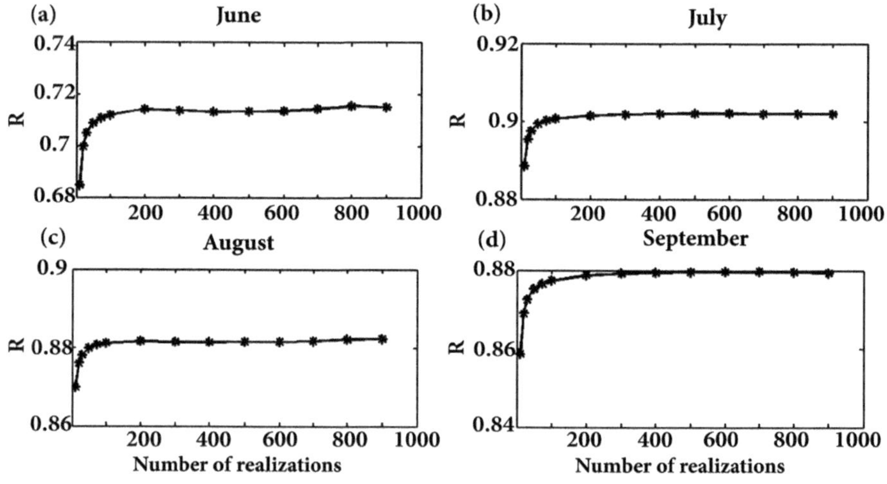

Fig. 7.1 Correlation (R) between observed versus disaggregated monthly intensity of rainfall for **a** June, **b** July, **c** August, and **d** September as a function of the number of stochastic realisations (1000) from 1971 to 2007

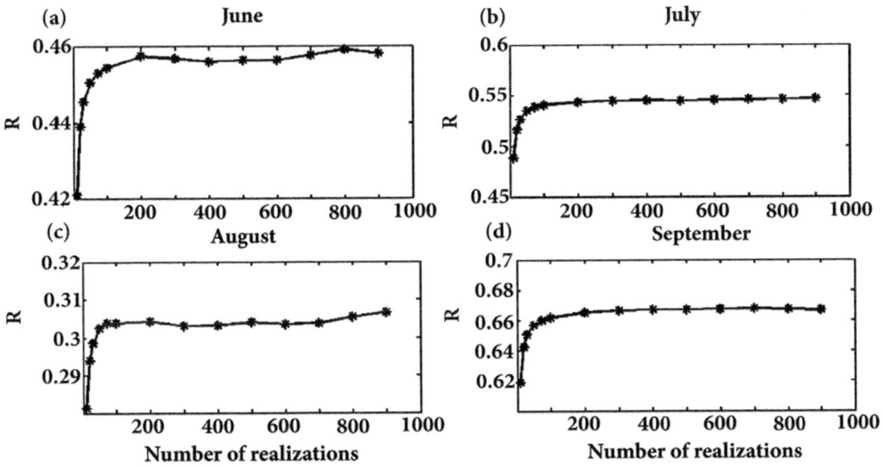

Fig. 7.2 Correlation (R) between observed versus disaggregated monthly frequency of rainfall for **a** June, **b** July, **c** August, and **d** September as a function of the number of stochastic realisations (1000) from 1971 to 2007

correlation values for frequency were reached at around 200 realisations, which are 0.46, 0.54, 0.29, and 0.66 for June, July, August, and September, respectively (Fig. 7.2).

The statistical skill scores of disaggregated realisations for seasonal frequency and intensity are evaluated and shown in Fig. 7.3. With the increase in the number of realisations, the correlation value for the whole season also increases (attains its maximum at around 200 realisations).

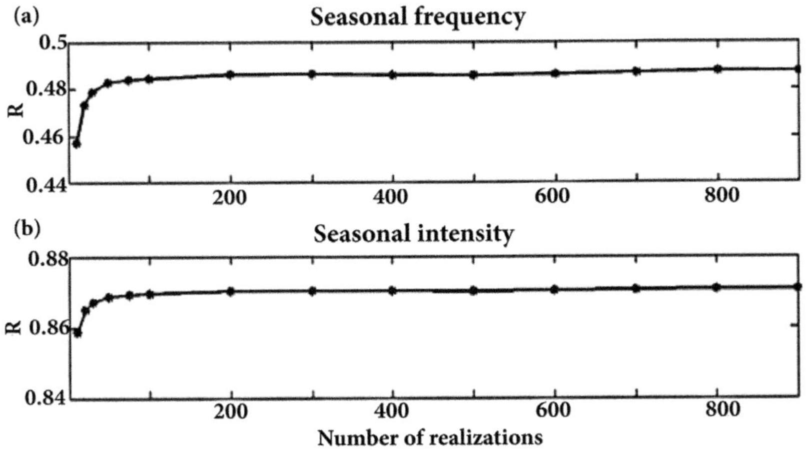

Fig. 7.3 Correlation (R) between observed versus disaggregated **a** seasonal frequency and **b** seasonal intensity of rainfall as a function of the number of stochastic realizations (1000) from 1971 to 2007

The correlation of seasonal frequency of rainfall approaches a maximum of 0.48, whereas that of seasonal intensity is nearly 0.85. Due to variability in rainfall distribution associated with the stochastic model, variation in correlations of the parameters among realisations occurs (Katz and Parlange 1993, 1998). The skills of the disaggregated forecasts are superior, and the results specify the usability of outputs from the stochastic disaggregation. Henceforth, the outputs of disaggregated monthly/seasonal forecast products can be considered as the input for the crop model.

The statistical skill of intensity and frequency parameters of rainfall at monthly and seasonal scales in terms of correlation shows saturation mostly at 200 realisations in this case study and it may vary for other regions or locations. Therefore, the number of realisations required to achieve reasonable accuracy in reproducing yields with a crop simulation model using observed as well as forecast rainfall data has been fixed at 200 realisations. This disaggregated observed rainfall is used as input in crop models, which may be considered reference yield data. The next chapter discusses the linkages of disaggregation of monthly and seasonal scale forecast products with crop simulation models for developing crop yield prediction systems.

Chapter 8
Crop Modelling and Simulation Concept

8.1 Introduction

Currently, about a billion people have very limited access to water and food, which makes maintaining a sustainable production system crucial to achieving global food security. Increased population and growing food demand are alarming trends that require the maintaining of a sustainable production system. It is estimated that by 2050, global food production will need to increase by 70–110% at a rate of 2.4% per year in order to meet the needs of 9 billion people (Ray et al. 2012; Zhou and Butterbach-Bahl 2014). Agricultural sectors are heavily dependent on the availability of resources and climate conditions. Therefore, optimising the resources is highly necessary in order to maximise crop productivity. Agricultural production is the mainstay of the rural Indian economy and plays a major role in the process of economic development of India (Naiyer and Md 2015). The agricultural sector in India's economy has declined by less than 15% over the past decade due to the expansion of the industrial and services sectors. Considering that almost three-fourths of Indian families rely on rural incomes, it becomes important to focus more on agriculture since approximately 70% of the country's poor reside in villages.

To meet the increased demands of the growing population, India relies heavily on the production of cereal crops, vegetables, fruits, milk, etc. Thus, the need for a well-diversified, productive and sustainable agriculture sector is increasing at a rapid pace to satisfy the needs of the growing population. India produces the most pulses, milk, and spices and has the world's largest herd of cattle. The country has the largest area under rice, wheat, and cotton. About 195 million hectares of land is under cultivation, of which 37% is irrigated and 63% is rainfed (www.worldbank.org). Climate change causes an increased variability in rainfall and temperature during the crop growing season, which negatively impacts agricultural production. Nevertheless, over 80% of the variance in productivity is caused

U. C. Mohanty et al., *Climate Risk Management in Agriculture*,
https://doi.org/10.1007/978-3-031-51862-1_8

by local weather and climate variations. The current changing climate makes it more important than ever to predict accurately the yield of any crop in advance, so adverse effects of bad weather conditions can be reduced to a minimum and good seasons are more productive, strengthening the economic standings of any nation.

Global food production is directly and indirectly affected by changing climate due to increased atmospheric temperatures, increased greenhouse gas concentrations, primarily (CO_2), and variability of rainfall (Kang et al. 2009). Higher temperature affects soil health and also affects the post-harvest processed yield of food grains. Increasing soil temperatures may accelerate the decomposition of organic matter and increase other chemical and biological processes that affect soil fertility. For the soil to be protected and crop productivity to be maximised, new innovative farming techniques must be utilised in place of traditional farming methods. Since groundwater levels have declined in India and water is used inefficiently, the per capita water resources have decreased over the years. With an increase in population, food demand will increase, and water availability will decrease along with it (Fig. 8.1). It is, therefore, imperative to develop technologies that will enable farmers to effectively utilize available resources. To minimize yield losses, farmers must get timely information on the outcomes of the crop growth prediction system so that they can adapt new technologies to minimize climate risk. Adapting to climate change through changes in farming practices, cropping patterns, and new technologies may help for a sustainable future.

The scientific community and stakeholders are concerned about the climate variability associated with the weather (floods, droughts, heat waves, and cold waves). As crop production involves a wide range of factors, including land type, water and nutrient management, pest control, cultural practices, genotype,

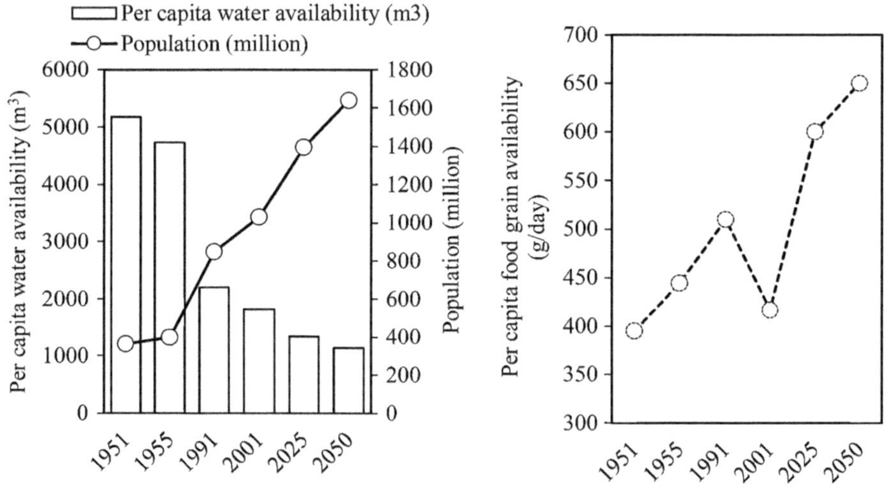

Fig. 8.1 Representation of per capita water and food grain availability and population in India by 2050. (*Source* Ministry of Agriculture, Government of India, Oct 2007)

and environment, it becomes increasingly difficult and time-consuming to conduct experiments to understand all these factors. With a validated crop simulation model, factors could be tested independently and in combination in a limited time and expense. These simulations can adequately describe relative trends in yields for varying agronomic management for quick evaluation of climate change adaptations (Penning de Vries and Laar 1988). The Wageningen group is the oldest working group for developing and applying crop models in the agroecological research program. The group developed and analysed several crop models with the first effort of C.T. de Witt in the 1960s. BACROS, LINTUL, SUCROS, ORYZA, WOFOST, INTERCOM, etc., are the most successful and popular crop models of the Wageningen group. Hoogenboom et al. (1994) developed a Decision Support System for Agro-technology Transfer (DSSAT), which integrates multiple crop simulation models for field crops with the weather, soil, and experimental databases. Recently, crop simulation models such as DSSAT have been used for decision-making and strategic adaptations for sustainable crop production and food security (Hoogenboom et al. 1994; Jones et al. 1998; Satapathy et al. 2014).

8.2 Description of Terms Used in Crop Modelling

The basic terminologies provide a general information frame for a better understanding of contexts. It plays an important role in understanding scientific texts more effectively. In crop modelling, the user needs to understand many scientific and specialized contexts. Any user must have a general idea about the major terms related to a model before starting to work in any model. In this section, some important terms related to a crop model are described.

8.2.1 Model

A model is a simplified representation of a process or a system. It can be defined as a simple description of a system used to perform the desired calculations. The process or system can be expressed in a formal mathematical statement or a series of statements known as Modelling. From simple sketches to sophisticated computer programs with tons of code lines, all are in the fields of models. The common thing in every model is to form and summarise the real situation in a scientific approach. There are many objectives of a model, depending on the user's need and specific problems, and include, in particular, (i) facilitating the scientific understanding of the system, (ii) assisting in decision making by simulating 'what if' scenarios, and (iii) monitoring and predicting.

8.2.2 Crop Model

Crop models are a quantitative description of different mechanisms that result in desired output like crop growth and yield during different stages of crop development. Crop growth and development are described as a function of soil conditions, weather conditions, and crop management. The biomass of crop components is estimated as the changes in a function of soil moisture, nutrients, and daily weather parameters over time. Crop models are tools for agricultural systems, which help in solving problems related to crop production (Bannayan et al. 2003).

8.2.3 Crop System

A system consists of various elements and their interactions with each other that work together to accomplish a specific goal. An intricate crop system is composed of all the factors that influence crop growth, including soil, plants, and weather. The elements or components for a cropping system are selected based on the objectives decided for the study.

8.2.4 Simulation

The process of model construction and management, model resolution and system analysis is known as Simulation. The simulation process includes the development of flowcharts, logical computer programs and codes, and the application of estimated computer codes to achieve the desired outputs. Simulation of crop systems comprises computer simulation and mathematical functions.

8.2.5 State Variables and Rate Variables

State variables represent the state or conditions of a system's components. Using the model type, these variables can be quantified or measured as changes over time. For crop models, these state variables change due to the interactions between the system components and environments. Rate variables represent the rates of different processes operating in a system. In a crop model, transpiration and photosynthesis rate are the main rate variables.

8.2.6 Parameters and Constants

When a model is applied to different experimental conditions, there are some quantities with accurate and reliable values that remain the same and are known as Constants. There are many examples of constants such as the number of hours per day and the gravitational constant. There are some quantities whose values are less certain but are assumed to be the same throughout the simulation process, known as parameters. For example, resistance to soil water flow and the number of degree-days from one growth stage to another growth stage of a crop is the parameters in the crop model. Constants are universal constants, whereas parameters remain constant for a particular simulation condition which changes with changing conditions.

8.2.7 Calibration

To calibrate a model, the parameters are adjusted so that simulated values match the measured or observed values obtained from field experiments. Initially, the model is run on as trial and error basis with any experimental data set and then calibrated correspondingly. The calibration process should be organised within the boundaries of a given data set.

8.2.8 Validation

In the next step, the calibrated model is validated using yet another experimental dataset to validate its ability to simulate under different conditions. Validation is the process of comparing simulated values to observed or real values, which have not been used in any calibration process. The main goal of validation is to compare the simulated state variables with observed values of state variables and to check the accuracy of the model.

8.3 Goal of Crop Modelling

Decision Support Systems (DSS) are a specific class of computer-aided information systems that supports entrepreneurial and organisational decision-making processes. A DSS is a software-based interactive system designed to help decision makers compile useful information from raw data, documents, personal knowledge, and/or business models to identify and resolve problems and make decisions. DSS are data-driven, as the entire process relies on the collection and availability of data to analyse using various tools, which use input data/information and produce output/decision after certain processing based on some set rules/criteria. DSS makes it easier for users to understand the effects of different factors/constraints on the system. With the advent of computers, DSS is being developed and used in almost every field. By using DSS, a farmer, planner, or decision-maker can prioritize the important activities with greater impact under existing constraints. DSS in agriculture are also used as convincing tools to explore, suggest, and disseminate technologies/strategies that can improve farmers' agricultural production, incomes, and livelihoods.

Crop simulation models as DSS aim to predict and estimate the effects of varying agronomic practices, soil types, environmental factors, etc., on crop production. The user needs to understand the objectives of the model in order to make better use of it. This modelling and simulation concept aims to provide an easier and more versatile means to generate knowledge and information to support the needs of scientists, engineers, decision-makers, and farmers. A few objectives of crop modelling and simulation concept are described below:

Integration and application of knowledge: The basic objective of crop modelling and simulation concepts is to implement the knowledge gained from different field experiments and analyse the results. The tendency to look for new methods to minimise cost and labour and to combat unfavourable conditions for crop growth.

Promoting versatile services: Crop simulation models provide a basic framework that promotes the versatility of the system components. It provides an independent platform to perform all the desired operations according to user requirements.

Providing user friendly and dynamic tools: The crop simulation model is simplified and managed to provide dynamic, user-friendly, and quantitative tools for predicting the output during crop cultivation. Since agriculture is a complex system, a dynamic and simple tool is needed. Evaluation of interactions among the system components can be done quickly, and the simulation process can easily find an identification of traits that have a significant impact on outputs.

Developing decision support system: The crop simulation model is developed to understand the problems and make appropriate decisions de-pending on the situation. It is essential to make appropriate decisions and provide the information required for the establishment of an effective crop management system.

Identifying the issue to be consigned and the data to be derived from the crop model is of utmost importance. In addition to identifying the problem, there should be a clear vision of the expected outcomes developed within the framework of a project. An inappropriate objective for a project or proposal may reduce the performance of the selected project. The model's development process includes the scientific and engineering goals as shown in Fig. 8.2. Before working on a model, a hypothesis is assumed and the model is calibrated, and a prediction is made. The prediction is then tested or validated which will further improve the understanding of the model. Otherwise, new research is carried out after testing the prediction, which will further improve user understanding and this complete model development process falls within the scientific goal. The model is used for

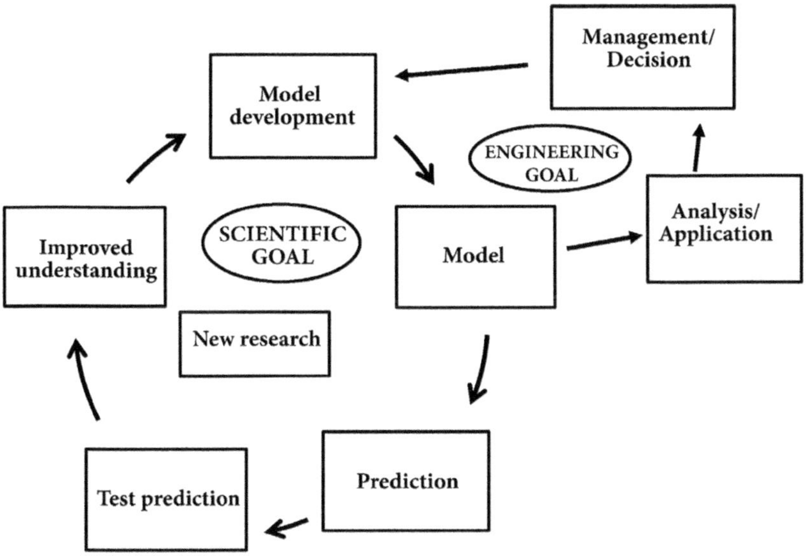

Fig. 8.2 Flow chart of steps involved in crop model development and prediction system

analysis or application; based on that, management and decisions are to be made. This complete process comes under the engineering goal, i.e., better predicting the system behaviour.

8.4 Crop System Model Development

In the face of a limited supply of farmland and an ever-growing population, there is an increase in complexity and pressure on natural resources. Again, these factors are integrated with climate change, which poses a number of challenges to food production. In order to plan agricultural production management and make good decisions, there is a need to develop new ideas and tools. Models of agricultural systems can generate huge amounts of data and information for enhancing the development of sustainable land management across many locations while taking up very little time and effort. Figure 8.3 presents an overview of the history of crop modelling, which includes a number of events that motivated scientists from different disciplines to develop and use models. The 1950s to 1960s were marked by the development of foundational science and early computational analysis of plants and soil processes. Identification of ecology and policy was introduced during 1960- 1970, including many new events such as pioneer water balance modelling, international biological programs and early crop modelling. From 1970 to the present day, the series of events has progressed as satellite technology and communication technology have been enhanced, systems models have been expanded,

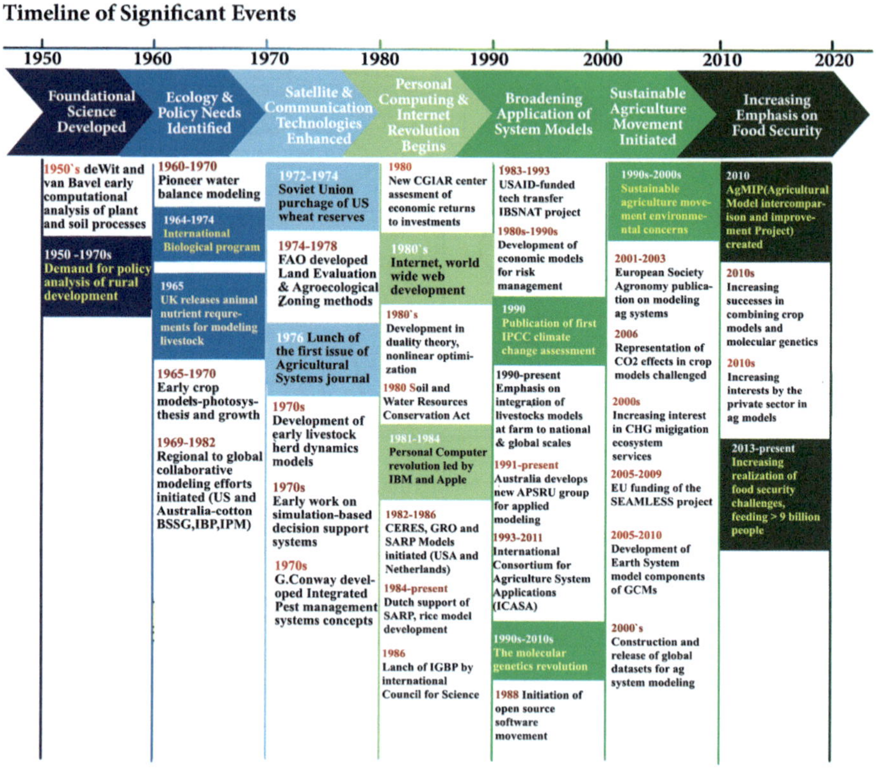

Fig. 8.3 Historical chronology of crop yield prediction system. (*Source* https://www.sciencedirect.com/science/article/pii/S0308521X16301585)

sustainable agriculture movements have been initiated, and an emphasis on food security. Different factors and management practices contribute to the complexity of the agricultural system. A flow diagram of crop model development is given in Fig. 8.4. The process of model development comprises the following steps:

1. ***Deciding appropriate goals***: The initial step to developing or understanding any crop model is deciding on well-defined goals. One should have a clear vision of the problem and requires proper knowledge of related areas. After deciding on specific objectives for the study, information from different fields is unified to develop a model.
2. ***Selection of system and boundaries***: Defining a proper system and its boundaries ensures an appropriate environment for the study and a well-established design for any system. In the case of crop models, the crop field is selected as a system.
3. ***Selection of key variables***: It is important to keep in mind the importance of key variables during the decision-making process for an agricultural system.

Fig. 8.4 Steps of crop system model development

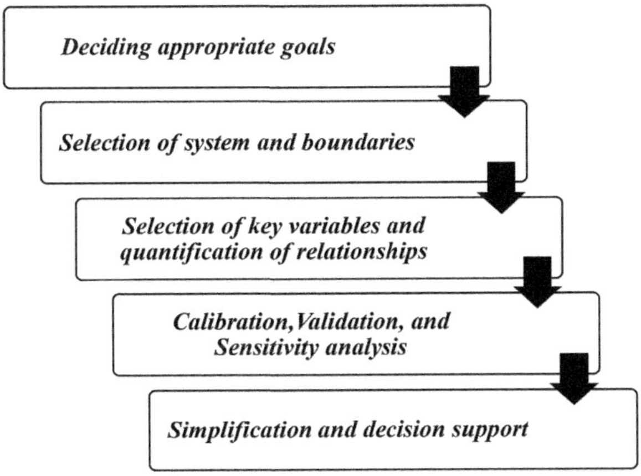

Crop systems differ based on state, driving, rate, and supplementary variables. State variables include crop biomass and soil moisture content. Driving variables are not part of the system but affect it, like rainfall, temperature, and solar radiation. Rate variables include transpiration and photosynthesis rates, and supplementary or auxiliary variables are like water stress at critical growth stages of the selected crop. Identification of all the variables is made in the crop field, and interaction or links among the selected variables is estimated to better understand the system.

4. ***Quantification of relationships***: After establishing a relationship among selected key variables, it is quantified by practicing different mathematical equations.
5. ***Calibration and validation of the model***: This step is crucial to check the accuracy and efficiency of the developed crop model. Field experiments' information is collected to run the developed model and calibrate it appropriately. The calibrated model is then validated with another dataset to check its simulation capacity under varying environments.
6. ***Sensitivity analysis of the model***: After validation, the developed model is checked and tested for sensitivity to different climatic and edaphic factors. This step is repeated to check the response of the developed model to changing factors.
7. ***Simplification of the developed model***: It is necessary to design a model as per the requirement and convenience of the user. In this step, the developed model is simplified to make it user friendly.
8. ***Using the developed model in decision support***: The final developed model is ready to be used in any decision support system for suitable decision-making and forecasting related to crop growth and management.

8.4.1 Standard Symbols Used in Crop System Models

A complex agricultural system can be easily represented as compartmentalized models, which are useful in conceptualising systems, focusing on the flow of variables and their storage. Forrester (1961) developed compartment diagrams to represent systems of equations for a variety of dynamic models in a visual manner. The mass flow between various compartments in agricultural systems is applicable for various applications. A compartment is defined to be a quantity of a state variable and is connected by lines that represent the rates of flow of the quantities between them. Figure 8.5 shows a set of symbols from the Forrester model. The level represents the state variable at a point in time and is shown by a rectangle. Sources and sinks represent the system's environment in the flow and the first can occur from a source into the system without affecting the environment. The rate symbol represents the flow from one compartment to another, from a source to a compartment, or from a compartment to a sink. Auxiliary variables are shown by circles and represent factors that influence the rates. Solid lines represent pathways for material flow whereas dashed lines represent the flow of information.

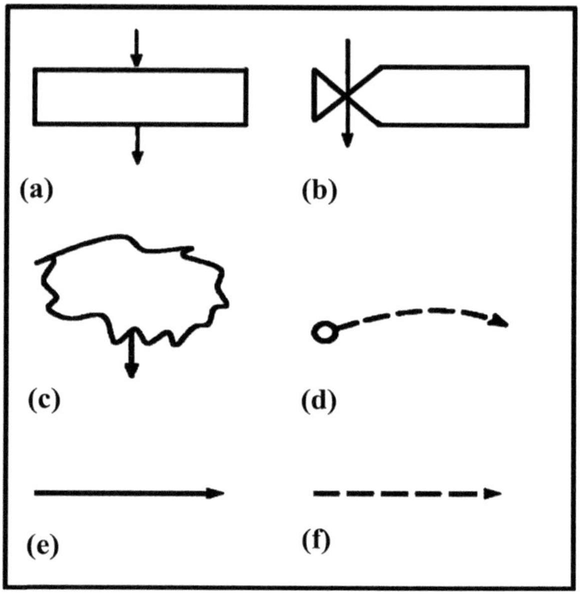

Fig. 8.5 Standard symbols used in crop system models (Forester model) **a** level **b** rate **c** source **d** auxiliary variable **e** pathway for material flow **f** pathway for information flow. (*Source* Agricultural Systems Modelling and Simulation by Robert M. Peart and R. Bruce Curry)

8.4.2 Crop Model for Different Production Situations

A crop simulation model's main components are three production levels: potential, attainable, and actual production. The characterisation of crop models is done for convenient application at field level. The actual, attainable, and potential yields are presented in Fig. 8.6 for easy understanding.

1. **Potential production**: Potential crop production is the yield obtained at optimum water, fertiliser, and pest and weed control conditions. Crop growth is dependent on genetic factors and weather factors such as solar radiation, temperature, CO_2 concentration, etc. It represents the first level to characterize a crop simulation model based on potential production, which includes basic crop growth processes such as respiration, photosynthesis, growth, and development. Models built for potential production assume that the crop is well protected from pests and diseases and does not suffer from water and nutrient stress.
2. **Attainable production**: In a particular soil and climate condition, nitrogen and water management plays an important role in achieving economic production. It represents the second level to characterise a crop model based on attainable production, which describes water and nitrogen limitations to crop yield. As the second level of characterising a crop model, it differentiates the crop yield limit due to water and nitrogen. Models built for attainable production assume that the crop is well protected from pests and diseases. Many crop models can interpret the growth and yield of the crop with these assumptions.

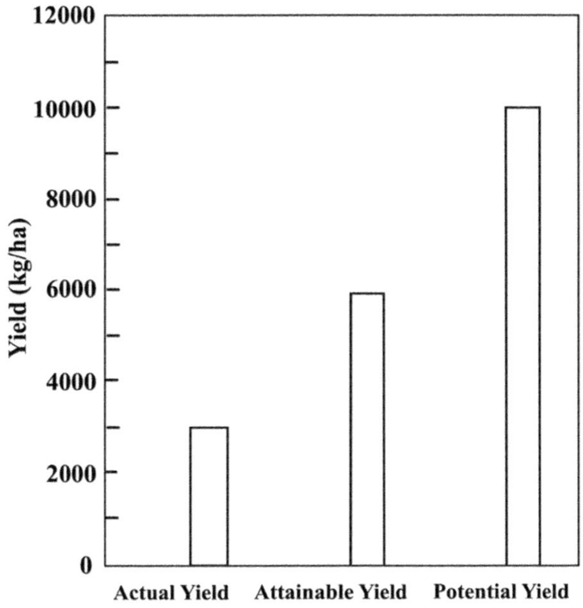

Fig. 8.6 Representation of actual, attainable, and potential yield

3. Actual production is the crop yield obtained in a farmer's field condition. It represents the third level to characterise a crop model on the basis of actual production, which attempts to explain all the factors that influence crop growth, including all the unfavourable conditions like pest damage, input management (water and nutrients), and climatic conditions. This level characterizes a model based on actual concepts, hence providing a practical pathway and conceptual framework for the proper application of the model.

Crop models have many choices of state variables. Figure 8.7 shows a Forrester diagram of a simple crop model with three state variables: stage of development (N, number of vegetative nodes), canopy biomass (Wc, g/m^2), and root biomass (Wr, g/m^2). Crop models with many state variables provide more details about crop behaviour, including lags in fruit growth and size distributions, but require much more time to develop and test and more time to solve in a computer.

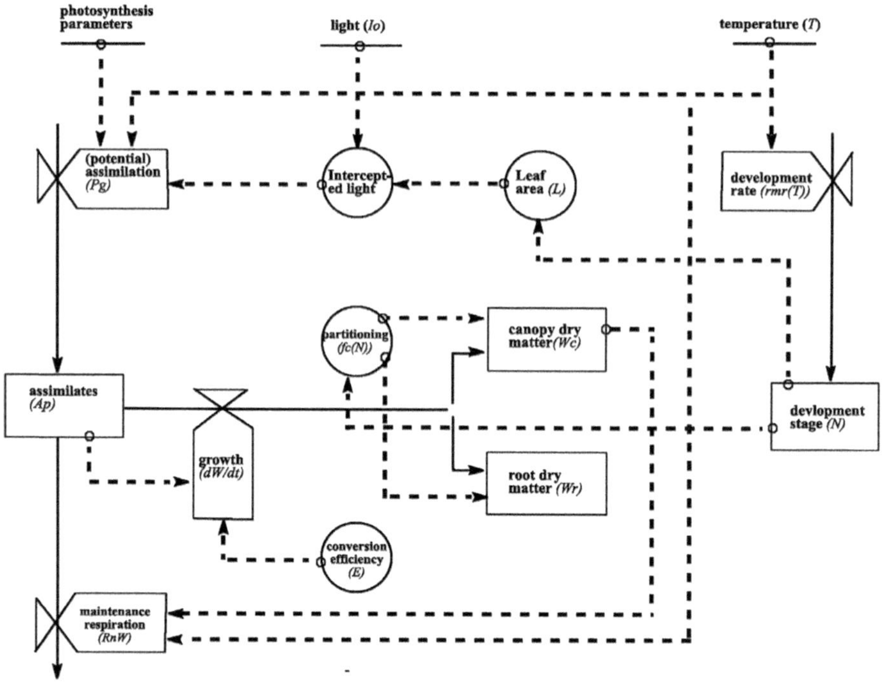

Fig. 8.7 Forrester diagram with three state variables at the potential production level. (*Source* Agricultural Systems Modelling and Simulation by Robert M. Peart and R. Bruce Curry)

8.4.3 Growth and Development Simulation Principles

Indirectly or directly, crop growth rate affects crop yield. Depending on the photosynthesis process, dry matter rate is of great importance in determining yield. A crop model simulates biomass and yield by utilising a variety of expressions and equations that describe the essential mechanisms during growth. Respiration and photosynthesis are two basic processes for predicting and simulating biomass. The crop biomass growth rate is divided into above-ground and root biomass as per the requirement of the crop model. An example of the model developed by Acock et al. (1978) describes canopy gross photosynthesis rate, temperature effect on photosynthesis, light flux density, and leaf area index for tomato crops.

For a three-state variable model (Fig. 8.7), N is the state variable for leaf number for a vegetative plant or number of nodes for plants. Taking the assumption, the leaf appearance rate (dN/dt) depends on the temperature, it can be calculated as

$$\frac{dN}{dt} = r_m r(T) \tag{8.1}$$

where, r_m is the maximum rate of leaf appearance (h^{-1}), $r(T)$ is a function of temperature. For the model taken as an example for the tomato crop, $r(T)$ is 0 when the temperature is lower than 8 °C or higher than 50 °C, $r(T)$ has the value 1.0 for optimum temperature in the range 30 to 35 °C. For temperature beyond the optimum value, the $r(T)$ is < 1.0.

The dry weight growth rate of the crop is calculated by existing expressions of the model which required many experimental data like plant dry weight, respiration rate, radiation intensity, etc. The equation for crop dry weight growth rate is

$$\frac{dW}{dt} = E\left(P_g - R_m W\right) \tag{8.2}$$

where dW/dt is rate of dry weight growth of crop (g tissue) / (m^2h), W is total plant dry weight (g/m^2), R_m is maintenance respiration rate (g CH_2O)/[(g tissue) h], E is the conversion efficiency of CH_2O to plant tissue (g tissue)/(g CH_2O), and P_g is canopy gross photosynthesis rate (g CH_2O)/(m^2 h). The total biomass growth rate is partitioned into the canopy and root biomass with the function $f_c(N)$. The equation is expressed as

$$\frac{dW_c}{dt} = \frac{dW}{dt} f_c(N) \tag{8.3}$$

$$\frac{dW_r}{dt} = \frac{dW}{dt}\left[1 - f_c(N)\right] \tag{8.4}$$

where dWc/dt is the canopy growth rate, dW/dt is the dry weight growth rate of the crop, $f_c(N)$ is assumed to be constant in this example as a function of crop development, and dWr/dt is root growth rate.

The biomass crop growth rate is further multiplied by the crop growing period to get the total biomass, then multiplied with the crop harvest index to get the desired crop yield. The harvest index is calculated as the ratio of economic yield (kg/ha) to biological yield (kg/ha).

8.5 Input Data

The input data requirement is set up by considering the objectives for the simulation effort. Crop simulation models require data related to weather, crop, soil, and management practices. The input data requirement is represented in Fig. 8.8.

8.5.1 Soil Data

Soil data includes layer-wise information on pH, electrical conductivity, available nutrients (NPK), texture (sand, silt, and clay content), organic carbon, soil moisture (saturation, field capacity, and wilting point moisture content), and bulk density. These data are the primary requirement for any crop model. We need to go through these general soil-related terminologies before going to work on any crop model. A few soil properties are described below:

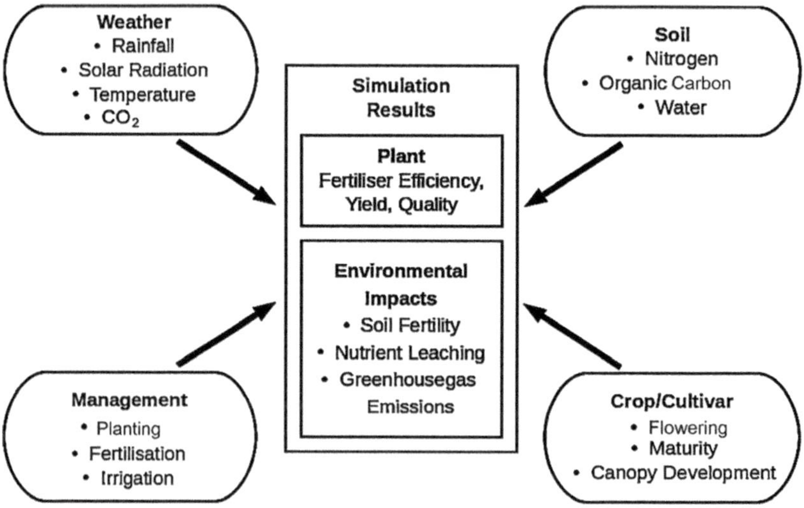

Fig. 8.8 Input data requirement of crop simulation model. (*Source* https://farmit.at/2018/01/smart-farming-with-crop-growth-models/)

- **pH**: Soil pH is a characteristic that describes the relative acidity or alkalinity of the soil. The technical definition of pH is described as a measure of the hydrogen ion concentration of any solution. If there is a high concentration of hydrogen ions in the soil solution, it is characterized as low pH soil and vice versa.
- **Electrical conductivity**: Soil can transmit an electrical current and is generally expressed as deciSiemens per meter (dS/m). The electrical conductivity of the soil varies depending on the moisture retention capacity of the soil particles. Sands have a low conductivity, silts have a medium conductivity, and clays have a high conductivity.
- **Soil texture**: The composition of the soil in terms of the percentage of clay, silt, and sand particles is referred to as soil texture. It represents the composition of the soil which plays a very important role in terms of crop production.
- **Field capacity**: The field capacity is the amount of water that remains in the soil in a few days after irrigation following cessation of free drainage. The matric potential at this soil moisture condition is approximately −0.1 to −0.3 bar. It is technically defined as the state of the soil when all the micro-pores are filled with water. The water travels in a downward movement inside the soil, and macro-pores are drained first due to gravity within two to three days of irrigation, after which the water remaining in the crop root zone is utilised by the crops. At the field capacity, the volumetric soil moisture content is about 15–25%, 35–45%, and 45–55% for sandy soils, loam soils, and clay soils, respectively.
- **Wilting point**: The moisture content of soil at which the plants get wilted and fail to recover their growth is termed a wilting point. The matric potential at this soil moisture condition is around −15 bar, and water present at this point cannot be extracted by the plants. The volumetric soil moisture content at the wilting point is about 5 to 10% for sandy soil, 10 to 15% in loam soil, and 15 to 20% in clay soil.
- **Bulk density**: Soils with a bulk density higher than 1.6 g/cm^3 are not favourable for crop growth as they restrict root growth. This property of soil increases with compaction and tends to increase with depth.

8.5.2 Crop Data

Crop data are collected from field experiments conducted in different regions of the world. The crop data are used for model calibration and validation. It includes the name of variety, duration, phenology, leaf area index, leaf biomass, stem biomass, above-ground biomass, 1000 grain weight, yield components, and selected crop yield. The data on both time series and single-stage events are collected for input to the crop model.

- **Crop phenology**: Crop phenology deals with the timing of the crop growth period, including all the biological events and their interactions with seasonal climate variations and other related factors.
- **Leaf area index**: It is a dimensionless quantity and is defined as leaf area per unit of ground surface area.
- **Above ground biomass**: Above ground biomass of a crop refers to the biomass above the ground surface, excluding the root growth inside the soil.

8.5.3 Weather Data

Weather data is the basic requirement for any crop model. It includes maximum and minimum temperature, rainfall, relative humidity, wind speed, and solar radiation. The data for all the desired weather parameters are required at the daily time step to estimate daily crop growth activities.

8.5.4 Management Data

The simulation process cannot be initiated in the absence of the sowing date of the crop. Usually, the date of sowing is taken as the initiation of the simulation process. Crop management data includes tillage operations, sowing depth, seed rate, irrigation and fertiliser application (amount, source, timing, method, etc.), manure application, crop residue, chemical application, etc.

8.6 Statistical Techniques for Model Evaluation

The agricultural system consists of many processes which generate many data sets. To study the relationships between various parameters, it is crucial to analyze the data recorded in the field statistically. The calibration and validation of the crop simulation model are performed by various statistical techniques, and details on several statistical verification methods for validation of the model products are given in Chap. 5 (Sect. 5.5). The simulated versus observed results are evaluated to check the accuracy and robustness of the calibrated model. In most of the crop simulation models, Root Mean Square Error (RMSE), D-index, and r^2 are vitally used for comparing the simulation values with the corresponding observed data collected from different field experiments (Willmott 1982; Willmott et al. 1985).

Root Mean Square Error (RMSE) has been widely used to analyse the degree of synchronism between simulated and measured values and taken as a benchmark for the evaluation of crop models. It measures how the simulated values have deviated from the observed values. RMSE value is always positive and is also considered as the measure of accuracy; a lower value indicates higher accuracy. RMSE is calculated by:

$$RMSE = \sqrt{\left\{\sum_{i=1}^{n} \frac{(s_i - o_i)}{n}\right\}} \tag{8.5}$$

where, S_i is simulated values, O_i is observed values obtained from the experiments, and n is the number of observations.

D-index (also known as Index of Agreement, IOA) is proposed to be a detailed measure, and it is both a relative and bounded measure that can be widely applied to make cross-comparisons between models. It is very important to know the major patterns or trends within a model and to know the nature of RMSE. The model is said to be good if the systematic difference is approaching towards zero. D-index is calculated as:

$$D = 1 - \frac{\sum_{i=1}^{n} (S_i - O_i)^2}{\sum_{i=1}^{n} \left(\left|S_i - O_{\text{avg}}\right| + \left|O_i - O_{\text{avg}}\right|\right)} \tag{8.6}$$

where, O_{avg} is the average of experimental observations. The R-square method needs mindful interpretation as higher values of r^2 coefficients in isolation will not show any indication of the accuracy of the developed model.

8.7 Crop Models

The use of crop simulation models helps predict and make decisions before sowing or planting a crop, as well as during its growing period; scientists, engineers, and decision-makers all over the world use these models. The major advantage of crop models is assimilating alternative adaptation strategies in case of unfavourable climatic conditions or late monsoon seasons. There are many types of crop simulation models used for forecasting and decision-making. A description of crop model characteristics is given below.

8.7.1 Crop Model Characteristics

- ***Dynamic models***: Crop simulation models are based on physiological parameters including photosynthesis, respiration, phenologic growth, etc.; soil fertility and water retention capacity including evaporation, infiltration, and drainage; and agronomic management. The variables in these models are time-dependent and give a dynamic response to weather and management inputs. The model has immense potential to evaluate varieties of crops, cropping patterns, and management practices for better decision-making in agricultural production.

- ***Deterministic models***: These simulation models are mainly used for distinct predictions for the desirable quantities without any association of variance, probability distribution, or random element. These models are inadequate in case of more uncertainties in a system.
- ***Mechanistic models***: These models describe the mechanism of the process along with the relationship between the weather parameters and crop yield. The nature or behaviour of the system can be explained by these models in terms of lower-level aspects. In these models, the operator can amend or add any extra parameters and variables based on the literature and knowledge gained to describe the yield.

Agricultural production is estimated using crop simulation models as a function of weather, soil conditions, and agronomic management (Tripathi and Kumar, 2014). Simulation models estimate the desired variables from planting to maturity using a series of equations. The major challenge is to increase the productivity while protecting natural resources to meet the rapidly growing demand. It is difficult to predict the growth of agricultural systems due to their complexity and dependence on climatic conditions. System components interact with one another, resulting in variations in the other components when any changes occur to one component. While performing a simulation study, the aim is to minimise undesirable factors and select desirable ones to achieve the target output. It is very difficult to manage these systems as weather conditions easily influence their components. So, system models or simulation models are used to manage the system's complexity, as models are based on engineering aspects. The crop simulation model combines various physiological and morphological features and analyses the desired output, such as crop yield with varying agronomic factors. For simulating the crop yield using different management scenarios, various crop models are developed, such as DSSAT (Jones et al. 2003; Hoogenboom et al. 2019, 2021), Agricultural Production Systems sIMulator (APSIM) (Keating et al. 2003), Info-Crop (Aggarwal et al. 2006), etc. These models can simulate the yield using given climatic conditions and management practices for a particular station. A detailed discussion on available crop models in general and DSSAT, in particular, are presented in this section.

8.7.2 Overview of DSSAT

DSSAT was developed through a collaboration of scientists from the University of Florida, the University of Georgia, University of Guelph, University of Hawaii, the International Center for Soil Fertility and Agricultural Development, Iowa State University, and other scientists associated with the International Consortium for Agricultural Systems Applications (ICASA). The DSSAT is a collection of independent programs that operate together. It integrates crop, soil, and weather data bases with crop models to simulate the outcomes of crop management policies. The software includes precision farm management, climate change impact assessment, variety selection, nitrogen balance, and soil organic carbon. The software comprises many crop models. It integrates the effects of soil, crop phenotype, weather, and management options that allow users to ask "what if" questions and simulates results by conducting trials in minutes on a desktop computer, experiments that would otherwise consume a significant part of an agronomist's career (Fig. 8.9).

The user can simulate not only single-year experiments but also multi-year outcomes of crop management strategies for different crops at any location in the world. DSSAT also provides for validation of crop model outputs, thus allowing users to compare simulated outcomes with observed results, which gives them confidence in the models or to determine if modifications are needed to improve accuracy (Uehara 1989; Jones et al. 1998). DSSAT was first released (v2.1) in 1989; additional releases were made in 1994 (v3.0) (Tsuji et al. 1994) and 1998 (v3.5) (Jones et al. 1998). The DSSAT has undergone major revisions in v3.5, not in its aim but in its design to make it more user friendly. The DSSAT v4.5 and 4.6 is an application software program that includes crop simulation models for more than 25 crops to make more reliable predictions (Jame and Cutforth 1996). DSSAT v4.5 and 4.6 include improved application programs for seasonal and sequence analyses that assess the economic risks and environmental impacts associated with irrigation, fertiliser and nutrient management, climate change, soil carbon sequestration, climate variability, and precision management. As a further advanced version, DSSAT v4.7 has been widely used; that includes crop simulation models for over 42 different crops and many tools that simplify the evaluation of the models for different functions (DSSAT.net).

CERES (Crop Environment REsource Synthesis) models in DSSAT are for cereal crops, including rice, barley, wheat, sorghum, millet, and maize.

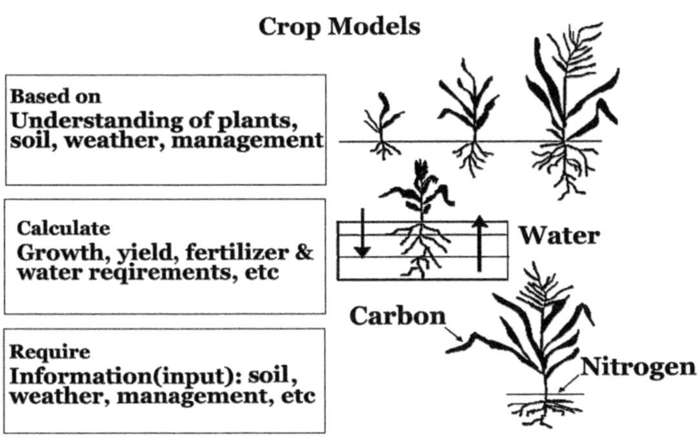

Fig. 8.9 Information about the crop model (Hoogenboom et al. 2021)

CROPGRO (CROP GROwth) models include chickpea, soybean, and peanut for legume crops. SUBSTOR (SUBterranean STORage) models are for root crops, including potatoes and cassava. DSSAT also included many other crop models for sunflower, tomato, sugarcane, and pastures. The objectives of the DSSAT crop simulation model are:

- Simulate crop production by considering the weather, genetics, soil properties (water, carbon, nitrogen, etc.), and agronomic management in single or multiple seasons and in crop rotations at any location where minimum inputs are provided,
- Provide a platform for easily incorporating modules for other abiotic and biotic factors, such as soil phosphorus and plant diseases,
- Provide a platform that allows one to easily compare alternative modules or specific components to facilitate model improvement, evolution, and documentation.
- Provide a capability for easily introducing the CSM into additional application programs in a modular, well-documented way.

The DSSAT model basically includes a main driver program, a land unit module, and primary modules. The main driver program controls the timing for each simulation, and the land unit module manages all simulation processes which affect a land unit. Primary modules are individually simulating the processes that affect the land unit, including weather, plant growth, soil processes, soil plant-atmosphere interface, and management practices. The major components of the DSSAT model are shown in Fig. 8.10.

Different Modules in DSSAT

The main program reads information from the DSSAT standard file that describes a particular experiment or situation to be simulated (Hunt et al. 2001) and establishes a series of variables to control a simulation execution. It initiates the simulation by setting the dynamic variable to initialize the execution of the run and calls the Land Unit module. It then starts a crop season time loop and calls the Land Unit module for initialising variables that must be set at the start of each season. After the seasonal loop initialisation, the program initiates a daily loop and calls the Land Unit module three times in sequence, first to calculate the rates, then to integrate, and finally to report daily outputs. After a crop season is completed, it calls the Land Unit module to produce season-end variables and create summary output files. The Land Unit module calls each of the primary cropping system modules each day. At the start of each new crop season, it obtains management information from the DSSAT input file. The Land Unit and Primary modules link to sub-modules and thus serve to aggregate processes and information describing successive components of the cropping system. For example, the Soil module has four sub-modules that integrate soil water, soil carbon and nitrogen, soil temperature, and soil dynamics processes. The plant module has sub-modules for different

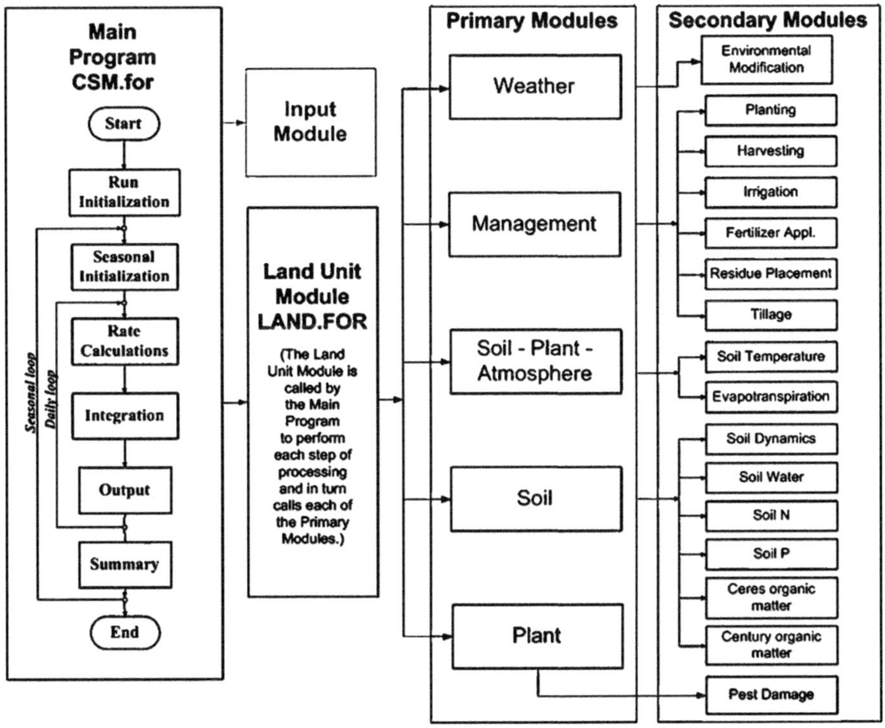

Fig. 8.10 Major components of the DSSAT model. (*Source* https://dssat.net/models-overview/components/) Jones et al. (2003)

crops. Table 8.1 presents a description of these modules and sub-modules, highlighting those for simulating plant growth and soil processes and their interactions.

A. **Weather Module**

The main function of the Weather module is to read or generate daily weather data. It reads daily weather values (maximum and minimum air temperatures, solar radiation, precipitation, relative humidity, and wind speed when available) from the daily weather file. The weather module is formed in such a way that it fills missing weather data or generates weather sequences using the WGEN (Richardson 1981, 1985) or SIMMETEO (Geng et al. 1986, 1988) weather generators. You can even modify weather variables to study climate or simulate experiments.

B. **Soil Module**

The soil in the land unit is represented as a one-dimensional profile; it is homogenous horizontally and consists of a series of vertical soil layers. The soil module integrates information from four sub-modules: soil water, soil temperature, soil

Table 8.1 Description of modules in the DSSAT crop simulation model

Primary modules	Sub modules	Behaviour
Main program (DSSAT-CSM)		Controls time loops, determine which modules to call based on user input switches, and controls print timing for all modules. Calls the input module (MINPT030.EXE) to read FILEX, soil file, and cultivar file and write the appropriate output information to a temporary input file (DSSAT40.INP)
Land unit		Provides a single interface between cropping system behaviour and applications that control the use of the cropping system. It serves as a collection point for all components interacting on a homogenous land area
Weather		Reads or generates daily weather parameters used by the model. Adjusts daily values if required and computes hourly values
Soil	Soil dynamics	Computes soil structure characteristics by layer. This module currently reads values from a file, but future versions can modify soil properties in response to tillage, etc
	Soil water module	Computes soil water processes, including snow accumulation and melt, runoff, infiltration, saturated flow, and water table depth. Volumetric soil water content is updated daily for all soil layers. The tipping bucket approach is used
	Soil nitrogen and carbon module	Computes soil nitrogen and carbon processes, including oiganic and inorganic fertilizer and residue placement, decomposition rates, and nutrient fluxes between various pools and soil layers. Soil nitrate and ammonium concentrations are updated on a daily basis for each layer. Resolves competition for resources in the soil–plant-atmosphere system. The current version computes the partitioning of energy and resolves eneigy balance processes for soil evaporation, transpiration, and root water extraction
Soil-Plant-Atmosphere (SPAM) CROPGRO Crop Template Module		Computes crop growth processes, including phenology, photosynthesis, plant nitrogen, and carbon demand, growth partitioning, and pest and disease damage for crops modeled using the CROPGRO model crop Template (soybean, peanut, dry bean, chickpea, cowpea, faba bean, tomato, Macuna, Brachiaria, Bahiagrass)
Individual Plant Growth Modules	CERES-Maize CERES-Wheat / Barley CERES-Rice CERES-Sorghum CERES-Millet SUBSTOR-Potato	Modules that simulate growth and yield for individual species. Each is a separate module that simulates phenology, daily growth and partitioning, plant nitrogen and carbon demands, senescence of plant material, etc

(continued)

Table 8.1 (continued)

Primary modules	Sub modules	Behaviour
Management operations	Planting	Determine planting date based on read-in value or simulated using an input planting window, soil, and weather conditions
Module	Harvesting	Determines harvest date based on maturity, read-in value or on a harvesting window, along with soil and weather conditions
	Irrigation	Determines daily irrigation based on read-in values or automatic applications based on soil water depletion
	Fertilizer	Determines fertiliser additions based on read-in values or automatic conditions
	Residue	Application of residues and other organic material (plant, animal) as read-in values or simulated in crop rotations

carbon and nitrogen, and soil dynamics. The soil dynamics module is designed to read in land unit's parameters and modify them according to tillage, long-term changes in soil carbon, or other field operations. Descriptions of the other three sub-modules of soil are given in the following subsections.

1. ***Soil water sub-module***: The soil water balance sub module was developed for CERES-Wheat by Ritchie (1985), and the same has been adapted for use by all of the DSSAT v4.5 and 4.6 crop models (Jones 1993; Ritchie 1998). This one-dimensional model computes the daily changes in soil water content by soil layer due to infiltration of rainfall and irrigation, vertical drainage, unsaturated flow, soil evaporation, and root water uptake processes. The model uses a 'tipping bucket' approach to calculate water drainage from the soil when the water content of the layer is above a drained upper limit parameter. Upward unsaturated flow is also computed using a conservative estimate of the soil water diffusivity and differences in volumetric soil water content of adjacent layers (Ritchie 1998).

   Soil water infiltration during one day is calculated by deducting the surface discharge of rainfall that occurs on that day. The SCS method (Soil Conservations Service, 1972) is used to partition rain-fall into runoff and infiltration, based on a 'curve number', which attempts to take into account texture, slope, and tillage. The modification to this method was developed by Williams et al. (1984) and used in the model; accounts for layered soils and soil water content at the time when rainfall occurs. When irrigation is applied, the amount applied is added to the amount of rainfall for the day to compute infiltration and runoff. Drainage of liquid water through the profile is first calculated on the basis of an overall soil drainage parameter assumed to be constant with depth. The amount of water passing through any layer is then compared with the saturated hydraulic conductivity of that layer if this parameter is specified. If the saturated hydraulic conductivity of any layer is less than the computed vertical drainage through that layer, actual drainage is limited to the conductivity value, and water accumulates above that layer. This function allows the model to simulate poorly drained soils and perched water tables. For example, the soil may have a layer with very low or no drainage at the bottom of the profile. Vertical drainage from the profile would not occur, or it would be very low, limited by the saturated hydraulic conductivity value of the bottom layer. Evaporation of water from the soil surface and root water uptake (transpiration) from each layer are computed in the SPAM and communicated to this soil water balance module. Each day, the soil water content of each layer is updated by adding or subtracting daily water flows to or from the layer due to each process.
2. ***Soil carbon and nitrogen balance sub-module***: The DSSAT CSM has two options to simulate the soil organic matter (SOM) and nitrogen balance. The original SOM model in DSSAT v4.5 (Godwin and Jones, 1991; Godwin and Singh 1998), based on the PAPRAN model of Seligman and Van Keulen (1980), was converted into a modular structure and retained in the new DSSAT CSM. Additionally, a SOM module developed by Gijsman et al. (2002),

based on the CENTURY model (Parton et al. 1988, 1994), is included. This CENTURY-based module was added to facilitate the simulation of soil organic sequestration potential for different crop rotations over long time periods after initialising soil C and other variables only once at the start of the simulation. The main differences are that the CENTURY-based module

(i) Divides the SOM into more fractions, each of which has a variable C: N ratio and can mineralise or immobilise nutrients,
(ii) It has a residue layer on top of the soil, and
(iii) The decomposition rate is texture-dependent.

In both SOM modules, organic matter decomposition depends on soil temperature and water content. Because of the widespread use of the CENTURY model and interest in its use in CSMs, we focus on this component in this section. This version is more appropriate for use in low-input agricultural systems, for example, those that use green manure where the surface layer is crucial. Gijsman et al. (2002) showed that this new component greatly improved the accuracy of simulating the long-term changes in soil carbon in the Rothamsted bare fallow experiment. The CENTURY-based module distinguishes three types of SOM:

(a) Easily decomposable (microbial) SOM1,
(b) Recalcitrant SOM2, which contains lignin and cell walls, and
(c) An almost inert SOM3.

At initialisation of the simulation, the fractional ratio of these three pools is set, with SOM1 of only about 2% of total SOM, while SOM2 and SOM3 vary with the management history of the soil (grassland or cultivated) and the degree of depletion. The improved SOM module also allows one to perform more realistic simulations on carbon sequestration, i.e., the buildup of soil organic C under different management systems. Most of the interface input variables to the soil carbon and nitrogen balance modules are soil properties and variables computed in the soil water and soil temperature sub-modules. Transport of N through the soil to deeper layers is based on water flux values obtained from the soil water module. The only interface variable from the Plant module is the array of plant mass being senesced and abscised onto the soil surface daily. The output variables are sent to other modules.

3. ***Soil temperature sub module***: The soil temperature model currently in the DSSAT CSM was originally derived from the EPIC model (Williams et al. 1984; Jones 1993) and is the same as the one in the CERES and CROPGRO models in DSSAT v4.5. Soil temperature is computed from air temperature, and a deep soil temperature boundary condition is calculated from the average annual air temperature and the amplitude of monthly mean temperatures and also uses a simple approach to calculate the impact of solar radiation and albedo on the soil surface temperature. The plant processes (emergence) and SOM decomposition are modified by using the soil temperature. In the description of the CERES-Maize model, additional details on this component are in Jones et al. (1986).

C. **Soil Plant Atmosphere Module**

This module estimates the daily soil evaporation and plant transpiration. The current version was originally developed by Ritchie (1972) and was used in all of the DSSAT v3.5 crop models as part of the soil water balance. This module brings together soil, plant, and atmosphere inputs and computes light interception by the canopy, potential evapotranspiration (ET) as well as actual soil evaporation and plant transpiration. It also computes the root water uptake of each soil layer. The main required inputs needed are the daily weather values as well as all soil properties and current soil water content by layer. The leaf area index (LAI) and root length density for each layer are also needed. The module initially computes daily net solar radiation, taking into account the combined soil and plant canopy albedo. It calculates potential ET using one of two current options. The default Priestley and Taylor (1972) method requires only daily solar radiation and temperature and was described in detail by Ritchie (1972, 1985). The Penman-FAO (Doorenbos 1977) method for computing potential ET can optionally be used to better account for arid or windy conditions, but weather data files must include wind and humidity data. We have also created options for using the Penman–Monteith (Monteith 1986) method for daily potential ET calculations and for using hourly energy balance (unpublished). The potential ET is partitioned into potential soil evaporation based on the fraction of solar energy reaching the soil surface, based on a negative exponential function of LAI, and potential plant transpiration. Actual soil evaporation is based on a two-stage process (Ritchie 1972). After the soil surface is first wetted due to either rainfall or irrigation, evaporation occurs at the potential rate until a cumulative soil evaporation amount since wetting is reached. Then, a soil limiting daily soil evaporation amount is computed as a square root function of time since stage one ended. Actual soil evaporation is the minimum of the potential and soil limiting calculations on a daily basis. If evaporation is less than potential soil evaporation, this difference is added back to potential plant transpiration to account for the increased heat load on the canopy when the soil surface is dry (Ritchie 1972). To determine whether the soil or atmosphere limits plant transpiration, the potential root water uptake is computed by calculating the maximum water flow to roots in each layer and summing these values (Ritchie 1985, 1998). These calculations account for root length density in each layer and the soil water content in the layer. The equation that computes potential root water uptake in each layer is an approximation to the radial flow equation, where assumptions are made about soil texture's effect on hydraulic conductivity, root diameter, and a maximum water potential difference between roots and the soil. The actual plant transpiration is then computed as the minimum of potential plant transpiration and the potential root water uptake. Thus, the atmosphere can limit transpiration by low solar radiation and cool temperatures, the canopy can limit it by low LAI, and the soil can limit it by low soil water content, low root length density, and their distributions relative to each other. This method for computing ET has provided an excellent functional approach for determining water stress in the plant without explicitly modelling water status in the plant component. The ratio of actual ET to potential ET, if less than 1.0, indicates that stomatal conductance would have had to be decreased sometimes during the day to prevent plant desiccation. This ratio is typically used in the Plant modules to reduce photosynthesis in proportion to relative decreases in transpiration. Similarly, a ratio of potential root water uptake and potential transpiration is used to reduce plant turgor and expansive growth of crops. The rationale for this is that as soil water becomes more limiting, turgor pressure in leaves decreases and affect leaf expansion before photosynthesis is reduced. In the current Plant modules, this ratio is set to 1.5 (Jones et al. 2003).

D. **Management Module**

The management module determines when field operations are performed by calling sub-modules. Currently, these operations are planting, harvesting, applying inorganic fertiliser, and irrigating crop residue and organic material. Users can specify these operations in the standard 'experiment' input file (Hunt et al. 2001). Users specify whether any or all of the operations are to be automatic or fixed based on input dates or days from planting. Conditions that cause automatic planting within the interval of time are soil water content averaged over a specified depth (i.e., 30 cm) and soil temperature at a specified depth to be between specified limits. Harvesting can occur on given dates, when the crop is mature, or when soil water conditions in the field are favourable for machine operation. Irrigation can be applied on specific dates with a specified irrigation amount or can be controlled by the plant's available water. If plant available water drops below a specified fraction of water holding capacity in an irrigation management depth, an irrigation event is triggered. The irrigation amount applied can be either fixed or refilled to the profile to the management depth. Similarly, fertiliser can be applied on fixed dates in specified amounts, or the applications can optionally be controlled by plant needs for nitrogen via the nitrogen stress variable from the Plant module. The crop residue and organic fertilisers, such as manure, are applied either at the start of a simulation, after harvesting the crop, or at fixed dates similar to the application of inorganic fertilisers. In addition to simulating experimentation previously conducted for model evaluation and improvement, these management options provide users with considerable flexibility in simulating an optional management system for different applications. Moreover, the management file allows the definition of multiple crops and crop rotation and sequencing strategies.

E. **Pest Module**

The Pest module was developed for the CROP-GRO models by Batchelor et al. (1993), following the approach described by Boote et al. (1983, 1993). It allows users to input field observations and scouting data on insect populations or damage to different plant parts, disease severity on different plant tissues, and physical damage to plants or plant components to simulate the effects of specified pests and diseases on growth and yield. Feedback on plant growth processes are through leaf area reduction, assimilate loss, loss of leaves, fruit, stems, or roots, and inactivation of the photosynthetic capacity of leaves (Boote et al. 1983). This feature has been used successfully for soybean, peanut, and tomato in the past (e.g., Boote et al. 1983, 1993; Batchelor et al. 1993).

8.7.3 Data Set Required for DSSAT

To simulate the yield using DSSAT for different crops, one needs the data on some parameters. The requirement for data is given below. The minimum data set (MDS) refers to a minimum set of data required ('*' indicate that it is compulsory required information) to run the crop models and validate the outputs. Based on availability, additional information (**) should be included for improved model performance.

A. **Location information**
*Name of station
*Cultivar type and year
*Climate class
*Latitude
*Longitude
*Altitude (from mean sea level), m

B. **Weather information**

*Daily solar radiation, MJ m^{-2} d^{-1}
*Daily maximum temperature, °C
*Daily minimum temperature, °C
*Daily rainfall, mm
**Reference height for weather measurements, m
**Reference height for wind speed measurements, m
**Temperature amplitude, monthly averages, °C
**Temperature average for the whole year, °C
**Daily dew point temperature, °C
**Daily wind speed (km/d)
**Daily photosynthetic radiation, moles/m^2/day
**Daily pan evaporation (mm/d)
**Daily maximum relative humidity, %
**Daily minimum relative humidity, %

Dry and wet bulb temperatures and wind speed may be included as optional data. The length of weather records for validation must, at minimum, cover the duration of the experiment and preferably should begin a few weeks before planting and continue a few weeks after harvest so that "what-if" type analyses may be performed.

C. **Soil Information**

Desired soil data include soil classification (SCS), surface slope, colour, permeability, and drainage class. Soil profile data by soil horizons include:

*Surface texture
*Soil depth, cm
*Water table depth, cm
*Bulk density, g/cm^3
*Organic carbon, %
*Total nitrogen, %
*Soil texture-Sand, silt and clay (%)
*Soil albedo
*pH in water
**Surface stones
**Drainage type
**Drainage rate
**Drain depth, cm
**Drain spacing, cm
**pH in buffer
**P (exchangeable), mg/kg
**K (exchangeable), cmol
**Stable organic carbon, %
**Saturated soil water
**Upper limit of stage-1 cumulative evaporation
**Runoff curve number
**Soil water in different layers
**Transpiration/evapotranspiration

D. **Experimental Information**

1. **Previous crop information**

*Previous crop
*Root weight, kg/ha
**Nodule weight, kg/ha
**Crop residue, kg/ha
*N % in residue
**P % in residue

2. **Plot Information**

**Gross plot area per replication, m^2
**Row per plot
**Plot length, m
**Plot relative to drains, degree
**Plot spacing, cm
**Plot layout

3. **Planting Information**

*Planting date
*Emergence date
*Planting depth, cm
*Row spacing, cm
*Plant population at seedling, plant/m^2
*Plant population at emergence, plant/m^2
*Planting method
*Planting distribution
**Row direction, degree from North

This data are needed for both model validation and strategy evaluation. In addition to site soil and weather data, experimental data include crop growth data, soil water and fertility measurements.

4. **Irrigation Information**

*Date of irrigation
*Depth of irrigation
*Method of irrigation

5. **Fertiliser Information**

*Date of fertiliser application
*Amount of fertiliser
*Type of fertiliser
*N:P:K ratio in fertiliser used
*Method of fertiliser application
*Depth of fertiliser application

6. **Genotype coefficients of the crop cultivar**

8.7.4 Experimental Output

A. **Growth Parameters**

Phenological events.
Anthesis days.
Maximum leaf area index.
Physiological maturity days.

B. **Yield and Yield Parameters**

Grain yield at maturity, kg/ha.
Weight of grain, mg/unit.
No. of grains/m^2
No. of grains/spike.

Tops weight at maturity, kg/ha.
Grain nitrogen (N) at maturity, %
Tops N at maturity, kg/ha.
No. of leaf per stem.
Leaf weight, g
Stem weight, g
Byproduct harvest, kg/ha.
Stem N at maturity, kg/ha.
Grain N at maturity, kg/ha.

C. **Soil Nitrogen**

Cumulative N mineralisation.
Cumulative nitrification.
Cumulative denitrification.
Cumulative N immobilisation.
Cumulative ammonia volatilisation.
Cumulative nitrate leached.

D. **Soil Water**

Cumulative runoff.
Cumulative drainage.
Cumulative precipitation.
Soil water profile.

E. **Evapotranspiration**

Cumulative evaporation.
Cumulative transpiration.
Cumulative soil evaporation.
Cumulative evapotranspiration.

8.7.5 Important Instructions to Avoid the Errors While Running the Model

Every DSSAT simulation consists of an Experiment file (FileX), which defines crop management for a particular 'experiment' (set of model runs or 'treatments') and references soil (FileS) and weather (FileW) files. These resources are separated in this way because soil definitions and weather data can be used in several simulations (even for different crops), whereas the experimental setup file is unique to a particular experiment.

Crop modelling has a complex nature. However, DSSAT is among the most user-friendly systems available for crop modelling and care must be taken to ensure that the simulations are reliably set up. The potential resources need to be defined—soil, weather, crop management, cultivar/genetic information and simulation options. Technical errors can creep within any of these resources. Scientific errors are not considered (i.e. the scientific bases for the model are assumed to be correct). DSSAT can, however, assist in detecting technical input errors, but this is beyond the scope of this document. These technical errors can be minimized, and more importantly, more easily detected, if an iterative process is followed when setting up a simulation. This process includes the following:

Define the management first—associate the treatments within an experiment with existing soil and weather data, existing cultivars, and standard simulation options. See if the model runs and check for any errors. Use of these 'placeholders' is recommended because they have been tested and shown in general to be free of technical errors.

- **Step 1:** Define the weather data; update the management file (FileX) such that these new weather data are used instead of the 'placeholder' weather data set up in step 1. Test to ensure that the simulation still runs and results make sense.
- **Step 2:** Set up any special simulation options. Run the simulation to check that it still works and makes sense.
- **Step 3:** Define the soil; update the management file (FileX) such that this new soil profile definition(s) is/are used instead of the 'placeholder' soil data set up in step 1. Test to ensure that the simulation still runs and results make sense.
- **Step 4:** Set up new cultivars by copying an existing one and updating the variables one-by-one; update the management file (FileX) such that these new cultivar definitions are used instead of the 'placeholder' cultivars set up in step 1. Test to ensure that the simulation still runs and results make sense.
- **Step 5:** The simulation setup is now complete. If the model ceases to operate at any step in this process, the user will know immediately where the error has occurred and remedial action can be taken without delay.

If such an iterative process is not followed, the user runs the risk of not knowing where an error has occurred—the model might simply 'crash', leaving the user with the potentially large task of tracking down the error(s).

An alternative error minimising approach is to first create each of the resources—soil, weather, cultivars, etc.—and then modify a known functioning simulation to use these. Again, an iterative approach is used—one change is made, the model is run, results checked, and then the next change is made. If, for example, the model ran fine with one soil and then crashed on the new soil profile, it could be inferred that the new soil definition has an error and should be checked.

8.7.6 *Crop Models in DSSAT*

8.7.6.1 CERES Models

For simulation processes to predict information about crops like crop growth rates and growth period, CERES models are widely used. Primary weather data and crop-related data are needed to input these models in order to get the desired

Table 8.2 Growth stages simulated by the DSSAT CERES-rice, maize, wheat barley, and maize models

Rice	Wheat	Barley	Maize
Transplant	Germination	Germination	Germination
End Juvenile	Emergence	Emergence	Emergence
Panicle initiation			End of Juvenile
Heading	Terminal Spikelet	Maximum Primordia	Floral Induction
Beginning grain filling	End ear growth	End ear growth	
End grain filling			75% Silking
	Beginning grain fill	Beginning grain fill	Beginning grain fill
Maturity	Maturity	Maturity	Maturity
Harvest	Harvest	Harvest	Harvest

output. These models are mainly used for biomass and yield simulation, which can be used to make decisions later on. Many types of CERES models are being used worldwide. A brief description of a few CERES models is explained below.

The CERES-Rice model is a process-oriented, management level model of rice crop growth and development that predicts the duration of growth, the average growth rates and the amount of assimilated partitioned to the economic yield components of the plant (Kaur and Hundal 1999). The simulation processes of the model are dynamic and are influenced by environmental and cultivar—specific factors. The duration of growth for a particular cultivar, however, is highly dependent on its thermal environment i.e. growing degree-days (GDD), which is computed based on the daily maximum and minimum temperatures. To some extent, the growth duration is influenced by the photoperiod during floral induction. Therefore, the model requires input data such as daily weather data, initial soil conditions, crop management and crop cultivar information. The daily weather data includes solar radiation, precipitation, maximum and minimum temperature. Initial soil conditions involve drainage and runoff coefficients, initial soil water, rooting preference factors, nitrogen and carbon contents. The output data for each model simulation run encompasses the results of simulated daily growth and development, grain yield, carbon balance, soil water balance, nitrogen balance and mineral nutrient aspects.

The CERES-Maize, Wheat and Barley models were modified for integration into the modular DSSAT CSM. For these CERES models, the plant life cycle is divided into several phases, which are similar among the crops (Table 8.2).

The rate of development is governed by thermal time, i.e. Growing Degree Days (GDD). The GDD required to progress from one growth stage to another is either defined as user input or is computed internally based on user inputs and assumptions about the duration of intermediate stages. Cultivar-specific inputs for all DSSAT CSM CERES models are presented in absolute terms for consistency, a convention change from that followed previously for wheat and barley for which relative values were used. The number of GDD occurring on a calendar day is a function of a triangular or trapezoidal function defined by a base temperature, one or two optimum temperatures, and a maximum temperature above which development does not occur. Day length may affect the total number of leaves formed by altering the duration of the floral induction phase, and thus, floral initiation. Day length sensitivity is cultivar-specific user input. Currently, only temperature and, in some cases, day length, drive the accumulation of GDD. During the vegetative phase, the emergence of new leaves is used to limit leaf area development until after a species-dependent number of leaves has appeared. Thereafter, vegetative branching can occur, and leaf area development depends on the availability of assimilates and specific leaf area. Leaf area expansion is modified by daily temperature GDD, and water and nitrogen stress.

Daily plant growth is computed by converting daily intercepted photo synthetically active radiation into plant dry matter using a crop-specific radiation use efficiency parameter. Light interception is computed as a function of LAI, plant population, and row spacing. The amount of new dry matter available for growth each day may also be modified by the most limiting of water or nitrogen stress, and temperature, and is sensitive to atmospheric CO_2 concentration. Above ground biomass has priority for carbohydrate, and at the end of each day, carbohydrate not used for above ground biomass is allocated to roots. However, roots must receive a specified stage dependent carbohydrate for growth. Empirical functions are being used to convert leaf area into new leaf weight and based on canopy weight, cultivar's genetic potential, the average rate of carbohydrate accumulation during flowering, and temperature, water and nitrogen stresses, kernel numbers per plant are computed during flowering.

The potential kernel number is a user-defined input for specific cultivars. Once the beginning of grain fill is reached, the model computes the daily grain growth rate based on a user-specified cultivar input defined as the potential kernel growth rate (mg/(kernel d)). The daily growth rate is modified by temperature and assimilate availability. If the daily pool of carbon is insufficient to allow growth at the potential rate, a fraction of carbon can be remobilised from the vegetative to reproductive sinks each day. Kernels are allowed to grow until physiological maturity is reached. If the plant runs out of resources, however, growth is terminated prior to physiological maturity. Likewise, if the grain growth rate is reduced below a threshold value for several days, growth is also terminated. Readers are referred to other papers for additional details on these CERES models (Jones 1986; Ritchie 1985, 1998). Genetic coefficients for CERES-Rice, Barley, Wheat, and Maize are presented in Table 8.3.

8.7.6.2 CROPGRO Models

The CROPGRO Crop Template module in DSSAT CSM is the same as that described by Boote et al. (1998), although its components were modified to fit the modular structure. CROPGRO was created from the combination of SOYGRO, PNUTGRO and BEANGRO (Hoogenboom et al. 1994). CROPGRO was developed as a generic approach for modelling crops in the sense that it has one

Table 8.3 Genetic coefficients used in CERES-barley, wheat, and maize models

Coefficients	Description
Rice	
P1	It is the time period expressed as growing degree days (°C-days) above a base temperature of 9 °C (basic vegetative phase)
P2O	It is in hours at which the development occurs at a maximum rate, (Critical photoperiod or the longest day length)
P2R	It is extent to which the phasic development leading to panicle initiation is delayed is expressed in °C-days (photoperiod sensitivity)
P5	It is the time period in °C-days from the beginning of grain filling to physiological maturity with a base temperature of 9 °C (grain filling phase)
G1	Potential spikelet number per panicle, no
G2	Single grain weight, g
G3	Tillering coefficients
G4	Temperature tolerance coefficient
Barley and Wheat	
P1D	Photoperiod sensitivity coefficient (% reduction/h near threshold)
P1V	Vernalization sensitivity coefficient (%/d of unfulfilled vernalization)
P5	Thermal time from the onset of linear fill to maturity (°C d)
G1	Kernel number per unit stem + spike weight at anthesis (#/g)
G2	Potential kernel growth rate (mg/(kernel d))
G3	Tiller death coefficient. Standard stem +/ spike weight when elongation ceases (g)
PHINT	Thermal time between the appearance of leaf tips (°C d)
Maize	
P1	Degree days (base 8 °C) from emergence to end of juvenile phase
P2	Photoperiod sensitivity coefficient (0–1.0)
P5	Degree days (base 8 °C) from silking to physiological Maturity
G2	Potential kernel number
G5	Potential kernel growth rate mg/(kernel d)
PHINT	Degree days required for a leaf tip to emerge (phyllochron interval) (°C d)

common source code, yet it can predict the growth of a number of different crops. Currently, it simulates ten crops; including seven grain legumes (soybean (Glycine max L. Merr.); peanut (Arachis hypogaea L.); dry bean (Phaseolus vulgaris L.): chickpea, cowpea, velvet bean and faba bean (Vicia faba L.)), and non-legumes such as tomato (Lycopersicom esculentum Mill.) (Scholberg et al. 1997; Boote et al. 1997, 1998). This versatility is achieved through input files that define species traits. Genetic coefficients used in the CROPGRO module are shown in Table 8.4.

An overview of the types of parameters contained in the species file is given in Table 8.5. Each species file contains information on base temperatures (T_b) and optimum temperatures (T_{opt}) for developmental processes (rate of emergence, rate of leaf appearance, and rate of progress toward flowering and maturity) and growth processes (photosynthesis, nodule growth, N_2-fixation, leaf expansion, pod

Table 8.4 Genetic coefficients used in the CROPGRO module

Trait	Definition of trait
ECO#	Code for ecotype to which this cultivar belongs (see *.eco file)
CSDL	Critical short day length below which reproductive development progresses with no day length effect (for short day plants) (h)
PPSEN	Slope of the relative response of development to photoperiod with time (positive for short day
EM-FL	Time between plant emergence and flower appearance (R1) (photothermal days)
FL-SH	Time between first flower and first pod (R3) (photothermal days)
FL-SD	Time between first flower and first seed (R5) (photothermal days)
SD-PM	Time between first seed (R5) and physiological maturity (R7) (photothermal days)
FL-LF	Time between first flower (R1) and end of leaf expansion (photothermal days)
LFMAX	Maximum leaf photosynthesis rate at 30 °C, 350 vpm CO_2, and high light (mg CO_2/(m^2 s)
SLAVR	Specific leaf area of cultivar under standard growth conditions (cm^2/g)
SIZELF	Maximum size of full leaf (three leaflets) (cm^2)
XFRT	Maximum fraction of daily growth that is partitioned to seed + shell
WTPSD	Maximum weight per seed (g)
SFDUR	Seed filling duration for pod cohort at standard growth conditions (photothermal days)
SDPDV	Average seed per pod under standard growing conditions
PODUR	Time required for cultivar to reach final pod load under optimal conditions (photothermal days)
R1PRO	Increase in daylength sensitivity after anthesis (CSDL decreases by this amount (h))
FL-VS	Time from first flower to last leaf on main stem (photothermal days)
THRESH	The maximum ratio of (seed/(seed + /shell)) at maturity causes seed to stop growing as their dry weight increases until shells are filled in a cohort
SDPRO	Fraction protein in seeds (g[protein]/g[seed])
SDLIP	Fraction oil in seeds (g[oil]/g[seed])

addition, seed growth, N mobilisation, etc.). The file also includes information on photosynthesis, N_2-fixation, tissue composition, growth and maintenance of respiration coefficients.

The CROPGRO Crop Template provides for ecotype and cultivar traits to be defined in read-in files. Cultivar differences are created by 15 ‘cultivar’ traits. The cultivar traits include two daylength sensitivity traits, five important life cycle ‘phase’ durations, light-saturated leaf photosynthesis, vegetative traits, and reproductive traits. There are 19 traits in the ecotype file that were proposed to vary less often, such as thermal time to emergence and first leaf stages, but some traits from this file have been used frequently to characterise cultivars.

Phenology is an important component of the CROPGRO Crop Template approach. This component uses information from the species file, which contains

Table 8.5 Summary of types of parameters used in the crop template approach

Section	Description
Photosynthesis	Canopy assimilation coefficients for effects of solar radiation and CO_2. Light extinction coefficient. Functions that define leaf N and temperature effects on photosynthesis
Respiration	Respiration parameters associated with various growth processes
Plant composition values	"Maximum", "normal growth", and "final" protein concentrations of leaf, stem, root, shell, seed, and nodule tissues. Carbohydrate-cellulose, lipid, lignin, organic acid concentration of leaf, stem, root, shell, seed, and nodule tissues. Effects of temperature on seed lipid concentration
Carbon and nitrogen mining parameters	Coefficients for carbohydrate reserves in stem tissue. Fraction of new leaf, stem, root and shell tissue growth that is available in carbohydrate. Mobilization rates of carbohydrate and protein from vegetative tissue
Nitrogen fixation parameters	Nodule growth and senescence parameters. Arrays that define the effects of temperature, soil water, and nodule age on nitrogen fixation and nodule growth
Plant growth and partitioning parameters	Dry matter partitioning to leaf, stem, and root as function of vegetative stage. Coefficients for partitioning at emergence, final growth stage, stem senescence, during water stress, and nodule growth. Parameters that define leaf expansion response to temperature and solar radiation. Initial root depth and length, root water uptake parameters. Relative effects of temperature on pod set, seed growth and relative change in partitioning. Relative effects of soil water content on peanut pegging and pod addition
Senescence factors	Senescence parameters related to vegetative stage, freeze damage, nitrogen mobilization, drought, canopy self shading
Phenology parameters	Curves that define temperature effect on vegetative, early reproductive, and late reproductive development. Parameters for each growth stage: preceding stage, photoperiod function, temperature function, temperature and water sensitivity, N and P sensitivity
Canopy height and width growth parameters	Internode length and canopy width increase as a function of plant vegetative stage. Internode elongation as a function of temperature and photosynthetic photon flux density

cardinal temperature values, as well as information from the cultivar and ecotype files, which contain physiological day durations for respective life cycle phases. Life cycle progress through any given phase depends on a physiological day accumulator as a function of temperature and day length, in many cases. A crop like soybean is sensitive to day length, but not peanut. When the physiological day accumulator reaches a value defined by a threshold given in the cultivar file, a new growth stage is triggered. A physiological day can be thought of as equivalent to one calendar day if temperatures are optimum 24 h per day and day length is below the critical short or long day length requirement, depending on species sensitivity. The species file also contains coefficients that indicate the effect of water or nitrogen deficit on the rate of life cycle progress. These coefficients may vary with the life cycle phase; for example, water deficit may slow the onset of reproductive growth, but accelerate reproductive growth after beginning seed fill. The species file also allows different cardinal temperatures for preanthesis development compared to postanthesis reproductive development. For additional information on phenology in CROPGRO, as studied by Boote et al. (1998), Grimm et al. (1993, 1994), Piper et al. (1996a, b), Mavromatis and Hansen (2001). Crop photosynthesis can be calculated by two options: (1) daily canopy photosynthesis, similar to radiation use efficiency models, or (2) hourly hedgerow light interception and leaf-level photo-synthesis. The daily canopy photosynthesis option, modified from the method used in SOYGRO V5.4 (Jones et al. 1989), predicts daily gross photo-synthesis as a function of daily irradiance for a full canopy, which is then multiplied by factors 0 to 1 for the light interception, temperature, leaf nitrogen status, and water deficit. There are additional adjustments for CO_2 concentration, specific leaf weight, row spacing, and cultivar. The hourly hedgerow photosynthesis light interception approach is described by Boote and Pickering (1994). On an hourly time step during each day, interception and absorption of direct and diffuse light components are computed based on canopy height and width, LAI, leaf angle, row direction, latitude, day of year, and time of day (Boote and Pickering 1994). Photosynthesis of sunlit and shaded leaves is computed hourly using the asymptotic exponential response equation, where quantum efficiency and light-saturated photosynthesis rate variables are dependent on CO_2 and temperature (Boote and Pickering 1994). Hourly canopy photosynthesis on a land area basis is computed from the sum of sunlit and shaded leaf contributions by multiplying sunlit and shaded leaf photosynthetic rates by their respective LAIs. The hourly time loop is handled completely by the subroutine that uses this approach; gross photosynthesis is integrated hourly to provide a daily total value for use by other subroutines in the CROPGRO module.

The growth of new tissues depends on daily available carbohydrates, partitioning to different tissues, and respiration costs of tissue synthesis. During vegetative growth, the model follows a partitioning pattern dependent on the vegetative growth stage but modified by water deficit and nitrogen deficiency. Partitioning coefficients for leaf, stem, and root are defined in the species Crop Template file. Beginning at flowering, cohorts of flowers, pods, and seeds are added daily. These cohorts have an explicit assimilate demand per day depending on genetic potential and temperature. Reproductive tissues have first priority for assimilation over vegetative tissues, up to a maximum reproductive partitioning factor. This factor may be less than 1.0 for indeterminate plants (such as peanuts and tomatoes) and 1.0 for determinate plants, indicating that reproductive tissue eventually can utilise 100% of the assimilate. Leaf area expansion depends on leaf weight growth and specific leaf area, whereas the latter depends on temperature, light, and water deficit. Leaf expansion during reproductive growth is terminated by the decrease of assimilate allocated to leaf growth and by reaching a phase that terminates leaf expansion. During seed fill, nitrogen is mobilised from vegetative tissues. As a result photosynthesis declines and leaf abscission increases. Protein and carbohydrates mobilised from vegetative tissue contribute to seed growth while photosynthesis declines. Growth respiration and conversion efficiency follow the approach of Penning de Vries and van Laar (1988) where the glucose cost for respiration and condensation is computed as a function of the composition of each tissue. The species file contains the glucose cost to synthesise protein, lipid, lignin, organic acid, cellulose-carbohydrate, and mineral fractions as well as the approximate composition of each tissue. Maintenance respiration depends on temperature as well as gross photosynthesis and total crop mass minus protein and oil in the seed. Maintenance respiration is subtracted from gross daily photosynthesis to give available carbohydrates for new tissue growth. Details on these relationships and sources of data used in their development have been published by various authors (Boote et al. 1986; Jones et al. 1989; Boote and Pickering 1994; Boote et al. 1997, 1998, 2002).

8.7.6.3 SUBSTOR Models

SUBSTOR (Simulation of Underground Bulking Storage Organs) models are for root crops including potatoes. The SUBSTOR-Potato model is one of the crop modules in the CSM framework of the DSSAT (Jones et al. 2003). This model can be used to simulate the growth and development of potatoes by capturing the effects of weather conditions, crop genotypes, soil properties and management practices. However, the effects of insect pests, diseases and natural calamities, such as wind and hailstorm damage are not accounted for in the model. The main features of the model are briefed here in, and more details are given elsewhere (Griffin et al. 1993).

SUBSTOR model takes into account several processes simultaneously to provide realistic information about the soil–plant-atmospheric system. The crop production function of SUBSTOR has two components such as soil and plant. The soil component includes evapotranspiration, water movement inside the soil, soil temperature, nitrogen transformation and movement and nitrogen release through the mineralisation of crop residue and organic matter. The plant component includes processes of leaf area expansion, net photosynthesis, root distribution, stage of tuber development, light interception, tuber dry matter production, the effect of nitrogen and water stress on development, and plant uptake of nitrogen.

Table 8.6 Genetic coefficients used in the SUBSTOR-potato model

Name	Description	Values
G2	Leaf area expansion rate in degree days ($cm^2\ m^{-2}\ d^{-1}$)	2000
G3	Potential tuber growth rate ($g\ m^{-2}\ d^{-1}$)	25.0
PD	Index that suppresses tuber growth during the period that immediately follows tuber induction	0.8
P2	Index that relates photoperiod responses to tuber initiation	0.4
T_C	Upper critical temperature for tuber initiation (°C)	15

Due to differences in genotype-specific coefficient depending upon temperature response, photoperiod, thermal time, and dry matter partitioning, the differences among the crop cultivars have been entrenched. The relative temperature function used for the model ranges from 0 to 1 and simulates the response of different processes over a wide range of temperatures. The relative temperature function increases from 0 at a base temperature of 2 °C to 5 °C to a plateau value of 1 at 15 °C to 25 °C, then decreases to 0 at temperatures of 33 °C to 35 °C. The biomass accumulation and partitioning in potatoes depend on temperature, intercepted radiation, and photoperiodicity.

Potato development is divided into five phenological stages within SUBSTOR-Potato viz., pre-planting, planting to sprouting, sprouting to emergence, emergence to tuber initiation and tuber initiation to maturity. When the observed emergence date is not an input, the model simulates bud sprouting and elongation. The emergence date also depends on planting depth. In SUBSTOR, the timing of tuber initiation is a function of cultivar response to both temperature and photoperiodicity. Genetic coefficients used in the SUBSTOR-potato model are presented in Table 8.6.

The model assumes that tuber initiation by early cultivars is less sensitive to high temperature and/ or long day lengths than tuber initiation by late cultivars. Therefore, the time of tuber initiation is based on the threshold temperature (T_C) above which tuber initiation is inhibited. For the day length factor, the dimensionless genetic coefficient (P2) links the effect of photoperiodicity to tuber initiation. The potential tuber growth rate is calculated by an equation that relates the maximum tuber growth rate (genetic coefficient G3) to temperature. The actual tuber growth rate is based on the potential tuber growth rate, soil water deficit and nitrogen deficit. Leaf growth is a function of soil water deficit and nitrogen deficit, the genetic coefficient for maximum leaf expansion rate (G2), the relative temperature and plant density. Stem growth is calculated as 75% of leaf growth and root growth is estimated by a function related to stem and leaf growth.

8.7.7 *Other Crop Models*

There are many crop models other than DSSAT which are used in many parts of the world. Input data for some crop models are shown in Table 8.7.

Table 8.7 Crop models and their input data requirement

Crop model	Input data
Infocrop	Weather data (rainfall, temperature and solar radiation) Crop data (phenology, physiology, and morphology) Agronomic data (seeds, fertilizer, farm yard manure, biopesticides, and irrigation) Soil data (salinity, fertility, sodicity and texture) Pest data (type and population of pest)
APSIM	Weather data (meteorological) Crop data (genotype, physiology) Soil data (soil pH, soil water, soil N and P, organic matter)
CROPWAT	Study area (location, topography Weather data (rainfall, temperature, humidity, wind velocity, sunshine hours) Crop data (crop name, planting date, harvest date, crop coefficient, rooting Depth, length of plant growth stages, critical depletion and yield response factor Soil data (available soil moisture, maximum rain infiltration rate, maximum rooting depth, initial soil moisture depletion and initial available soil moisture

8.7.7.1 Infocrop

InfoCrop is a user-friendly DSS developed by the National Agricultural Technology Project (NATP) by the Indian Council of Agricultural Research (ICAR) (Aggarwal et al. 2006). The system comprises many models having similar structures and is used for the simulation of soil, weather, agronomic management, and major pest effects on the growth and yield of the crop. The models are designed with a target of increasing applications in the field of research with easily available input requirements. InfoCrop system has been developed for 12 crops, namely, wheat, rice, millet, sorghum, sugarcane, pigeon pea, chickpea, mustard, cotton, potato, groundnut, and maize.

The input data required includes weather data (rainfall, temperature and solar radiation), crop data (phenology, physiology, and morphology), agronomic data (seeds, fertiliser, farm yard manure, biopesticides, and irrigation), soil data (salinity, fertility, sodicity and texture), and pest data (type and population of pest). The model provides several outputs relating to water use, growth and development, soil carbon, N uptake, greenhouse gas emissions and loss of crop yield due to pest attacks. The model helps in estimating yield gaps, evaluating experimental data, and decision making on the basis of further assessment of climate change and variability.

8.7.7.2 Agricultural Production Systems Simulator

Agricultural Production Systems Simulator (APSIM) is a modelling dais used for the simulation of biophysical processes in cropping systems (Holzworth et al. 2014). This tool provides accurate predictions for crop production based on soil, climate, genotype, and other management factors and helps in providing knowledge related to natural resource management affairs. APSIM is a custom-built modelling framework comprising many modules which explain the processes related to soil, climate and management. These include a distinct range of crops, soil processes (N and P transformations, water balance, and soil pH) and an extensible range of management controls. APSIM framework has the capability to integrate models which enables the research work from one domain to be carried to the aid of some other domain.

8.7.7.3 CROPWAT

Models such as CROPWAT, developed by Smith of the Food and Agricultural Organization of the United Nations (FAO, 1992) for water management, are used extensively across the world. In essence, it performs standard calculations for crop water requirements, reference evapotranspiration and more specifically, designing irrigation schemes and manages them. Input data requirement for this model includes details of the study area, climate, soil, rainfall, and crop data. As a result, irrigation recommendations can be generated, irrigation scheduling can be adjusted for different water sources, and production can be analysed.

In the agricultural production management process, crop simulation models can be used to forecast yields and make decisions. Understanding the behaviour of biological systems is facilitated by the incorporation model at the fundamental level. While there are no such universal models that can solve every problem, crop models can be developed and adapted as per the need and for unfavourable conditions. The system modelling increases the efficiency of agricultural research with better management decisions to achieve global food security in a sustainable way. The model offers dynamic and quantitative tools to interpret the complexity of agricultural systems and promote research.

It is very important to know the basic behaviuor of a crop model before working on it for easy operation. For a better understanding of the user, this chapter provides a clear and concise picture of crop modelling and simulations. The knowledge of those can assist in the preparation of climate risk matrices (CRM) in association with the weather and climate pattern. The following chapter shows the use of disaggregated monthly and seasonal climate forecasts (into daily sequences) in crop models and the preparation of CRM. The CRM tool accounts for crop weather calender, and pest and disease information including the response of yields on weather patterns through crop models. The details of the preparation of CRM for management decisions in agricultural operations have been discussed in the next chapter.

Chapter 9
Generation of Climate Risk Matrices

9.1 Introduction

Climate is one of the key factors that determines the success or failure of agricultural enterprises, particularly, crop production and livestock. It manifests its influence on agricultural operations and farm production through its effect on the soil and plant growth. The location-specific cropping patterns in various regions largely depend on the prevailing climatic conditions of those regions. So, higher crop yields can be achieved by adjusting cropping patterns and following suitable agronomic practices to mitigate the adverse impacts of the climate.

Generation of appropriate climate risk matrices (CRM) for a specific crop over a particular region benefits the farmers because it provides effective guidelines for saving a lot of resources like land, labour, capital, etc.; thereby decreasing the cost of production in the agricultural sector. For example, during periods of anticipated heavy rains, irrigation from the wells can be avoided by which both water and electricity are saved. The threshing of the harvested produce can be done before the rains, thereby avoiding crop losses. The losses in seed, fuel, fertilisers, labour, time, etc., can be avoided by postponing the sowing of the crops if anticipated weather is not suitable for crop management operations. Therefore, a sound knowledge of the climatic factors and their interactions with the crops is essential for successful agriculture. This chapter deals with the study of different kinds of aberrant climatic conditions faced during crop production like erratic monsoon events, droughts, floods, etc. and the contingent planning methods to minimise the stresses on crops and generation of the CRM for prosperous growth in agricultural production. In order to generate the CRM for better agriculture practices, one has to develop a crop weather calendar for each crop in its particular region based on the actual scenario with the growing period of that crop.

U. C. Mohanty et al., *Climate Risk Management in Agriculture*,
https://doi.org/10.1007/978-3-031-51862-1_9

9.2 Crop Weather Calendar

A crop weather calendar (CWC) consists of a specific crop's typical chronological life history, from sowing through vegetative growth, flowering, and grain filling to a period of maturity. It provides information on crop growth stages, normal weather for crop growth, warnings to be issued based on prevailing weather conditions, the water requirement of crops during their various phytophases, and meteorological conditions favourable for the development of crop pests and diseases. This advanced information is useful for crop planning, irrigation scheduling, and plant protection measures, which are vital for persuasive crop planning and maximising and stabilising food production in the country. Thus, it is indeed essential to have a CWC for a particular crop in a specific region. The following section will provide a useful direction on the preparation of CWC which is inevitable for proper agro-management systems to minimise the loss and maximize production.

9.2.1 Preparation of Crop Weather Calendar

A CWC is basically the pictorial representation of detailed information for a crop concerning the sowing period and duration of important phenological stages in its life cycle, the optima of climatic requirements during different stages of the crop, and the actual and normal weather for that station/location. So, each CWC is specific to each crop for a particular location from which one can get the (A) weather and climate information (site information), (B) information about the optimum climatic conditions suitable for the growth and development of the crop in different stages, and (C) information about the diseases and pests which reduce the yield of that crop. Therefore, confining and acquiring the knowledge on the above-mentioned sections, one can prepare CWC for targeted crops which will contain all the climatic/weather information necessary for the cultivation and management of that crop for a specific region and its neighborhood where the life cycle/phenology of that crop is similar.

Thus, it is imperative to understand each of the three parts mentioned above (A, B, and C) in detail for developing a CWC. These three parts will concisely provide instructions for generating the CWC of any crop.

Part A (Weather and Climate Information): The first part of the calendar contains climatological information of various meteorological variables for different months as well as the respective standard meteorological weeks at the location/station for the entire growth period of the crop. The different meteorological parameters such as maximum and minimum surface air temperatures, rainfall amount, number of rainy days (if rainfall on the day is ≥ 2.5 mm, it is a rainy day), relative humidity (both morning and evening), solar radiation/sunshine hours etc. are recorded, which can be computed from long-term (at least 30 years period of data) averages of the observational data-sets. The months and standard meteorological weeks (1 year = 12 months = 52 meteorological weeks) are marked at the top of the calendar (Table 9.1). Crop growth mainly depends on climatic conditions and optimum weather conditions (temperature, rainfall, relative humidity, solar radiation), which are necessary for the proper growth and development. The maximum temperature (T_{max}) is the temperature above which crop growth does not occur, and the minimum temperature (T_{min}) is the one below which no crop growth fails to occur. Therefore, the crop grows well when the current atmospheric temperature synchronises with the optimum temperature (T_{opt}) needed for the growth of the crop, which falls between T_{min} and T_{max}. For example, for *kharif* rice in India, the T_{max}, T_{min}, and T_{opt} are 45–50 °C, 15–20 °C, and 30–38 °C, respectively; but for *rabi* potato, these values are 30–38 °C, 0–5 °C, and 25–30 °C, respectively. The amount, timings, and precipitation distribution decide the crop choice for agriculture in a particular location. Relative humidity (RH) of around 60–75% is optimum for the growth of most crops. When RH is too high then, it increases the succulence of the crops, which in turn leads to more attacks of diseases and pests on the crops, and when RH is too low then, it leads to desiccation of the crops; thereby decreasing the crop yields in both the cases. Solar radiation is very important for the photosynthesis of crops, which is the basic process of food production in the crops. A minimum amount of 10–12 h of sunlight in a 24-h period is needed to get a higher yield of the crops. A 12-h photoperiod consists of 12 h of light and 12 h of darkness in 24 h. So, less amount of bright sunshine (BSS) hours decreases the rate of photosynthesis in the crops, and higher bright sunshine hours increase the rate of photorespiration, thereby decreasing the yield of the crops in both cases. The climatic data of the past 30 years are taken into consideration as specified by the World Meteorological Organization (WMO), Geneva, Switzerland to prepare the CWC.

Part B (Crop Information): The second part of the calendar indicates (Table 9.1) the important growth phases like sowing, germination/emergence, transplanting (in the case of rice), vegetative growth, flowering, grain formation and maturity period, etc. within typical life history of a specific crop species. This part of the crop weather calendar also informs the favourable meteorological conditions for the crop (stage-wise or whole crop growth period), which will lead towards a high

Table 9.1 Crop weather calendar (AICRP on Agrometeorology, CRIDA, Hyderabad, 2011)

Months		Part A↓
Standard meteorological weeks		
Name of the meteorological Parameter	Climatic normal	
Name and pictures of phenological stages of crop		Part B↓
Stage-wise climatic normals for high yield of crop		
Climatic normals for diseases or insect pests		Part C↓
Name of diseases or insect pests of crops	Climatic normals conducive for major diseases of the crop along with susceptible crop phenological stages	

yield of the crop. For example, the rice crop (120 days crop) has many growth phases, and each phase requires a specific climatic condition for its proper development. These growth stages are the germination phase, vegetative phase, reproductive phase, and ripening phase. These stages for rice are discussed below.

Germination phase: When the first shoots and roots start to emerge from the seed, and the plant begins to grow, germination occurs in rice crops. This phase begins with the appearance of the young shoot and roots through the seed coat at one end of the seed. The optimum temperature range for the germination of rice is 10–40 °C. The seeds need to absorb a certain amount of moisture for proper germination. Germination and emergence occur almost simultaneously in rice crops after the seed is sown in the soil.

Vegetative phase: The vegetative phase in the rice crop is described by the development of tillers and the number of leaves and a continuous increase in plant height. Depending on the cultivar, the vegetative phase of rice varies from crop to crop but generally ranges between 55 and 85 days. The number of active tillers in this phase has a direct effect on rice yield, and the maximum tillering stage is also included in this phase.

Reproductive phase: The booting stage is the sign when the rice plant gets ready to enter its reproductive phase. Booting means a bulging of the leaf stems that conceals the developing panicle, then its tip emerges from the stem. When the panicle is fully visible, rice is said to be at the heading stage, and after completing a day of heading, flowering begins. As the flowers open, they shed their pollen on each other so that pollination can occur. The flowering stage continues for 7 days.

Ripening phase: This phase usually takes 30 days of the crop growing period and starts with the initiation of flowering and ends with the maturation & harvest of the rice crop. This phase is affected by the duration of climatic variables. The low temperatures may lengthen the ripening phase, while warm days may shorten it. The last three stages of growth i.e., the milking stage, soft dough and hard dough stages, make up the ripening phase.

Part C (Pests and Diseases Information): The third part of the calendar consists of meteorological conditions conducive to the incidence of pests and diseases and the nature of the weather warnings that can be given. It also indicates the susceptible periods of the crop during which disease/pest incidence may occur under conducive weather. According to this information, an agro-meteorologist can issue a weather-based agro-met advisory for the control of diseases/pests in the crop. The diseases and pests incidence cause considerable losses in crop production. Therefore, the cataloging of the pests and diseases which are known to be weather-sensitive in their proliferation and determination of the meteorological conditions conducive to their growth and spread can help in locating the likely regions and periods of their occurrence and initiation of control measures on the basis of the weather warnings. So, this part of CWC mainly focusses on the climatic factors which cause the susceptibility of the crops to certain diseases and pests.

The diseases/pests of the cropping attack it during a specific period of the growing season of the crop. Temperature and RH are the main climatic factors that induce crop susceptibility to diseases/pests. For example, in rice, the Yellow Stem Borer attacks the crop when the current atmospheric temperature is 19–33 °C and RH around 90%, while Blast disease occurs when the temperature is 25–27 °C and RH 86–98% (Banerjee 1956).

Table 9.1 depicts the representation of different parameters of any CWC and Table 9.2 represents the required elements in preparing the CWC for rice, potato, and groundnut crops of a particular region (Kharagpur, West Bengal) of India. In addition to it, different weather parameters like T_{max}, T_{min}, RH-I (morning RH), RH-II (Evening RH), and Rainfall are also given for different growth stages of crops and for different pests and diseases that are most likely attacking the crops.

9.2.2 Climatic Normals for Specific Crops

Water and solar energy are necessary for plant growth. A systematic appraisal of our natural resources, including climate, soil, fauna, and flora is essential to meet the need for optimum food production on a sustainable basis to support the increasing population demands. Such appraisals may help the country in more rational planning and optimising the use of resources for the present and preserving them for the future. The common elements that determine climate are rainfall, atmospheric temperature, relative humidity, solar radiation, and wind. All these have their own significance for the life of crop plants. The importance of an individual element depends to a certain extent on the intensity of the other element. Therefore, the options for cropping systems in most parts of the globe and the management largely depend on the climate, particularly over the tropical belt. In India, a number of crops are grown like rice, wheat, maize, millets, chickpea, arhar, pea, green gram, black gram, groundnut, soybean, sesamum, jute, cotton, tobacco, tea, coffee, etc., which come under the main crop productions in the country. Detailed information about the climatic normal for different crops in India has been described in textbook by Varshneya et al. (2003). The user may refer to this book for the normal climatic conditions for the other crops in India. The climatic requirements of some of the important crops are given below:

(a) **Cereal Crop**

Rice (Oryza Sativa Indica)

The rice crop is mostly grown in *kharif* season (June to September) and requires high temperature, ample water supply, and high atmospheric humidity during the growth period. It tolerates up to 40 °C, provided water is not a limiting factor. A mean temperature of 22 °C is required for the entire growing period. If night temperature drops lower than 15 °C during the entire vegetative phase, the rice yield is greatly reduced by the formation of sterile spikelets, which are induced by low temperature. The period during which low temperature is most critical is about 10 to 14 days before heading.

Table 9.2 Crop weather calendar of rice, potato, and groundnut for Kharagpur location, West Bengal, India

		Kharif (June–September)		Rabi (October–February)	Zaid (March–May)
Site information	T_{max} (°C)	31–34		24–30	34–38
	T_{min} (°C)	25–27		11–22	20–27
	$T_{avg.}$ (°C)	28–30		18–26	27–32
	Rainfall (mm.)	1080–1240		75–110	230–310
	RH (%)	67–88		52–74	50–60
Crop information:	T_{max} *(°C)*	T_{min} *(°C)*	*Optimum Temp. (°C)*	*Rainfall (mm.)*	RH (%)
(a) *Rice (kharif)*					
Germination stage	37	23	29	30	76
Seedling stage	37	24	28.5	50	80
Tillering stage	38	27	31	200	84
Panicle Initiation stage	31	26	33.5	200	89
Flowering stage	29.5	26.5	33	80	81
Milking stage	28	21	26	60	77
Dough stage	29	20	25	50	75
(b) *Potato (rabi)*					
Germination stage	29.5	16.6	18	10	82
Vegetative stage	26.6	14.4	20.5	20	77
Tuber initiation stage	29	13	21.5	5	68
Tuber bulking stage	30	21	24	15	65
Maturation stage	34	20	26	20	64

(continued)

Table 9.2 (continued)

		Kharif (June–September)		Rabi (October–February)	Zaid (March–May)
(c) *Groundnut* (*Zaid*)					
Germination stage	29	22	25	20	68
Vegetative stage	30	25	26.1	28	65
Flowering stage	32	25.5	28	42	60
Pegging stage	30	25	26.1	40	60
Pod development stage	32	26	27	38	57
Maturity stage	34	25	31	40	57
Diseases & pests information:	T_{max} (°C)	T_{min} (°C)	*Optimum Temp. (°C)*	*Rainfall (mm.)*	*RH (%)*
(a) *Rice*					
Blast	28	20	23.5	110	93
Bacterial leaf blight	26	21	21.5	90	92
Rice stem borer	33	19	26	110	90
Gall midge	35.2	19.8	25.6	90	89
(b) *Potato*					
Late blight	21	10	18	15	91
Colorado beetle	27	16	25	30	75
(c) *Groundnut*					
Tikka	29	20	25	90	90
Aphids	31	24	26	80	90

Low sunshine hours during the vegetative stage have a slightly ill effect on grain production but such a situation prevailing during the reproductive stage has an effect on spikelet number and their development, thus reducing crop yield. A combination of low daily mean temperature and high solar radiation during the reproductive phase gives a higher yield.

There should be a monthly rainfall of 200 mm to grow lowland rice and 100 mm to grow upland rice successfully. The base of the plant should remain submerged in water from the time of sprouting to the milking stage of the grain. At the ripening stage, no standing water in the field is required.

Wheat (Triticum Spp.)

The crop can withstand intense cold conditions. The optimal range of temperature for the germination and vegetative growth of wheat is 15–20 °C (Venkataraman 1992). High temperature during rapid growth and tillering periods results in poor tillering, fewer effective tillers, poor growth rate, short shoot height, lower grain yield, etc. The crop shows considerable resistance to drought. Bright sunny days with dryness and cooler nights during the ripening period give better-sized quality grains. The optimum water requirement of the crop during the entire growing season is 500 mm. Daily consumptive use of the crop varies from 1.7 to 8.7 mm from emergence to grain filling and the highest value during the booting to dough stage.

(b) **Pulse Crop**

Chickpea (Cicer Arietinum)

A sufficiently cool and humid climate is ideal for chickpea cultivation. The seeds of the crop can germinate over a wide range of temperatures, from 10 to 45 °C. A temperature of around 15–20 °C is optimum for its growth. The root zone soil temperature of 15–25 °C is highly beneficial for early and adequate nodulation. Nodulation is adversely affected when the root temperature exceeds 30 °C (Venkataraman 1992).

Excessive rains after sowing and flowering are harmful. Excessive relative humidity appears to have a negative influence on seed set. The highest pod formation has been recorded at relative humidity ranging from 20 to 40%. The crop is moderately tolerant to soil salinity.

(c) **Oilseed Crop**

Groundnut (Arachis Hypogaea)

It is normally a warm and moderately moist climate crop requiring abundant sunshine and moderate rainfall (Venkataraman 1992). A longer season with moist soil conditions from flowering to pod development, a warm climate (24–34 °C temperature), and plenty of sunshine are highly favourable for the crop. Temperatures above 44 °C and below 4 °C are not desirable. The optimum seed germination temperature is 14–16 °C. A well-distributed rainfall of 50–70 cm is generally adequate for the crop; however, rain at maturity reduces the crop yield. Moist soil conditions at pegging and early pod development are highly critical. Relative humidity in the range 80 to 90% stimulates greater rates of flowering, resulting in increased peg setting.

Soybean (Glycine Max)

It is a crop of warm and humid temperate regions. Its cultivation has now been extended to semi-arid tropics with daily maximum temperatures below 40 °C and fairly assured rainfall conditions due to the availability of different varieties (Peters et al. 1971). The optimum temperature for seed germination is 30 °C. The freezing temperature can result in the complete destruction of the leaves. The high temperature during flowering and ripening increases fat but reduces proteins. Nodulation and nitrogen fixation in soybean is greatly affected by soil temperature. Soybean growth is limited by temperatures over 33 °C. The seasonal crop water requirement of the crop is 45 to 75 cm.

(d) **Horticultural Crop**

Potato (Solanum Tuberosum)

The potato crop is grown in *rabi* season (October to March), and the climatic factor plays a major role in the production of potatoes. The optimum planting time temperature needs to be around 18–20 °C. Higher temperature damages the tubers and too low-temperature damages sprouting. For tuber formation and development, lower temperatures are desirable. A decrease in the tuber formation and production at a constant temperature above 25 °C and complete inhibition at 29 °C is observed. Plants are damaged even by a mild frost.

The water requirement of the crop is around 40 to 60 cm. Heavy rain, alternating with dry and wet spells, is harmful to the crop. The optimum soil moisture for the crop is 65–80% of the field capacity.

(e) **Commercial and Fibre Crop**

Cotton (Gossypium Spp.)

The crop is heat-loving and is the best adapted to the sub-tropical climate. It is sensitive to temperature and soil water conditions. Cool night and low daytime temperatures encourage vegetative growth. Usually, the crop requires abundant sunshine with adequate moisture, fairly high temperature, and around 200 days of frost-free period (Venkataraman 1992).

A mean annual temperature of more than 16 °C is essential for higher yields, and a temperature above 39 °C is detrimental to crop growth. After emergence, the surface soil moisture should be maintained for the establishment of the seedling. Cloudiness prolongs vegetative growth. During the later boll development period, low night temperatures result in shorter fibers and bad boll opening. The overall development rate is maximised at a temperature range of 25–30 °C. The minimum rainfall for cotton is 50–65 cm. Heavy rainfall during sowing and the early stage of the crop is undesirable. Similarly, excessive rainfall at a later stage may cause the shedding of leaves, blooms, and bolls.

Tobacco (Nicotiana Spp.)

The crop is grown under a variety of soil and climatic conditions. Although it is a tropical plant, it can be grown in varied environments with a temperature range of 10–32 °C (Venkataraman 1992). The temperature beyond 35 °C results in leaf burning, with 24–27 °C being the optimum temperature for most varieties. It needs low temperatures for germination. Warm and moist weather is the best for its growth and cool, dry weather for ripening. Bright sunshine for 2–3 weeks after planting is harmful. High humidity during leaf curing is necessary. High rainfall during its early growth stages is harmful. The crop requires a well-distributed rainfall of about 80–100 cm during the growing period. The maintenance of soil moisture at around 50% of available moisture holding capacity is found to provide a higher crop yield. It is very sensitive to wet soil conditions, and hence poor soil drainage, particularly during the early growth period, affects the production level.

9.2.3 *Identification of Important Pests of Crops*

Important field crops like rice, wheat, maize, potato, cotton, etc., are heavily influenced by pests and diseases. The infestation goes beyond the economic threshold level (ETL) in a particular season, causing a significant loss in production. Some of the important pests of major crops are discussed Fig. 9.1 in relation to their climatic requirements.

(a) **Rice Stem Borer (*Scirpophaga Incertulas*)**
Pest infestation is negatively correlated with minimum temperature but positively with maximum temperature and humidity. The stem borer moths are active between the temperature range of 19 and 33 °C, and the maximum number of eggs are laid at 29–30 °C and 90% RH (Banerjee 1956).

(b) **Rice Gall Midge (*Orseolia Oryzae*)**
Favourable conditions for infestation are a minimum temperature of 19.8 °C, a maximum temperature of 35.2 °C with a relative humidity of 90%, and mean rainfall of 4.5 to 62.5 mm (per 5 day period). High humidity, less sunshine hours, and high temperature are conducive for the more infestation of gall midge on rice (Banerjee 1956).

(c) **Colorado Potato Beetle (*Epilachna Philippinensis*)**
They are very active at temperatures ranging between 16 and 27 °C. Annual rainfall of 600–1500 mm is favourable for its optimum development. Egg laying is maximum at 25 °C, and considerable mortality occurs at 30 °C. The incubation period is 5 days at 30 °C and 19 days at 12 °C. Beetles go for hibernation when the temperature falls below 12 °C and can tolerate the lowest temperature up to −4 °C (Bourke and Robertson 1955).

Fig. 9.1 Image of some important pests of crops. *Source* http://www.crida.in, http://www.knowledgebank.irri.org, https://extension.wvu.edu, http://agropedia.iitk.ac.in

(d) **Cotton Spotted Bollworm (*Earias Vitella*)**
During the boll formation stage (in July/August), the infestation occurs at a rapid rate. The RH higher than 90% with low sunshine hours (<7 h) increase infestation in the crops. In October/November, the larval activities are more when T_{min} is 19–20 °C and T_{max} is 30–32 °C (Venkataraman 1992).

9.2.4 *Identification of Important Diseases of Crops*

The major field crops like rice, wheat, maize, groundnut, potato, etc., are affected by various diseases at different growth stages. Some diseases that significantly effect important crops are discussed (Fig. 9.2) below in relation to their climatic requirements.

(a) **Rice Blast Disease (*Pyricularia Oryzae*)**
It is a fungal disease. Environmental factors such as humidity, moisture, and temperature influence the development of the disease. The temperature of

Fig. 9.2 Image of some important disease of crops. *Source* https://www.forestryimages.org, http://www.knowledgebank.irri.org, https://microcosmos.foldscope.com

around 20 °C during the night, alternating with a day temperature of around 30 °C with daylight for about 14 h, has been found to predispose the crop to infection. The fungus invades the rice leaves most rapidly at 24–28 °C. High relative humidity of 92% and above and free water are required for conidial germination and infection. Cloudy, overcast weather is reported to encourage the blast spread. The peak concentration of spore dispersal occurs around 4 am favoured by a night temperature of 25–27 °C and relative humidity of 86–98% (Govindaswamy 1964).

(b) **Bacterial Blight of Rice (*Xanthomonas Oryzae Pv. Oryzae*)**

A combination of rainy weather, dull windy days, and atmospheric temperatures of 22–26 °C are conducive to its development. Plants in the shade and close-planted crops supplied with high doses of nitrogen show more disease incidence. Plant infection occurs when relative humidity is higher than 90%. The disease is most likely to develop in areas that have weed growth and stubbles of infected plants. It is commonly observed when strong winds and continuous heavy rains occur, allowing the disease-causing bacteria to easily spread through ooze droplets on lesions of the infected plants. It is commonly called as 'kresek' disease of rice (Padmanabhan et al. 1971).

(c) **Late Blight of Potato (*Phytophthora Infestans*)**
It is a fungal disease and is a serious disease in potatoes. This disease was the main reason behind the Irish famine in the 1840s. The late blight fungus is especially adapted for growth under conditions where water is present, and cool temperature persists. Significant crop loss can result from tuber infection. During the rains, late blight spores may wash down from the infected above-ground parts (foliage and stem) of the plant into the ground and infect the underground growing tubers. The following climatic conditions prevailing in a fortnight form an outbreak of potato blight:

- At least four hours of dew in the night
- Minimum temperature during night below 10 °C
- Rainfall of at least 2.5 mm in the 24 h following the dew night

(d) **Tikka Disease of Groundnut (*Cercospora Personata*)**
Heavy precipitation ensures free water on the leaves, the RH about 90%, and the temperature around 20 °C for 6–7 days during the growing period of groundnut favours the onset of tikka disease. An average maximum temperature of 33 °C, minimum temperature of 20 °C, and monthly rainfall of 88 mm were found favourable for the outbreak of this disease.

9.3 Contingency Crop Planning

Rainfall behaviour in dry farming areas is erratic and uncertain. The adverse effects of these rainfall aberrations on crop growth vary with the degree of deviation and the crop growth stage at which such deviations occur. Suitable manipulations in crop management practices are needed to minimize such adverse effects of abnormal rainfall behaviour. These management decisions constitute contingency planning. So, contingency crop planning refers to mitigating any unexpected, unusual, unfavourable, and hence unwanted accidental weather situations occurring at any time without prior knowledge before the crops are sown or even after the crops are sown. The contingency crop plan is the change-over cropping in place of a normally sown crop which otherwise would have been a highly suitable and profitable crop of the tract, but for aberrant weather situations. It is highly location-specific due to variations in the amount and distribution of rainfall. The selection of crops for contingent sowing is equally important. The choice of crops should be such that the duration of the crop coincides with the length of the growing season. Management practices are also extremely important in contingent crop planning. As a greater portion of agricultural areas in India depends on dry land agriculture (around 67%), so rainfall is the key factor for higher growth and productivity of the crops. The amount, duration, and distribution of monsoon rainfall (June–September) decide the cropping pattern of the particular area. Here, monsoon rainfall is chosen as it accounts for nearly 75% of the total rainfall received by any region in India. The monsoon rainfall is beneficial mainly for the *kharif* season crops like rice, maize, soybean, pigeon pea, etc. The *rabi* (wheat, barley, chickpea, etc.) and *zaid* crops (watermelon, groundnut, cucumber, etc.) are mainly grown with the residual moisture from the monsoon rainfall under partially/fully irrigated conditions. Dry spells are very commonly seen in the monsoon season in tropical belts of the world. A dry spell is a duration of days between one rainfall event and the succeeding rainfall event. Generally, the duration of a dry spell in heavy-textured soils is about 10 days and in light-textured soils is about 15 days. So, an abnormality in quantity, duration, and distribution of the monsoon rainfall adversely affects the cropping pattern of any region and decreases the yield of the crops. In general, the monsoon rainfall aberrations are:

A. Late onset of monsoon
B. Dry spell immediately after sowing
C. Dry spells during the crop growing period
D. Early withdrawal of monsoon
E. Extended monsoon

The contingency planning owing to different situations of the monsoon activities stated above are briefly addressed here:

(A) **Late Onset of Monsoon**: During some years, the onset of the southwest monsoon gets delayed, which brings a challenge for the timely sowing of regular crops/varieties. The delayed sowing of the crops can lead to reduced and even uneconomical crop yields. Under these circumstances, management options available are

(1) **Transplantation**
Transplanting of rice has been found to be advantageous, especially under delayed sowing conditions. This is the surest way of compensating for the delayed commencement of sowing rains, since transplantation is a labour intensive operation. The transplanted rice crop gives a higher grain yield than the direct-seeded rice. As a thumb rule, the seedling could be raised in the nursery for a week for every month of its life period. So, for four months of rice crop, 28 days old seedlings are used for transplanting. The staggered transplanting method is used for rice crops in regions with chronic water stress. In this method, the transplanting of seedlings is done in batches with a minimum 1-week interval from the start of initial transplanting (10th July) with receipt of a good amount of rainfall.

(2) **Land Preparation**
Deep summer ploughing is done in the 1st week of June to increase the rainwater infiltration and hold the rainwater in the furrow opening, as a good practice for in-situ water conservation. This water can be used for puddling operations in rice cropping during the late onset of the monsoon.

(3) **Alternate Crops/Varieties**
Certain crops/varieties can perform better even if sown late in the season. Depending on the receipt of rain, such crops and varieties can be selected. Castor (Aruna) is more remunerative in red soils of Telangana (AP), India, than pigeon pea under late sown conditions. Sunflower, because of its shorter duration, performs better than groundnut when sowings are delayed beyond September's first fortnight in relatively deeper red soils. In western Rajasthan, short-duration green gram and cowpea are more remunerative than pearl millet when sown late. The crops and varieties that can be sown depending upon the date of sowing have been identified for dry farming tracts of India. In eastern India, summer crops should be grown like sesamum, groundnut, cowpea, fodder crops, etc. In case of the delay in monsoon up to August, instead of rice, other crops like maize, soybean, pigeon pea, etc., are grown in *kharif,* followed by sesamum/linseed. Sunflower crops can also be grown in both *kharif* and *rabi* seasons continuously.

(B) **Dry Spell Immediately After Sowing**: Dry spells are commonly found in dryland agriculture. It might result in poor germination due to soil crusting, withering of seedlings, and poor establishment of crop stand due to the dry spell after the sowing of the crop. The adverse effect of moisture stress should be mitigated to avoid total crop failure. It is necessary to maintain a proper plant stand to ensure a better yield. So, growing rice crops is avoided in this situation, and crops like pigeon peas, maize, soybean, etc. are preferred.

(C) **Dry Spells During the Crop Growing Period**
After the sowing of crops and in early growth, any unexpected or unfavourable weather abnormality like a long dry spell of 3 to 6 weeks, drought, or early cessation of monsoon may occur. To mitigate the damaging effects of droughts after sowing of the crop following mid-season corrections are recommended:

(1) **Mulching**: If the break in monsoon is very brief, soil mulching was found to be a tool in extending the water storage period in the soil profile due to reduced heat load on the soil and hence the evaporation losses, which in turn leads to extended periods of water availability. Soil mulch minimises deep cracking leading to reduced evaporation losses. Moisture can be conserved by using organic surface mulches at 5 tonnes/ha.

(2) **Thinning**: Thinning is recommended if a drought occurs during the pre-flowering stage. Every third plant or alternate row may be removed to reduce the soil moisture loss and preserve it. Moisture stress of 35–50 days can be lessened by such thinning.

(3) **Weed control**: A plant that is considered undesirable in a particular situation is called a weed. Weed control is essential to save the crop from the onset of drought as weeds rob soil moisture and nutrients.

(4) **Rainwater harvesting and recycling**: In-situ water harvesting and/or runoff recycling are potent measures for saving crop life during periods of moisture stress. There is no control over the receipt of rainwater, but it can be managed in a way better than what is being done at present after it is received. Supplementary irrigation provided to the crop through rainwater harvesting and recycling, particularly during the grain formation stage will be very much rewarding. Run-off recycling holds immense prospects in deep black soil where the seepage losses are very less. High-intensity rains are quite common during the rainy season contributing 75% of seasonal rainfall in India. The inevitable runoff (10–40% of the total rainfall) could be stored in dugout ponds of appropriate sizes and recycled to the donor area in the event of severe moisture stress.

(5) **Intercropping and risk distribution**: Application of meteorological information in terms of the frequency and probability of breaks in monsoon rains can be made to select a combination of crops of different durations in such a way that there is a time lag in the occurrence of their growth for appropriate intercropping systems. Intercropping systems like maize-green gram, maize-cowpea, arhar-finger millet, etc. are chosen to avoid the risks of crop failure.

(6) **Choice of crops and rice planting technique**: The low water requires crops like millets, pulses, and oilseeds chosen to grow in these areas. If the transplanted rice crop is grown, then sufficient care is taken to conserve the moisture in the fields. The bund height of the rice fields is increased to avoid drainage of the surface water, thereby securing the proper depth of standing water for the crop. However, under prolonged water stress, direct-seeded rice is chosen over the puddled transplanted rice.

(7) **Crop sowing technique**: The broad bed and furrow (BBF) method of sowing is well suitable for dry farming areas. It is developed at the International Crops Research Institute for the Semi-arid Tropics (ICRISAT) in 1981. The recommended system of BBF consists of broad beds of about 100 cm in width separated by sunken furrows of about 50 cm in width. The preferred slope along the furrow is between 0.4 and 0.8%. It is mainly used in the deep vertisols in India to grow cotton. Other crops like wheat, groundnut, soybean, etc., also can be grown in the BBF system. In this method of sowing, 3–5 rows of the crops are grown on the broad beds, and the furrow irrigation method is employed. So, irrigation water for all the crop rows is fulfilled by seepage of the applied water from the furrows. This method of sowing increases the water use efficiency and yield of the crops as compared to the conventional flat bed method.

(8) **Protective irrigation**: If possible protective irrigation may be given. Usually, protective irrigation is given at 55–60 days of growth. However, due to the early withdrawal of monsoon, the same may be applied at 35–40 days of growth.

(9) **Increased frequency of Inter-cultivation**: Early stoppage of monsoon results in early cracking in the soil. To prevent cracking and loss of moisture, the frequency of inter-cultivation may be increased. The intercultural operations can be increased to 5 or 6, which acts as a dust mulch to decrease the evaporation loss.

(D) **Early Withdrawal of Monsoon**

All the suggested measures to mitigate the adverse effect of prolonged dry spells during the crop period are necessary to overcome the soil moisture stress during the early withdrawal of the monsoon.

(E) **Extended Monsoon**

Extended monsoon is seldom a problem in dryland agriculture. Such a situation is rarely experienced. Drainage facilities should be made available to drain out excess water from the field. Sowings of *rabi* crops must be well planned, and summer crops should be taken to utilize the excess water. In the case of rice, long-duration varieties are chosen so that the grain filling stage is not affected by excess rainfall. Rainwater harvesting structures like farm ponds and percolation tanks should be employed to store the excess water, providing life-saving crop irrigation.

Table 9.3 depicts the potential cropping systems to be followed in different regions of India based on parameters like rainfall, soil type, length of the growing season, profile storage capacity, etc. Here, some of the cropping systems which may be followed in different aberrant conditions are also suggested. In Table 9.4, different intercropping systems which may be followed in the dry land areas of India are depicted.

9.4 Agromet Advisory Service in India

Agro-meteorological Advisory Service (AAS) rendered by the Indian Meteorological Department (IMD), Ministry of Earth Sciences (MoES) in active collaboration with the Indian Council of Agricultural Research (ICAR), Indian Institute of Technology (IIT), State Agriculture Universities, State Department of Agriculture, etc. is a step to contribute to weather information based crop/livestock management strategies and operations dedicated to enhancing crop production and food security. The main emphasis of the existing AAS system is to collect and organize climate/weather, soil, and crop information and to amalgamate them with the weather forecast to assist farmers in making management decisions.

Table 9.3 Potential cropping system for different regions of India based on rainfall and soil characteristics (Reddy et al. 2017)

Rainfall (mm)	*Soil type*	*Growing season (weeks)*	*Profile storage capacity (mm)*	*Suggested cropping system*
350–600	Alfisols, shallow Vertisols, Deep Aridisols, Entisols (alluvium)	20	100	Single rainy season cropping of Sorghum/Maize/ Soybean
350–600	Deep Vertisols	20	100	Single post-rainy season cropping of Sorghum/chickpea
600–750	Alfisols, Vertisols, Entisols	20–30	150	Intercropping of Sorgum + Pigeon pea/ Cotton + Black gram
750–900	Entisols, Deep Vertisols, Deep Alfisols, Inceptisols	30	200	Double cropping with Maize-Safflower/ Soybean-Chick pea/ Groundnut-Horse gram
900–1400	Entisols, Deep Vertisols, Deep Alfisols, Inceptisols	>30	>200	Assured double cropping of Maize-Chick pea/ Soybean-Safflower
>1400	Alfisols, Entisols	>30	>200	Assured double cropping of Rice-Chick pea/ Maize-Lentil/ Soybean-Mustard/ Rice-Mustard

Table 9.4 Intercropping system suitable for drylands of India with crop geometry and duration (Reddy et al. 2017)

Crops	*Geometry (crop rows)*	*Base crop duration, days*	*Intercrop duration, days*
Sorghum + Lablab bean	6–8:2	100–120	150–180
Sorghum + Pigeon pea	6–8:1	100–120	180
Sorghum + Cowpea	2:1	100–120	80
Cotton + Black gram	2:1	150–185	65–75
Groundnut + Pigeon pea	6–8:1	105	180
Groundnut + Castor	6–8:1	105	150–180
Maize + Cowpea	2:1	100–110	75–80

9.4.1 The Information Support Systems: The Following Information Are Available Under AAS

Provision of weather, climate, crop/soil, and pest disease data to identify biotic and abiotic stress for on-farm strategic and tactical decisions.

Provide district and block level weather forecast (rainfall, cloudiness, maximum/minimum temperature, wind speed, wind direction, relative humidity (both morning and evening)) for up to 5 days.

Translate weather and climate information into farm advisories using existing research knowledge on making more efficient use of climate and soil resources through applications of the medium-range weather forecast. A broad spectrum of advisories includes weather-sensitive farm operations such as sowing/transplanting of crops, fertiliser application based on wind conditions and intensity of rain, pest and disease control, intercultural operations, quantum and timing of irrigation using meteorological threshold, and advisories for the timely harvest of crops.

- Introduction of state-of-art technologies such as remote sensing, GIS, and crop simulation model-based decision support systems for agro-meteorologists to adapt agricultural production systems to weather and climate variability and the increasing scarcity of natural resources.
- Develop an effective mechanism for on-time dissemination of agro-met advisories to the farmers.
- Effective training, education, and extension on all aspects of agricultural meteorology.

9.4.2 Agro-Meteorological Support for Farm Management

Weather-based farm advisories as a support system have been organised after the characterisation of agro-climate, including the length of crop growing period, moisture availability period, distribution of rainfall and evaporative demand of the regions, weather requirements of cultivars, and weather sensitivity of farm input applications. All this knowledge is used as background information. Following are the ingredients of a typical Agromet Advisory Bulletin to reap the benefits of benevolent weather and minimize or mitigate adverse weather impacts.

i. District/county-specific weather forecast in quantitative terms for the next 5 days of rainfall, cloud, max/min temperature, wind speed/direction, and relative humidity, including forewarning of hazardous weather events likely to cause stress on standing crops and suggestions to protect the crop from the weather variability.
ii. Weather forecast-based information on soil moisture status and guidance for land preparation, seeding, application of irrigation, fertiliser, herbicides, etc.

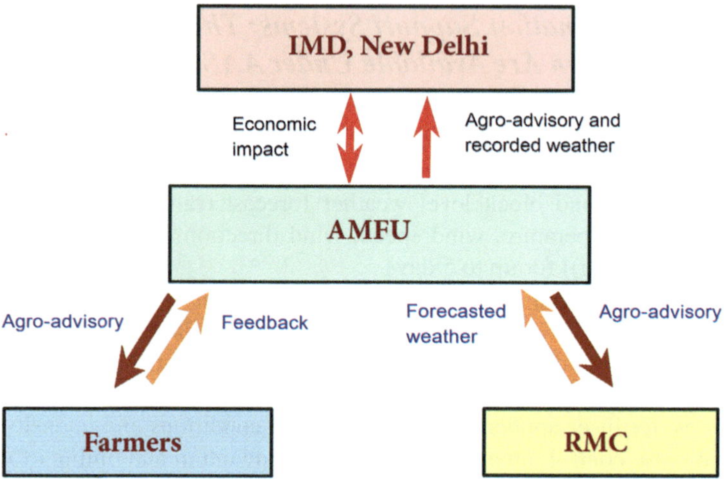

Fig. 9.3 Flow chart of transfer of agromet advisory information for AMFU, India

iii. The advisories on dates of sowing/planting and suitability of carrying out intercultural operations covering the entire crop spectrum from pre-sowing to post-harvest guide the farmer in his day-to-day cultural operations.
iv. Weather forecast-based forewarning system for major pests and diseases of principal crops and advice on plant protection measures.
v. Dissimilation of techniques for manipulation of crop's microclimate e.g. shading, mulching, other surface modification, shelterbelt, frost protection, etc. to protect crops under stressed conditions.
vi. Advisory on contingency plan under extreme weather situations, late on-set of monsoon, long dry spells during the crop growing period, etc.
vii. They are reducing the share of the agricultural production system to global warming and environmental degradation through precise management of land, water, and farm inputs, particularly pesticides, herbicides, and fertilisers.
viii. Advisory for livestock on health, shelter, and nutrition.

The flow chart of transfer of Agromet advisory information for any Agro-meteorological Field Unit (AMFU), India, is given in Fig. 9.3. It depicts the channel of the flow of information between the AMFU, Regional Meteorological Centre (RMC), IMD, and the farmers. The RMC sends weather data to the AMFU, which are formatted into the form of the agricultural weather bulletin as an Agro-advisory based on the forecasted weather by the AMFU. The AMFU sends the Agro-advisory to IMD, RMC, and the farmers. In the chain, the farmers send feed-back about the utility of the received advisory to the AMFU, which helps in further improvement of the advisory system through an integrated assessment.

9.5 Climate Change Impacts and Adaptation Metrics for Agriculture

CRM is a growing concept to deal with the changing climatic scenario of the world. CRM is the use of climate information to cope with the possible impacts of climate change on development and resource management. CRM deals with the management processes to reduce risks by considering the given points like (a) analysing the impacts of climate change; (b) assessing risks and uncertainties for informed decision-making; (c) planning, sharing, and pooling of risks, and (c) taking adaptive measures to reduce risks. This consists of several elements like an early response, strategic diversification, dynamic resource allocation, financial instrumentation, infrastructure design, and capacity-building systems. The CRM concept has been popularised by WMO, Geneva, and it focusses on the best management practices that are already in use in different parts of the world, especially in the water and agriculture sectors (Selvaraju et al. 2012). The quantity, duration, and distribution of rainfall are very important factors in determining the Kharif season characteristics, cropping systems, and livestock production. The seasonal variability of rainfall is the main barrier to crop production in the tropical belts of the world. Unreliable rainfall combined with high evaporative demand and soils with low water-holding capacity and high run-off potential results in a high risk of water deficit at any stage of crop growth (Muchow et al. 1996). The temperature during the cropping season often exceeds the optimum range for physiological processes such as phenology, leaf area development, assimilate accumulation, and grain filling. The high air temperature around flowering can reduce pollen viability and grain set in major cereals of the tropics (rice, maize, sorghum, etc.). The incidence of high soil temperature during crop establishment is also a threat in semi-arid and arid environments.

9.5.1 Principles of Good Metrics in Agriculture

There are several key considerations in choosing metrics. Metrics must be well defined, meaningful, measurable, motivational, and easy to understand and communicate to all stakeholders, including farmers, policymakers, business executives, and consumers.

A concise set of universal indicators will provide broad, general information and should be supplemented by metrics tailored to national and local challenges. Because a central objective of sustainable development is to ensure social inclusion, metrics should be disaggregated by gender, geography, socio-economic status, disability, ethnicity, age, and other dimensions wherever possible.

There are infinite ways to measure progress; therefore, there must be a conscious limiting of the number of metrics. Indicators associated with well-tested methodologies and guidelines should be chosen; proxies can be used that cover several areas of interest simultaneously. Moreover, there will be trade-offs between metrics in terms of precision, scale, and cost that will require a clear vision of measurement objectives at the start.

9.5.2 Managing Climate Risks

CRM aims to manage the climate variability currently affecting the farming community. The integration of climate information in risk management and adaptation planning is one of the important aspects of sustainable agriculture. The climate risk assessment and management approach is mainly used for weather monitoring, climate data analysis, crop–weather relationships, seasonal forecasting, and economic modelling.

Researchers are currently aiming at the inclusion of the following elements into CRM:

(i) Modern methods/tools for climate data sourcing and analysis, automatic meteorological measurements (rainfall, temperature, wind, etc.) at the local level on a near real-time basis, and seasonal forecasts.
(ii) Analysis of climate risks and assessment of climate impacts using crop–weather interactions.
(iii) Integration of economic models, linear and non-linear optimisation methods, AI/ML techniques, and risk perception by farmers.
(iv) Preparation of agro-meteorological advisory for farmers and communication of the advisory and actionable production technologies in farm fields.

9.5.3 Data and Information Needed for Generating CRM

The climate risk assessment emphasises the comparison of current quantitative data with historical information on trends and impacts. Daily rainfall records are essential in order to determine rainfall onset dates and relate rainfall probabilities to crop growth stages. Decisions can be taken on the current weather risk after its comparison with historical weather. While preparing this climate risk matrix, one should always keep this point in mind. What is the main crop or livestock is at risk, and what other options one has, review of existing climate data availability and quality; information synthesis related to cropping pattern, soil types, natural resources, and management practices; assessment of the suitability of sources of real-time weather and climate-forecast products and modern data sourcing (e.g. remote sensing)?

The preparation of the climate risk matrix is an important component of the CRM system. A variety of data and information are required for generating a climate risk matrix for a specific crop in a specific region/locality. The broadly grouped data types needed to develop the climate risk matrices are given below:

(i) **Climatic data**: This includes the historical data related to climatic parameters like rainfall, evapotranspiration, maximum, minimum, and optimum temperatures, relative humidity (both morning and evening), etc., at least for the past 30 years of that particular region/area.
(ii) **Local conditions data**: It includes data related to topography, soil type and depths, land slopes, surface drainage, water holding capacity, nutrient retention capacity, soil fertility status, problem soils (acidity, salinity, alkalinity, nutrients toxicity, etc.), agro-climatic risks, etc. of that particular geographical area.
(iii) **Farming and cropping systems data**: It includes data related to cropping systems, integrated farming systems, livestock production systems, cropping patterns, diversification of different crop species and their varieties, etc. of that particular geographical area.
(iv) **Socio-economic data**: This includes data related to socio-economic conditions of the farming community residing in that particular geographical area under study availability of credits to the farmers, presence or absence of credit funding agencies like banks in that area, market availability, household strategies, presence or absence of technical support providing institutions, etc.

9.5.4 Methodology for Preparation of Climate Risk Management Matrix

The climate risk assessment emphasises the comparison of current quantitative data available regarding climate data availability and quality; cropping pattern, soil types, natural resources, and management practices; assessment of the suitability of sources of real-time weather, and climate-forecast products; modern data sourcing (e.g. remote sensing), etc. For the preparation of the CRM matrix, one needs the data and information on various parameters like climate data, cropping systems, knowledge of local environmental conditions and agricultural practices, costs of inputs, market price, etc. The steps for the preparation of the CRM matrix are given below.

1. Choice of the main crop in the particular region.
2. Determination of the effect of climatic parameters of that region on the crop and its yield variability in the past 30 years.
3. Documentation of agronomic management systems based on the variability of the yield of the crop to counteract the yield-limiting factors due to various climatic parameters.
4. Application of Decision Support System Technology, Crop Models, Pests & Pathogens models, etc. to choose Best Management Practices (BMPs) for the crop and also to verify that the agronomic practices documented in the previous step are applicable or not using different simulations.

5. These BMPs are employed in that particular region, and a CRM matrix is developed, keeping in mind the yield-limiting factors like climatic aberrations, pests and disease attacks on the crop, socio-economic status of the farmers of that region, etc.
6. Finally, an evaluation of the change in the yield of the crop is done after employing those strategic measures and the CRM matrix. From this evaluation process, the number of climatic risks was reduced, and an increase in the yield of the crop due to the application of the CRM matrix is obtained by the researchers. So the matrix mainly aims at reducing climatic risks, reducing the chances of crop failures, increasing the crop yield, enhancing adoption levels, and implementing efficiency of new technologies by the farmers, etc. Table 9.5 describes the criteria for developing agricultural matrices for the marginal and small farmers of India based on different biophysical indicators,

Table 9.5 Agricultural matrix developed for marginal and small farmers of India

Categories	*Vulnerability criteria*	*Indices*
Biophysical indicators	Exposure	• Aberrant weather (e.g. heat wave, cold wave, flash flood, drought, etc.) • Problematic Soils (e.g. salinity, acidity, alkalinity, etc.) • Faulty Crop Management practices like burning of crop residues, monoculture, no seed treatments, etc. • Acute/Chronic water shortages leading to temporary/permanent droughts • Increased pest, disease, and weed infestation • Low yield of crops due to aberrant weather, and/or faulty management practices, insect pests or diseases attack on the crops etc.
Socio-economic data	Sensitivity	• Lack of knowledge about the crop insurance schemes provided by the Govt. of India and private sectors • Higher percentage of total agricultural lands under subsistence farming • Lower wages to landless tenants • Lack of commercialization of agriculture
Agricultural system characteristics	Adaptive capacity	• Lack of knowledge about advanced agricultural technologies and climate-smart agriculture • A higher percentage of lands under rainfed crops and lack of intensive agriculture • Poor adoption of resource conservation technique • Lack of training and demonstration programs to the farmers by the extension officials
Climate risk mitigation programs	Risk reduction	• Organizing of Afforestation programs like social forestry and JFM (Joint Forest Management), • Putting a ban on 'slash-and-burn' agriculture, • Promotion of organic agriculture and Natural Farming • Reduced flooding of rice field • Minimum tillage production techniques

socio-economic data, agricultural system characteristics, and climate risk mitigation programs. By observing this table, one can understand the crucial factors which are taken into account for the development of any agricultural matrix. Table 9.6 depicts the methodology to evolve CRM options for region-wise focal species based on the usage of different spatial scales, production technologies, linkage with weather variability, etc.

9.5.5 CRM Matrix for Maize Crop Grown in India

The climate risk matrix of the maize crop is given below as an example of the preparation of climate risk matrices for different crops. In this way, the researchers can prepare climate risk matrices for different crops to mitigate the adverse effects of climate change. The CRM matrix for maize crops is prepared for Mahbubnagar district, Telangana, India, after discussion with expert scientists and farmers. Mahbubnagar is the most vulnerable drought-prone district in Telangana state, which receives rainfall of less than 600 mm with a deviation of 10 to 45%.

Maize is generally sown between 25th May and 2nd August in Mahbubnagar and takes around 100–105 days for harvest. According to the farmer's and expert scientists' suggestions, three different sowing periods (Tables 9.7, 9.8) are considered as per the soil type and rainfall of the district.

Now considering these dates of sowing (Table 9.8), the CRM matrix is prepared for maize by considering the commonly grown cultivar. Maize crop yield is affected not only by abiotic factors (temperature and rainfall) but also by biotic factors (pests and diseases). Abiotic and biotic stress management are considered while preparing the CRM for the maize crop.

Abiotic Stress

1. For early sown maize in light soil, summer ploughing is recommended for increasing the infiltration of rainwater into the soil. This will increase soil moisture content, which helps in the germination of the maize crop. Hence, during the germination period, water stress impact is not observed.
2. For heavy textured soil during the monsoon season (22nd June to 19th July), the maize crop is sensitive to germination for rainfall more than 6 cm causing waterlogging. Hence it is suggested to postpone the sowing time after the rainfall ceases.
3. During the vegetative growth period, hoeing is recommended to loosen the top soil to remove weeds and reduce loss of water due to evaporation.
4. Tasseling and silking stages are highly sensitive to water stress; hence, supplementary irrigation is suggested from surface water storage structures.
5. The silking stage is highly sensitive to fungal disease if a wet spell continues; hence, the fungicidal spray is suggested.

Table 9.6 Methodology to evolve the CRM options for region wise focal species

Sr. No	*Spatial scale*	*Activity required*	*Details of task 1*	*Details of task 2*	*Details of task 3*
1	Study region, State and Country	Decide on the focal crop/tree/livestock for the study area-Criteria and methodology	Importance of the crop to the farmers in the region	Area of cultivation over time-Minimum of 25 years look at other competing crops also	Relative cultivation areas in rest of the growing region
			Influence of climate in the cultivation of species	Productivity over time-Minimum of 25 years	Relative productivity to the rest of growing region
2	Study region	Focal species-Determine the importance and quantum of climate component of yield variability (variation with time)	Reasons for variation in productivity with Time	Climate related—e.g.- Rainfall distribution, critical growth periods, pest and diseases	Others like Cropping pattern related
			How much of variation due to climate?	Discussion with farmers	Discussion with scientists
3	Study region	Determine the cultivation technologies based on Spatial variability and intraseasonal climate variability-major land and crop management factors causing large variation of yield	Reasons for variation in productivity with Space	Climate related—e.g.- Rainfall spatial variability, critical growth periods, pest and diseases	Others like soil and cropping pattern related
			Spatial variability impact on yield and crop management in the study region	Discussion to determine the quantum of such variability	If found significant; distinguish appropriate technologies
			Intra-seasonal climate variability impact on yield and crop management in the study region	Discussion to determine the quantum of such variability	If found significant; distinguish appropriate technologies

(continued)

Table 9.6 (continued)

Sr. No	*Spatial scale*	*Activity required*	*Details of task 1*	*Details of task 2*	*Details of task 3*
4	Each cultivation technology	For each technology in 3-Identifying decisions that depend on climate information and basket of options (choices) available for each decision	Decisions prior to planting	Soil and moisture conservation measures, Manure application	Choice of variety, inter crops, planting designs, cropping sequence
			Decisions at planting	Sowing window	Inputs at sowing—e.g. Basal dose of fertilizer
			Decisions after planting till harvest	Schedule of operations from sowing to harvest	Will include—(a) quantity and timing of inputs (b) timing of farming operations (c) technique of operation
			Decisions after harvest till sale of produce there has to be a link to climate	Post-harvest processing, storage and transport	Both timing and choice of—(a) post-harvest processing, (b) storage (c) sale
			Decision regarding credit and input supply		
5	Each cultivation technology	For each such decision in 4, choose amongst options based on farm economics of cost, projected range of benefit and penalty (for wrong choice)	Work out typical costs (range of costs depending on quantity and also interest rate of credit) and penalties	By discussion with growers and scientists	
			Work out range of projected benefits	The range will depend on rain till harvest and input management of the farm plot	Use of appropriate crop and input management that are validated for the region
			Pragmatic choice, Transaction cost depending on various factors		

(continued)

Table 9.6 (continued)

Sr. No	*Spatial scale*	*Activity required*	*Details of task 1*	*Details of task 2*	*Details of task 3*
6	Each cultivation technology	For each choice of options in 5, work out requirement & availability of—(a) climate information needed like lead time, skill of forecast (b) tools like crop models, remote sensing, pest, pathogen models (c) Data analysis, etc.	Lead time needed	by discussion with farmers, scientists	
			Skill of forecasts	As inputs from Climate scientists, local agro-met scientists	
			Tools and other data	Local agro-met scientists and other researchers	
7	Study region, State and Country	For each option chosen in 5 & 6 (considering all cultivation technologies);—How to improve the adoption levels and implementation efficiency by growers of the focal species by (a) by current opportunities (b) new options?	To implement this option—Other than climate information what else (inputs, advice, credit) is required?		
			For each of this requirement in the current system how do farmers access?		
			What are the problems faced by farmers in access to these requirements?		
			Institutional improvements-including harmonizing and utilizing current options		

Table 9.7 Dates of sowing for maize crop in Mahbubnagar district

Soil type	*Sowing window*
Light soils (Located along the higher contours in the watershed)	**Early**: 25th May to 21st June (27 days)
Heavy soils (Located along the lower contours in the watershed)	**Timely**: 22nd June to 19th July (27 days) **Late**: 20th July to 2nd August (27)

Table 9.8 Appearance of phenological stages of maize in Mahbubnagar district

Phenological stage	*Period*	
	From	To
Pre-sowing	03-May	21-Jun
Sowing	15-Jun	22-Jun
Germination	22-Jun	01-Jul
Vegetative growth	02-Jul	01-Aug
Tasseling	06-Aug	21-Aug
Silking	21-Aug	15-Sep
Physiological maturity	15-Sep	30-Sep
Harvest	30-Sep	05-Oct

Biotic Stress

Maize is highly sensitive to pests during the vegetative and tasselling stage; hence proper control of pests and diseases is recommended.

Wet spells with high temperatures can lead to pest attacks during the vegetative stage, and hence right control measures are suggested.

According to the crop weather calendar based on the appearance of phenological stages (Table 9.8), risk factors associated with each phenology stage should be identified, and accordingly, management options may be suggested with cost and benefit analysis. As an example, the CRM matrix for early, timely, and late sown maize in the Mahbubnagar district is shown in Tables 9.9, 9.10, and 9.11, respectively. (This CRM is adopted from the National Innovations in Climate Resilient Agriculture, ICAR-Central Research Institute. Dryland Agriculture, Hyderabad).

The team led by Prof. DK Swain of IIT Kharagpur, India, worked on the development of CRM tools for making the decision on agricultural management for rice and peanut crops in the Kharagpur location. They used crop simulation models of DSSAT to develop the CRM tools for the years 2030 and 2050 under RCP4.5 climate scenario. In view of decreasing water availability, the most sensitive crop growth stages were flowering and panicle initiation in rice, flowering, and pod development in peanut. Late planting of any crop suffered due to unfavourable weather, hence the yield loss. The CRM matrices were prepared as decision support and agro advisory for farmers to minimize the adverse impact of climate change on agricultural production (Tables 9.12 and 9.13). The above tables are guidelines for preparing a CRM of a specific crop over a particular region. The readers can consider the above example as a reference for preparing

Table 9.9 The CRM matrix for early sown maize in Mahbubnagar district

Early sowing: 25th May to 21st June (27 days)

Days after (before) sowing	Cropping time line	Abiotic risk-water stress impact	Good climate event with low risk outcome	Adverse climate event with high risk outcome	Risk Management Options (RM0) available	Cost	Range of benefit	Penalty
−50 to 0	Before and during Land preparation				Soil amendments, Watershed treatment	For soil amendments—the cost is Rs. 2400 to 6000. At 4 to 10 cart loads per acre—it is 2 to 5 tons. At Rs. 600 per cart load (including transport and spreading)	The benefit range is Rs. 800 to 4000. The impact will last for 3 years. Increase in yield can be from 10 to 50% over control (non application). Benefit increases with (a) higher application rate (b) lack of moisture stress during crop growth (c) Medium level of moisture stress during critical growth stages like silking stage. With a typical yield of 1000 kg /acre in the region and Market price of Rs. 8 per kg	Just the cost in most years. But, 10–30% of yield of control. As benefits persist over 3 years we should only take a fraction of the cost as penalty

(continued)

Table 9.9 (continued)

Early sowing: 25th May to 21st June (27 days)								
Days after (before) sowing	Cropping time line	Abiotic risk-water stress impact	Good climate event with low risk outcome	Adverse climate event with high risk outcome	Risk Management Options (RM0) available	Cost	Range of benefit	Penalty
0	Sowing							
0–9	Germination	None						
Early sowing: 25th May to 21st June (27 days)								
10–40	Vegetative growth	Medium			Inter cultivation and Hand weeding instead of weedicide application.Take up an additional intercultivation if needed	Cost, Rs. 200 to 600 per acre. Added cost compared to herbicide use will be Rs. 200 per acre. Additional intercultivation will be Rs. 400/ acre	Benefit range is Rs. 400 to 1200. Increase in yield can be from 5 to 15% over control (non-application).Benefit depends on the distribution of dry spells and wet spells during the season. With a typical yield of 1000 kg /acre in the region and Market price of Rs. 8 per kg	Just the cost in most years. But, 10% of yield of control

(continued)

Table 9.9 (continued)

Early sowing: 25th May to 21st June (27 days)

Days after (before) sowing	Cropping time line	Abiotic risk-water stress impact	Good climate event with low risk outcome	Adverse climate event with high risk outcome	Risk Management Options (RM0) available	Cost	Range of benefit	Penalty
45–60	Tasseling stage	High						
60–85	Silking stage	Very high	15 cm and more rain and Minimum of 2 cm per week	Rain less than 5 cm or dry spell of 15 days	1–2 supplementary irrigation from Surface water storage structures	Cost is 0 to Rs. 1000/ irrigation. (At Rs. 50 per hour from a bore well for 20 h). It is 0 if it is from CPR surface water and Rs. 1000 if a storage structure is created within the farm	The benefit range is Rs 1600 to 4800. Increase in yield can be from 20 to 60% over control (non-application). Benefit increases with (a) higher application rate (b) lack of moisture stress during crop growth (c) Medium level of moisture stress during critical growth stages like silking stage. With a typical yield of 1000 kg/ acre in the region and Market price of Rs. 8 per kg	just the cost

Table 9.10 The CRM matrix for timely sown maize in the Mahbubnagar district

Timely sowing: 22nd June to 19th July (27 days)

Days after (before) sowing	*Cropping time line*	*Abiotic risk-water stress impact*	*Good climate event with low risk outcome*	*Adverse climate event with high risk outcome*	*Risk Management Options (RMO) available*	*Cost*	*Range of benefit*	*Penalty*
−50 to 0	Before and during Land preparation							
0	Sowing							
0–9	Germination	Very high	Rain less than 3 cm	Rain of 6 cm or more	Postpone sowing	0	Benefit range is Rs. 750 to 3000. Increase in yield can be from 5 to 20% over control (not postponed). Benefit depends on the distribution wet spells and clay content. With a typical yield of 1500 kg /acre in the region and Market price of Rs 8 per kg	0 to 30% of control. In case of long wet spells or high rainfall; the next opportunity may take several days. There is a risk of missing the optimal sowing window
10–40	Vegetative growth	Low			Field drains	Cost is Rs. 600/ acre	Benefit range is Rs. 750 to 3000. Increase in yeild can be from 5 to 20% over control (not making). Benefit depends on the distribution wet spells and clay content. With a typical yield of 1500 kg /acre in the region and Market price of Rs. 8 per kg	Just the cost

(continued)

Table 9.10 (continued)

Timely sowing: 22nd June to 19th July (27 days)

Days after (before) sowing	*Cropping time line*	*Abiotic risk-water stress impact*	*Good climate event with low risk outcome*	*Adverse climate event with high risk outcome*	*Risk Management Options (RMO) available*	*Cost*	*Range of benefit*	*Penalty*
		Moderate	Minimum temperature of less than 20 degree C; for 7 consecutive days or more. Low minimum temperatures reduce fecundity (number of eggs / female) and their further survival	Coupled events of Wet spell of 5 days with dry spells of 15 days or more. More such iterations will increase loss. High temperature and humidity increase incidence. Both do not co-occur in the growing season of the district. Hence, alternative periods of high temp and high humidity will help the pest. The pest will have multiple generations in region with extensive and sustained mono cropping like Mahbubnagar	1–2 Sprays systemic insecticides depending on the severity	Cost of Rs. 750 to 1500. Per spray the chemical cost is Rs. 500. Labour and equipment will cost of about Rs. 250	Benefit range is Rs. 1500 to 7500. Increase in yield can be from 10 to 50% over control (noapplication). Benefit increases with (a) higher infestation rate (b) lack of moisture stress during crop growth. With a typical yield of 1500 kg / acre in the region and Market price of Rs. 8 per kg	Just the cost
45–60	Tasseling stage	Moderate		Wet spell of 7 days or more or Rain of 10cm/week	Fungicide sprays			
60–85	Silking stage	High-increase in Vulnerability to fungal diseases						

Table 9.11 The CRM matrix for late sown maize crop in the Mahbubnagar district

Late sowing: 20th July to 2nd August (27 days)

Days after (before) sowing	*Cropping time line*	*Abiotic risk-water stress impact*	*Good climate event with low risk outcome*	*Adverse climate event with high risk outcome*	*Risk management options (RMO) are available*	*Cost*	*Range of benefit*	*Penalty*
−50 to 0	Before and during land preparation	Low			Apply compost, tank silt in large quantities	(same as for Moisture stress). The same application will provide both the benefits	Same as for moisture stress	None
0	Sowing							
0–9	Germination-							
10–40	Vegetative growth	Moderate	Rain of 2 cm or more per week		First top dressing with 30 kg urea	Cost Rs. 200	Benefit from both top dressings are considered together benefit range is Rs. 1600 to 4000. Increase in yeild 20–50% over control (non application). Benefit increases with (a) higher application rate (b) lack of moisture stress. during crop growth. With a typical yield of 1500 kg /acre in the region and Market price of Rs. 8 per kg	Just the cost

(continued)

Table 9.11 (continued)

Late sowing: 20th July to 2nd August (27 days)

Days after (before) sowing	*Cropping time line*	*Abiotic risk-water stress impact*	*Good climate event with low risk outcome*	*Adverse climate event with high risk outcome*	*Risk management options (RMO) are available*	*Cost*	*Range of benefit*	*Penalty*
5–60	Tasseling stage	High	15 cm and more rain. Minimum of 2 cm per week				Second Top dressing 30 kg urea and 10 kg potash per acre	Cost Rs. 300
60–85	Silking stage	Very high						

Table 9.12 The CRM matrix for best planting dates of rice varieties under rainfed and irrigated conditions for the years 2030 and 2050 at Kharagpur, India

Rice varieties	*2030*		*2050*	
	Rainfed	*Irrigated*	*Rainfed*	*Irrigated*
IR 36 (transplanted)	2nd fortnight of July	1st fortnight of August	2nd fortnight of July	1st fortnight of August
Swarna (transplanted)	2nd fortnight of June	1st fortnight of August	2nd fortnight of July	1st fortnight of August
Naveen (direct seeded)	2nd week of June	2nd week of June	3rd week of June	1st fortnight of July

Table 9.13 The CRM matrix to find critical irrigation stages in order of their irrigation importance for early and late planting of rice varieties at Kharagpur, India

Rice varieties	*2030*		*2050*	
	Single irrigation	*Two irrigations*	*Single irrigation*	*Two irrigations*
IR 36, Swarna	FL	PI + FL	FL	PI + FL
Naveen	FL	TI + FL	FL	TI + FL

the CRM for the other crops in their targeted region. One needs to consider all the table components (denoted in each column) and prepare the CRM, which will be useful for agriculture practices. A CRM based on farmers' experience and agricultural scientists' advice will be helpful to the farmers and decision makers for field-level implementation, which in turn minimizes the loss and maximizes the profit through increased production. A few examples on the utilizations of agromet advisory based on CRM tools to the farmers and their feedback are provided in the next. These examples are from different corners of the world for various major crops, and the reports will be helpful in understanding the benefits of the use of CRM.

Chapter 10
Lessons from the Farmers on the Application of Weather Forecasts

10.1 Introduction

Agriculture is a vital sector for supplying food and contributes precisely to the national economy, but agricultural production is highly influenced by weather variables. Agriculture mainly includes crops, livestock, and fisheries, and weather parameters have variable effects on these enterprises in different ways. Every year, calamities like floods, droughts, severe storms, heatwaves, wildfires, etc., cause severe damage and significant losses to agricultural products worldwide, and climate change manifests an increasing trend of climate extremes. In general, the weather changes, which lead to high temperatures, hampers the growth and development of crops, further reducing crop yield. However, it does not show that all crops are only affected by high temperatures. The crops could be adversely affected by the extremely cold temperatures, making them vulnerable. Changes in weather can have both beneficial and detrimental effects on crops and the practices associated with growing them. Extended periods of hot weather can result in drought, which has a detrimental impact on the growth of crops and the well-being of livestock. Having a thorough knowledge of the ideal weather conditions for crops prior to planting can significantly improve their productivity.

Climate changes, such as fluctuations in monsoon rainfall and temperature variations within a season, have a significant impact on food production in India. The quality of vegetables, fruits, tea, coffee, medicinal and aromatic plants, etc., is greatly influenced by the subtle variations in precipitation and temperature. Other weather conditions also have an impact on agricultural and related industries, such as reduced milk production from dairy cows and a decrease in fish reproduction, migration, and catches. According to various reports, it is predicted that there will be a potential decrease of 10–40% in global crop production by the year 2100 (Mahato 2014).

U. C. Mohanty et al., *Climate Risk Management in Agriculture*,
https://doi.org/10.1007/978-3-031-51862-1_10

The majority of precipitation in the Indian region is brought by the southwest summer monsoon, which has a significant influence on the climate. According to Mahato (2014), the potential decrease in net revenue at the farm level is projected to be between 9 and 25% if the temperature increases from 2 °C to 3.5 °C. The limited availability of complementary inputs and institutional support systems, as well as the heavy reliance on natural factors with varying climatic conditions, greatly hampers the adaptability of farmers in India. According to the scientists' calculations, if there is a 7% increase in average rainfall and a 2 °C increase in average temperature, the country's overall net revenues would decrease by 12.3%. (https://www.manage.gov.in/studymaterial/CCA-E.pdf).

Weather forecasting has proven to be a dependable tool for numerous farmers in enhancing their farm management practices to mitigate losses caused by unfavourable weather conditions. Farmers often attribute the decline in crop yields primarily to shifts in weather patterns, according to various studies examining their experiences with weather variability. Although farmers used to trust the historical weather patterns, such as the starting and end of the rainy season, they are not as clear as they used to be earlier. The weather patterns are not the same as in earlier days, so there is a need for a weather forecasting system to help the farming community. Farmers desire to know the exact date of starting the rainy season, especially for the rainfed agricultural system. Many farming activities can be optimised during the growing season if farmers received localised and timely weather forecasts. Forecasting helps optimise the timing of fertiliser application during the crop growing period, which can be shifted to dry periods to prevent the loss of excess fertiliser. It helps increase the efficiency of fertiliser use and reduces the cost of inputs. Early knowledge about rainfall can help the farmers to protect the crops by making a drainage system in case of excessive rainfall forecast. The excess rainwater can also be harvested and preserved for further use on dry days. In addition, the occurrence of pests and diseases is heavily influenced by weather conditions, and numerous small-scale farmers do not have the resources to afford pesticides. In this scenario, a prompt notification can remind farmers to manually eliminate pests from their fields in a timely manner. Furthermore during the critical stage of crop growth, receiving timely notifications about adverse weather conditions can help prevent potential damage to the crops. These warnings and the corresponding measures are effective in minimizing crop damage, maximizing harvest yields, reducing expenses, and optimizing the entire cultivation period. Providing location-specific weather information to small-scale farmers is a crucial advancement in enabling them to make timely and informed decisions regarding their farm management.

10.2 Weather Forecasting for Agricultural Application

The quality of the forecast provides information about the probability of the respective weather forecast arriving. The credibility and reliability of forecasts define the quality and value of any weather information. To maintain the quality and authenticity of the weather forecasting system, we need to understand and take the necessary steps towards bettering the system. Its critical factors monitor and manage any system's success, which affects the information system. Nevertheless, weather predictions are inherently uncertain, and acknowledging this uncertainty is the initial step in managing it. The details of different time and space scale forecasts categories are explained in Chap. 3. However, a very brief is summarised here for a quick view of the readers.

The weather forecast is divided into different types based on time and space or temporal coverage of projections. The types of weather forecasts are defined below.

10.2.1 Now-casting

The atmospheric state estimates for the next few hours (mostly 2–3 h) over a particular region are known as now-casting. Now-casting is mainly utilised to understand any extreme weather events in a short period (within a few hours) to provide warnings in case of any unwanted weather conditions for saving lives and damage to properties. It has minimal application in the agricultural field. The now-casting forecast is confined to a few hours; thus, it is difficult to disseminate it using a broadcasting system. The time scale of now-casting is only about 2–3 h, which may not give sufficient scope for use as the routine farming operation starts early in the morning regularly.

10.2.2 Short Range Forecast

The short-range forecast usually predicts weather events in a time scale from 12 h to 3 days. This kind of forecast can be effectively broadcasted through several media such as newspapers, television, web, etc. This type of forecast has potential use in the agricultural fields for intercultural operations such as fertiliser and pesticide application, irrigation water management, hoeing, weeding, etc. Skill full adaptive strategies are needed to be developed for the farmers to use this forecast more.

10.2.3 Medium Range Forecast

The medium-range forecast usually predicts weather events on a time scale of 4–10 days. It is an essential requirement for a farming operation to decide on sowing time to ensure seed germinating soil moisture and the proper seeding technique in general for uniform seed establishment. The forecast is used explicitly in intercultural operations for efficient use of resources which helps in increasing farmers' economic grain while reducing the production cost. Developing useful adaptive strategies by holding seminar for farmers and capacity building for more extensive and intensive use of this forecast is needed.

10.2.4 Extended Range Forecast

The extended range forecast usually predicts the average weather events in a time scale of 10–30 days. It is mostly used for deciding the right sowing time for crops so that the crop should not suffer from any stress during sensitive stages. Farmers have to face huge economic losses if the crops are affected by weather-related damage during seedling. This forecast helps farmers prepare for intra-seasonal water management to avoid yield loss due to drought at critical growth stages of crops. It is necessary to develop useful adaptive strategies and educate the farmers through training for using this forecast in their farming operations.

10.2.5 Long Range Forecast/Seasonal Forecast

The long-range forecast mostly provides an expected outlook of the average weather on a time scale from one month to season. The seasonal forecast is mainly used for policy-making and risk management of different sectors such as agriculture, water management, hydropower, etc. This is used to select crops, varieties, and sowing windows as adaptive strategies to minimise the adverse effect of weather abnormality on agricultural production. Useful adaptive strategy and effective policy are required to make this forecast useful.

10.3 Success Stories of the Weather Forecast

10.3.1 Information and Technology Customised to User Needs

The weather information through forecasts should be easily broadcast and accessible to the farmers. Early (at least one week before) precipitation forecasts are mainly of utmost priority for most farmers for proper scheduling of farming activities. It is well known that crops are weather-sensitive depending on their varieties and species; their growth and development depend on many weather parameters like temperature, humidity, and wind. An instance of this is when Coffee is extremely susceptible to heat waves, whereas the maturation of sesame is highly vulnerable to powerful gusts of wind. The specific weather information needed for a particular crop is determined by the differences in behaviour among different types of crops. Customizing the information provided to farmers based on the specific crop they are growing and aligning farm-management practices to their needs is crucial. Along with the type of information, farmers require information in the local language to facilitate comprehension, which can be difficult in countries with multiple spoken languages. Translating messages from English to local languages can be a time-consuming task, as the meaning of phrases can vary across different regions. This can lead to miscommunication or misunderstandings within the farming community. Therefore, it is essential to have accurate translations provided by local experts who are knowledgeable about the language or can guide the farmers in that specific area.

10.3.2 Location-Specific and Authentic Forecasts

The second important factor for the success of any forecast is its reliability and credibility. Providing accurate and high-quality forecasts is not only essential but also a challenge. The issue of uncertainty in weather forecasts is widely acknowledged, and the primary objective is to comprehend, assess, and minimize it to the greatest extent feasible, while also providing farmers with the necessary knowledge to navigate these uncertainties. By utilizing the most accurate ensemble weather forecasting models and implementing suitable data techniques and statistical analysis, we can significantly minimize the level of uncertainty in the forecasts. Another method to decrease uncertainty is by offering a forecast that is tailored to a specific location. Service providers can ensure that farmers in a particular region receive the most precise and relevant information by concentrating on a specific location. In countries like India, not all farmers are educated enough to avail the weather forecast facility. It is not always possible to get the exact location of the farm if needed through telephone GPS locations if the farmer is illiterate. The location-specific forecast can be more reliable if some extension workers are appointed to go to the field or area to register the farmers for better communication. Although the groundwork process requires a lot of labour, it is worth it because it leads to a more accurate local forecast.

10.3.3 Approvement of the Service by National Weather Services

There are many local authorities in every country to provide weather alerts and related information. There should be a central entity or national weather service to provide excellent and accurate weather forecasts and reduce any unauthorised entity's false reports. If there is any private source of weather services, it should be allowed after approval by the federal entity to set up a trustworthy and sustainable weather service. The leading cause of approval is to help farmers get reliable service and prevent confusion among them due to different weather information sources. These national weather services also have a distinct responsibility in monitoring and storing these weather observations.

10.3.4 Service Ingrained in Current Local Networks

The experts need to monitor the services embedded with existing local networks wisely. The trust between the users and the service provider is important to maintain, which is only possible if the agricultural experts or extension workers are appropriately engaged in providing services. These experts take the lead in actively involving farmers in the service and providing ongoing support to ensure their continued participation. Another approach is to establish community-based organizations consisting of individuals who share the same language and culture, which facilitates the more effective exchange of knowledge. The unions and cooperatives have the ability to promote services to their members, as well as to local farmer networks. Furthermore, it is essential to organize workshops that will educate farmers about the service and outline the steps involved in the local service process. Once the service is up and running, the extension officers will be available to assist farmers in understanding the messages and making informed decisions based on the weather information.

10.4 Success Stories and Farmers' Feedback

10.4.1 India

The usability of extended-range forecasts has been evaluated by the IMD based on feedback received from various Agro-met Field Units (AMFS) and farmers across the country (Chattopadhyay et al. 2018) given below.

10.4.1.1 Feedback from various Agro-met Field Units

(i) **AMFU, Hyderabad**: Using Extended Range Weather Forecast, AMFU, Hyderabad, mentioned that the Extended Range Weather Forecast has the potential to be utilized in the creation of agro-met advisories for a range of agricultural activities, including land preparation, sowing, harvesting, fertilizer application, irrigation scheduling, and insecticides/pesticides usage.

(ii) **AMFU, New Delhi**: The AMFU, Delhi, mentioned, the Extended Range Weather Forecast proves to be beneficial in effectively managing input and strategically planning agricultural operations.

(iii) **AMFU, Coimbatore**: The AMFU Coimbatore relies on the Extended Range Weather Forecast to make informed decisions regarding the timing of sowing, selection of crop varieties, cultivation methods, management of inputs such as seeds, fertilizers, and bio-inoculants, as well as the harvesting and storage of output. The AMFU has emphasized the importance of incorporating this forecast into crop models to make informed decisions regarding management strategies. Additionally, they have highlighted the significance of analyzing the economic implications of SCF information.

(iv) **AMFU, Pantnagar**: They have stated that the Extended Range Weather Forecast was found to be helpful in the decision on various farm operations like the selection of land/crops, planning irrigation, and adopting plant protection measures. Extended range forecast is found to be of great importance for agriculture planning (sowing, harvesting, etc.), which can enable tactical adjustments to the strategic decisions that are made based on the longer lead seasonal forecasts and also help in timely review of the ongoing monsoon conditions for providing outlooks to farmers.

(v) **AMFU, Lamphelpat**: The extended weather forecast provided by IMD has proven to be extremely valuable in assisting us with the preparation of agro-met advisories for farmers, as it offers valuable guidance and knowledge. The accuracy in giving weather forecasts has increased with the use of the additional information from the extended weather forecast, especially on extreme weather, temperatures, rainfall, and the past day's weather information.

(vi) **AMFU, Ananthapuram**: The medium-range weather forecast information received from IMD for the next 5 days use to provide to farmers every Tuesday and Friday. In recent years, the precision of weather forecasts by IMD has been found to increase due to downscaling the forecast to the block level from the district level. Besides weather forecasts, weather-based agro advisories are disseminated to farmers in the form of bulletins that help the farming community in making timely correct decisions at the farm level and execution of agricultural operations like time of sowing, weeding, fertiliser application, irrigation, spraying, harvesting, etc. on day to day basis thereby reducing the risks associated with the weather during the crop growing season.

(vii) **AMFU, IIT Kharagpur**: The AMFU located at Indian Institute of Technology Kharagpur, India, is primarily situated in the Red and Laterite Agroclimatic Zone of West Bengal, encompassing five districts including Paschim Medinipur, Bankura, Jhargram, Purulia, and Birbhum. The medium-range weather forecast for the next five days and extended-range forecast are helpful in preparing district as well as block-level agro-advisories every Tuesday and Friday and special weather advisories both in Bengali and English languages. These advisories are directly disseminated to farmers and the community via WhatsApp and SMS to make timely decisions for various agricultural operations from land preparation to harvesting and also in the post-harvest operations to mitigate weather-related risks during the growing season.

10.4.1.2 Feedback from Farmers

(i) **Farmer: Mr. Matti, Dharwad, North Interior Karnataka**

The IMD extended range forecast, released on September 16th, 2015, predicted that there would be a revival of rainfall in the last week of September. Farmers were advised to prepare for the harvest of their groundnut crops after the anticipated rainfall. The revised extended range forecast, shared in the third week of September, predicted a revival of rainfall starting from October 1st. This four-day window provided an opportunity for farmers to prepare their fields by plowing them in preparation for sowing *rabi* crops.

(ii) **Farmer: Mr. Anand Joshi, Vijayapura, Bijapur, North Interior Karnataka**

According to IMD's extended range forecast on August 15, 2015, there was a prediction of rainfall during the last week of August, prompting the farmer to make plans for sowing sunflowers. The farmer decided to cancel the sowing of sunflowers after the modified extended-range forecast on August 22, 2015, indicated that the expected rainfall would occur between September 4 and September 13, rather than at the end of August as previously anticipated. As a result, land preparation for sunflower cultivation was also halted. He made changes to the plan for planting sorghum and chickpea, and on September 5th, there was rainfall, so he proceeded to sow the sorghum and chickpea crops. According to the extended range forecast on September 16, 2015, there was a prediction of rainfall in the last week of September and early October for horticulture crops as well. The farmer made the decision to avoid pruning the grapes in September in order to prevent the occurrence of downy mildew disease. It has been observed that approximately 80% of grape orchards that are pruned in September end up being affected by the disease, primarily due to rainfall that occurs in the first week of October. The KVK in Vijayapura (Bijapur) provided an extended range forecast on August 28th, 2015, indicating that there was an expectation of rainfall between September 4th and 13th. Farmers made preparations for the cultivation of sorghum, chickpea, and safflower seeds, among other crops. The land was prepared and seeds were obtained and treated. Sowing was carried out between the 10th and 24th of September, following rainfall on the 5th of September 2015. Sorghum was sown on 10 acres (3000 ha), chickpea on 12 acres (5000 ha), and safflower (A-1 variety) on 500 ha. There was no rainfall from the second week of September until the first week of October 2015.

(iii) **Farmer: Mr. G. Nageswara Reddy, Yagantipalli village, Banaganapalle mandal, Kurnool district, Andhra Pradesh**

The farmer has taken up land preparation utilising the rainfall (62.4 mm) received during the first week of June 2021. The redgram crop was sown during the last week of June 2021 as per the rainfall forecast. Later, during July and August months, the fertilisers application and weeding operations were taken up on rain free days as per the agro-advisories, thereby facilitating the farmer to apply fertiliser at the right time in the right dose at the right place to meet the crop nutrient requirement. During September month, the farmer prepared drainage channels in view of continuous rains as per extended-range forecast. As forecasted, a rainfall of 85 mm was received during 7 rainy days during the month and the farmer was able to drain out excess water from fields through drainage channels and prevented crop damage from waterlogging. Later, during II and III weeks of October, the farmer was given forewarning about the incidence of the spotted pod borer as the prevailing weather conditions were congenial for the pest development, the farmer followed this agro-advisory and controlled the pest effectively during the early stages itself by spraying pesticide twice at seven days interval and protected the crop from the pest menace. Further, the farmer harvested the crop as per the forecast and prevented the crop loss by timely harvesting, which resulted in higher pod yields besides good quality seed. By adhering to the agricultural recommendations, the farmer was able to efficiently organize their farming activities, resulting in a seed yield of 1400 kg/ha. This led to a net income of Rs.76,000/ha and a B:C ratio of 3.68. On the other hand, if the farmer had not followed the agro-advisories, the seed yield would have been 1000 kg/ha, resulting in a net income of Rs.41,500/ha and a B:C ratio of 2.23.

(iv) **Farmer: Mr. Arup Mandal, Nishchinta Village, Narayangarh Block, Paschim Midnapore District, West Bengal**

Mr. Arup Mandal cultivates Aman paddy (2–10 ha), potato (2–10 ha), and strawberry (0–2 ha) in *kharif* and *rabi* seasons with an irrigation facility. He prepared the land under Aman paddy using the rainfall received in the third week of June 2021 and following the agro-advisories issued from AMFU Kharagpur. Receiving the rain forecast and agro-advisory timely, he saved the paddy seedlings by pumping out the excess rainwater from the seedbed. After 21–22 days, he completed the mainland preparation. During planting, he applied fertiliser on the mainland following the weather forecast and advisory. Subsequently, in July and August, as per the agro-advisories, fertiliser application and weeding activities were undertaken on rainy days, which helped to apply the fertiliser in the right quantity at right time to meet the need of the crop. Necessary advice was given to control the pests and disease when the weather was favourable for the infestation of insects (brown sucking insects, beetles, caterpillars, etc.) and the prevalence of diseases (brown spot, blight, blast, etc.). Mr. Mandal followed the advisory properly, hence was able to protect the rice crop from pests and disease infestation. In November, in a rain-free environment, following the agro-advisory bulletin, he harvested the paddy. According to him, the nature of rainfall is the most important factor in the cultivation of Aman paddy. He added that the weather-based agro-advisory of AMFU Kharagpur was very helpful in seedbed preparation, planting, fertiliser management, pests and disease control, and harvesting of paddy grown in *kharif* season. By adhering to the weather-based agro-advisory bulletin, Mr. Mandal was able to achieve an extra 150 kg of paddy yield per bigha during that particular year. Following the weather-based agro-advisory bulletin, he earned an additional Rs. 750 per bigha from paddy cultivation and Rs. 615 per bigha (1 bigha = 1/3 acre) from potato cultivation. Overall, Mr. Mandal had an additional income of about Rs. 15,000 from his entire farmland in that year. Following the weather-based agro-advisory bulletin, he realised that crop damage could be reduced to a large extent. He showed his positivity about Meghdoot mobile app and WhatsApp messages being helpful in weather-based agro-advisory service for better crop planning and management.

10.4.2 World

10.4.2.1 Example 1: Early Warning of Winter Seasons for Greenhouse Crops in China

According to Stigter et al. (2008), the adoption of plastic greenhouses in northern China was accompanied by a lack of available services for greenhouse farming. Vegetable growers inside greenhouses have been experiencing difficulties due to low temperatures and limited sunshine hours, which are hindering their production. Implementing an early warning system could greatly benefit these farmers by alerting them to unfavourable conditions in advance. The fully climate-controlled greenhouse typically includes temperature, humidity, and light controls, but it is more expensive than basic greenhouses. On the other hand, simple plastic greenhouses are largely influenced by external weather conditions, with limited human intervention. As a result, these greenhouses experience high humidity, reduced sunlight exposure, and inadequate ventilation during the winter season. The challenges that arise in cultivating indoor fruits and vegetables are directly influenced by the availability of heat and light. If there is a prolonged period of low temperatures and insufficient sunlight, the temperature will decrease and humidity levels may rise, which is not ideal for the optimal growth of vegetables. These extreme conditions also create a favourable environment for pests and diseases, which can further impact the health and development of the vegetables. The weather forecast services offer valuable information to both the government and farmers regarding days with minimal sunshine and potential weather disasters. This forecast data is accessible to farmers through various delivery channels, such as V.T.V. weather forecast programs, telephones, and meteorological websites. The relevant warning details and service offerings regarding an adverse incident are promptly shared with agricultural management entities and agricultural websites, which are then disseminated by the agrarian authorities through the internet. This helps in providing guidance to farmers on how to prevent and minimize the consequences of the event.

10.4.2.2 Example 2: Agrometeorological Service for Irrigation Advice in Cuba

The research conducted by Stigter (2010) examines the provision of irrigation advice through agrometeorological services in Cuba. These services were developed based on input from farming communities and irrigation specialists from agricultural organizations in Villa Clara. The aim of the advisory service is to promote responsible water resource management and encourage farmers to optimize water usage according to the specific needs of their crops at different stages of growth. The agrometeorological service for irrigation plays a crucial role in assisting farmers with timely crop irrigation. Information regarding plant water consumption is provided every ten days, and this valuable service is transmitted through local radio broadcasts at specific times. The forecast for the upcoming ten days includes predictions for evapotranspiration based on the evaluation of the previous ten-day period's evapotranspiration. The FAO-recommended Penman/Monteith formula was utilized to calculate evapotranspiration, taking into account the resistance values and default coefficients specific to Cuba's standardized conditions. The Provincial Government Meteorological Service provided policy support for this capacity building initiative, which aimed to develop participatory irrigation advice as an agrometeorological service.

10.4.2.3 Example 3: Opportunities and Constraints for Farmers of West Africa to Use Seasonal Precipitation Forecasts with Burkina Faso

The study conducted in Burkina Faso's three agro-ecozones aimed to assess the effectiveness of seasonal precipitation forecasts for farmers and their potential impact on crop production and resource management. The study aimed to determine farmers' interest in and ability to use these forecasts, as well as identify the need for effective dissemination, interpretation, and application strategies to further improve agricultural practices. The study focused on three different agro-ecozones: the cotton-based system in the Sudan area of southwest Burkina Faso with relatively high rainfall, the sorghum and millet-based system in the central plateau with low rainfall, and the cattle-based system in the Sahel area in the north with very low rainfall. The farmers in the central plateau are expected to benefit the most from the study, while the cattle herders in the Sahel area are expected to benefit the least. The farmers in the selected agro-ecozones showed a strong desire to receive forecasts for seasonal precipitation. Their main interest was in knowing the timing and duration of rainfall, as well as any potential interruptions. To effectively disseminate these forecasts, it is recommended to include them as part of an extension package that also covers the probabilistic nature of forecasts, possible response strategies, and risk management. Additionally, it may be necessary to provide farmers with better access to basic agricultural technologies, such as improved crop varieties, plows, and fertilizers, in order for them to fully benefit from the rainfall forecasts.

10.4.2.4 Example 4: Farmer Forecasts: Impacts of Seasonal Rainfall Expectations on Agricultural Decision-Making in Sub-Saharan Africa

The fluctuating seasonal climate patterns in agricultural systems have an impact on crop yields, leading to reduced income for farmers. As a result, farmers are compelled to make decisions that are influenced by the climate, such as choosing early or late-maturing seed varieties, determining the optimal sowing time, and selecting a diverse range of seed varieties for planting. In order to investigate this phenomenon, a study was conducted by Guido et al. (2020), which involved both in-person visits and phone surveys of 501 smallholder farmers in central Kenya. The study focused on the 2018 growing seasons and the farmers' expectations for the upcoming March-April-May growing season in 2019. The findings of the report indicated that farmers' predictions regarding the upcoming rainfall for the season have a notable impact on the number of maturing varieties they plan to plant and the types of seed maturity varieties they choose. However, the influence on the timing of seed planting is not as significant. Additionally, the study revealed that approximately 79% of farmers develop expectations about future climate conditions, with around two-thirds of them relying on past weather patterns to make these predictions. However, a concerning discovery was made that one-third of farmers base their assumptions about rainfall timing solely on the previous season. Based on our current knowledge of climate variability, there is no evidence to suggest a correlation between El Niño events and seasonal rainfall. This lack of correlation poses challenges for farmers who rely on anticipating rainfall patterns for crop diversification and selecting drought-resistant cultivars. It emphasizes the importance of farmers' expectations and understanding of upcoming seasonal climate in making informed decisions. To support the farming community, it is crucial to have a comprehensive understanding of all aspects of climate variability in order to develop effective decision-making systems.

10.4.3 India's Integrated Agro-Meteorological Advisory Service Program from a Farmer's Perspective

The report by Venkatasubramanian (2014) encapsulated the results of the Consultative Group on International Agricultural Research (CGIAR) program on Climate Change, Agriculture, and Food Security (CCAFS) that commissioned an evaluation of India's Integrated Agrometeorological Advisory Services (AAS). The evaluation of India's Integrated Agrometeorological Advisory Services was conducted by the International Crops Research Institute for the Semi-Arid Tropics (ICRISAT), Climate change, Agriculture and Food Security (CCAFS), and India Meteorological Department (IMD) in June-July of 2012. The assessment involved gathering information from 132 individuals in 18 villages across six states through focus groups and individual interviews. The purpose was to understand how the AAS advisories were utilized, identify any issues or gaps, and gather suggestions for improvement. The assessment revealed several important findings, but the most significant factor in most cases was the varied approach to communication. Villages that utilized multiple channels to share AAS information, including meetings with agricultural extension officers, SMS and voice messaging, farmers clubs, local knowledge centers, and regular announcements or warnings over microphones, were well-informed about the AAS advisories and found them useful. Farmers particularly appreciated the training and discussions with extension officers as their preferred method of receiving information. For shorter discussions, it is easier to address the matter at hand. However, for larger issues, it is crucial to ensure and clarify broad representation in the discussion. The villages that saw the most utility and benefits were those where women were fully engaged in receiving and disseminating advisory information. In general, women had less awareness of AAS compared to men, highlighting the need to target women and provide information that meets their specific needs in communication efforts. The farmers also suggested implementing specific training and discussion programs for women farmers in their villages to improve interaction. Increasing membership in farmers' groups could also help expand awareness of the information provided by AAS. To maximize the impact of AAS information, it would be beneficial to target existing local groups for extension services.

10.5 Specific Lessons from Farmers in India

10.5.1 Grape Cultivation in Vijayapura District of Karnataka

According to Bal and Minhas (2017), Shri Gurupadappa Shivalingappa Biradar, a small farmer in the Vijayapura district of Karnataka, decided to cultivate grapes on 2 acres of land. He invested around Rs. 1.45 lakhs to construct a telephone trellis system for the grapevines. Initially, he expected to harvest approximately 20 tonnes of grapes and earn a net income of about Rs. 1.5 lakhs. However, due to the cold waves during the winter season, he experienced satisfactory yield and net income in subsequent years. The crops suffered from downy mildew disease due to the cold temperatures, resulting in significant losses in yield and increased expenses for plant protection measures. The farmer managed to harvest 16.6 tonnes of crops from 2 acres of land, but it cost them approximately Rs. 1.49 lakhs. The farmer became aware of the Agromet Advisory Services (AAS) provided by the All India Coordinated Research Project on Agrometeorology (AICRPAM) center in Vijayapura through its networking centers, ICAR-CRIDA and IMD. He learned about the service from his fellow farmers and registered his mobile number to receive the AAS bulletin and access the service. In the following season, he received advance information about the expected drop in temperature through the Agrometeorological Advisory Service (AAS) and learned how to maintain the temperature in his field to protect his crops from cold and frost conditions. Thanks to the AAS service and early weather information, Gurupadappa achieved a yield of around 19 tonnes in the next season while reducing his cultivation costs. The successful implementation of the AAS service inspired many other farmers to adopt it for their own benefit. As a result, almost all farmers in the village now provide their mobile numbers to access the agro-met advisory service.

10.5.2 Farmers Feedback About the Agro-Met Advisory Services at Mahasamund District of Chhattisgarh

The study conducted by Gandhi et al. (2018) aimed to assess the effectiveness and usefulness of Agrometeorological Advisory Service (AAS) in relation to Climate Change. The research was conducted in specific villages of the All India Coordinated Research Project on Agrometeorology-National Innovations on Climate Resilient Agriculture (AICRPAM-NICRA) Project, located in the Mahasamund district of Chhattisgarh. Initially, the farmers were asked about the medium and source of weather information they received. The farmers were divided into two groups, AAS and non-AAS, based on their access to weather information. AAS farmers used mobile phones and newspapers to stay informed, while non-AAS farmers relied on TV, newspapers, and personal contact with other farmers. According to the feedback from farmers, 46% rated the NICRA services as excellent, while 20% rated them as very good. Overall, farmers expressed satisfaction with AAS in terms of fertiliser application, plant protection, and timing of field operations, but felt that advisories on water management could be improved.

10.5.3 Usability of Extended Range and Seasonal Weather Forecast in Indian Agriculture

Weather forecasts play a crucial role in the agricultural system of India, as they enable farmers to effectively manage their crops and increase productivity. Providing advanced information on weather conditions can help farmers make informed decisions and take appropriate measures to protect their crops. A comprehensive study conducted by Chattopadhyay et al. (2018) highlights the importance of extended range and seasonal weather forecasts in assisting farmers. Additionally, the study explores the application of field measures and operational tools to mitigate weather-related uncertainties through agro-meteorological applications. The India Meteorological Department (IMD), under the Ministry of Earth Sciences (MoES), already offers Agro-meteorological Advisory Services that utilize Medium-Range Weather Forecasts (MRWF) to assist farmers in optimizing the use of natural resources. To adapt the agricultural system to increased weather variability, it is crucial to provide farmers with climatological information combined with extended range and seasonal weather forecasts. In collaboration with the Indian Institute of Tropical Meteorology (IITM), Pune, the IMD has begun preparing extended range and seasonal weather forecasts. These forecasts are used to create Agro-met Advisories, which are provided fortnightly and for the entire season to farmers across the country. The feedback received from farmers emphasizes the necessity of prior information on weather forecasts at extended and seasonal scales to effectively plan their farming operations. This feedback highlights the importance of these forecasts in making informed agriculture decisions.

10.5.4 Farmers Feedback About the Agro-Met Advisory Bulletin: A Farm Broadcast on All India Radio, Kolhapur, Maharashtra

The All India Radio in Kolhapur began producing and airing their own agricultural program on April 1, 2005. A team of scientists from the National Agricultural Research Project (NARP) in Kolhapur created a weather forecasting message, which was sent to All India Radio and broadcasted during the "Kisanwani" program every Wednesday at 7.30 pm. A study conducted in the Kolhapur district, based on feedback from farmers, found that 94.78% of respondents were receiving weather information from the radio. However, the majority of them rated the Agro-met Advisory Bulletin as satisfactory to good. Therefore, it is important to enhance the radio's capabilities, including the Agro-met Advisory Bulletin, as a reliable source of weather information and improve its rating to excellent. To achieve this, the Agro-met Advisory Bulletin should be broadcasted twice a week and repeated 2 to 3 times a day for better implementation.

10.6 Benefits of the Use of the Weather Forecasts

10.6.1 Benefits of Weather and Seasonal Forecasts to Millet Growers in Niger

The study conducted by Roudier et al. (2016) aimed to evaluate the effects of 10-day and seasonal forecasts on cropping practices and income for millet farmers in Niger. They utilized the SARRA-H crop model, along with an economic model that simulates the selection of cropping strategies. The study considered risk aversion and focused on two types of farmers with varying adaptation capacities. The results indicated that both the 10-day and seasonal forecasts were advantageous for Nigerian farmers, particularly those with more adaptation strategies. However, the farmers' limited knowledge and skills regarding the forecast system remained a significant obstacle. As a result, the current abilities of Nigerian farmers led to losses in approximately 20% of cases, despite the positive impact of weather forecasts.

10.6.2 Economic Impact of the Agrometeorological Advisory Service of India

A study conducted by Maini and Rathore (2011) assessed the economic impact of weather forecast-based advisories issued to 15 out of 127 AAS units of the Ministry of Earth Sciences (MoES) under the Government of India. The study analyzed six seasons, including three *kharif* and three *rabi*, from 2003 to 2007. The selected crops for the study included food grains, cash crops, oilseeds, fruit, and vegetable crops. The sample-set consisted of 80 farmers, with 40 responding and 40 non-responding farmers. The main objective of the study was to examine the percentage increase or decrease in yield and net return resulting from the AAS. The findings indicated that AAS farmers experienced a net benefit of 10–15% in overall yield and a reduction of 2–5% in cultivation costs compared to non-AAS farmers.

10.6.3 Direct Use of Weather Information for Agricultural Decision-Making

Study conducted by Frisvold and Murugesan (2013) utilized data from a subset of the National Agricultural, Food, and Public Policy Preference Survey to evaluate the utilization of weather data in agricultural decision-making. The study analyzed responses from 284 farmers and ranchers in Arizona. Through the development of a demand model for weather information, count data and discrete choice multivariate regression models were employed. The findings of the study indicated that the significance of weather data was higher among diversified producers, who were more inclined to utilize the data for determining the optimal timing of planting, cultivation, and harvest. Conversely, producers with a greater reliance on off-farm income exhibited lower usage of weather data. The study also revealed that access to satellite television had a positive impact on data utilization, while access to the internet did not contribute to an increase in the use of weather data.

10.7 Limitations of the Weather Forecasts

- A participatory approach is crucial in ensuring that farmers have ownership of the processes related to weather and climate information, which is essential for effectively linking forecasts to farm decisions. This approach allows farmers to actively engage in the decision-making process and facilitates the integration of weather information into their decision-making processes.
- There is a need for the development of suitable interdisciplinary systems that can effectively connect weather and agronomic information with farm management.
- Pre-seasonal forecasts must have a validity period of at least 10 days and not less than a week, as it takes some time to organise and execute any farming strategy for making high investment decisions. The condition of the atmosphere must be known at a specific time with sufficient accuracy.
- It is important to consider the provision of climate change scenarios and trend information to assist in making long-term strategic decisions for farm management, particularly in areas with farming zones that are susceptible to vulnerability.
- To counter the effects of forecasted hazardous pests, diseases, weather, etc. The control measures in the field are time taking. Some of the measures like control operations against pests and diseases, pre-seasonal agronomic corrections, and supplementary irrigation, will be high-cost decisions. Hence, mid-seasonal forecasts must preferably be broadcasted five days in advance.
- Training farmers in agricultural skills is crucial for them to effectively understand and utilize weather and climate forecasts. It is important to prioritize the development of farmers' abilities to adapt to unpredictable conditions by enhancing the versatility of existing resources.

10.8 Take Home Note: Elective Bridge Between Farmers Need and Weather Advisory

Agriculture is a wide range of sectors covering different sub-sectors like crop varieties, seed varieties, livestock, fisheries, etc. The requirement of farmers varies from region to region worldwide. Overall, we can summarise the farmers' issues regarding their expectations and needs from weather forecast systems. It is more convenient to improve the weather forecast system according to the farmers' use as they are the actual users and producers. Farmers are constantly making decisions based on weather conditions, as the primary components of the farming system are heavily influenced by the weather.

Crop growth/irrigation: Farmers rely on crop yield as it directly impacts their income in the market. The growth of crops is crucial in determining the yield, making it vital to address the various factors that influence crop growth. Farmers heavily depend on accurate and reliable weather data to make informed decisions about their crops. This includes historical, real-time, and forecast weather information, which helps them track the growth stage and make important choices regarding irrigation timing and quantity. Having access to precise weather data is essential for farmers as it guides them in making significant and potentially costly decisions that can impact their overall crop yield and financial success.

Pest and disease control: The growth and development of pests and diseases are mainly encouraged by specific weather conditions, destroying crops. Farmers are facing lots of issues while controlling these, and sometimes it becomes uncontrollable. The integration of weather forecast guidance into pest and disease modelling enables the identification of the optimal timing and dosage of insecticides or pesticides for effective control.

Fertiliser timing and delivery: Incorrect application of fertilizer due to weather conditions can lead to the loss of the entire field's profits. Utilizing precise weather forecasts can help ensure that fertilizer is applied under suitable conditions, whereby it is neither washed away nor left on dry soil.

The weather forecast system plays a crucial role in assisting farmers in making timely decisions for their crops. However, there is a need for a slight adjustment in the system to enhance the communication between agriculture extension officers and farmers. A study conducted in the Southeast USA aimed to determine the economic benefits of climate forecasts for farm management decisions. The study found that having perfect knowledge of daily weather for the upcoming season could potentially result in an expected value of approximately US$ 190 per hectare. However, they observed a number of obstacles in achieving these advantages, as these challenges are typically associated with the unpredictability of climate predictions and the complexities of agricultural systems (Jones et al. 2000). Having accurate forecasts that extend 3–6 months into the future can enable farmers to make informed decisions to mitigate adverse effects or capitalize on expected favourable weather conditions.

Changing both the farmers' and societal beliefs and perceptions of forecasts will have a significant impact on their use and influence, as observed by Artikov et al. (2006). This suggests that forecast originators should prioritize altering farmers' perception of forecast accuracy by providing easily understandable forecasts through widely accessible and trustworthy media in a timely manner. Farmers' primary goal is to increase their farm income and improve crop production to ensure the financial viability of their farms. To make climate forecasts more influential and beneficial for farmers, it is important to develop practical frameworks that can be applied in specific locations. By providing early forecasts before the crop growing season, farmers can plan their agricultural decisions, such as selecting the appropriate crop and seed varieties. This planned behaviour can help enhance the utilization and impact of climate forecasts in agricultural decision-making. Enhancing women's access to climate information services and building their capacity to manage climate risks is crucial for improving the effectiveness of seasonal forecasts in supporting farmers' decision-making, particularly in acquiring crop insurance policies. The study conducted by Raj et al. (2020) aimed to strengthen the institutional processes of capacity-building, specifically targeting women, to ensure the integration of climate information into agro-advisories and mitigate risks in agricultural production systems. The research proposed strategies to enhance the utilization of climate information and decision-making among women farmers. Providing them with tailored news and agricultural advice, adopting a group-based approach, training local communicators, and offering support to implement the information could assist women farmers. By gaining knowledge and understanding of advisories, women farmers can comprehend the scientific basis for their actions. The research suggests that capacity-building programs at various levels of the information chain could be an effective strategy to address gender-related challenges in agriculture. These programs would specifically target women farmers, who play a significant role in the agricultural sector. By equipping them with the necessary knowledge and skills, they can make informed decisions and adopt climate-smart practices in the face of increasing climate variability and change. This would ultimately lead to improved farming outcomes and resilience in the agricultural sector.

Chapter 11
Hands-On Practice

11.1 Introduction

The present book focusses on the meteorological forecast system that is aimed at providing monthly and seasonal scale forecasts of rainfall and surface air temperatures (hereafter referred to as only temperature), one-month in advance (discussed in Chap. 6) over the targeted regions for climate risk management during agricultural purposes (discussed in Chap. 9). Foremost, it is suggested to examine the raw products of each GCM output. For this, climatological knowledge of the targeted regions is useful. For example, consider the observed climatological seasonal mean temperature over a particular region is 7 °C during the winter season with a standard deviation of 2.1 °C and a model temperature climatology and standard deviation are 9.5 and 1.8 °C, respectively for the winter season over the same region. Now, if that model is predicting seasonal mean temperature for the forthcoming winter season of about 23 °C, then it is essential to verify that model products (it may be owing to a data downloading problem, interpolation problem, or a problem in model dynamics; for the first two, one need to cross verify the results).

For generating reliable monthly and seasonal scale forecasts, different statistical approaches are needed to be employed at several stages on GCMs products. It is necessary to examine the efficacy of GCMs and estimate the systematic bias of each GCM. The details of a few bias correction techniques that are standard and effective, and widely used have been discussed in Chap. 6. It is strongly suggested to the user that each bias correction technique should be applied to remove the systematic bias, and inter-comparisons among the bias-corrected results will provide guidance to choose a suitable bias correction technique for the prediction of rainfall and temperatures for the targeted region. In the next, various multi-model statistical methods can be applied to the selected bias corrected GCMs outputs to improve the monthly and seasonal scale forecasts. An integrated method combines all the products from different statistical approaches to generate a single forecast.

U. C. Mohanty et al., *Climate Risk Management in Agriculture*,
https://doi.org/10.1007/978-3-031-51862-1_11

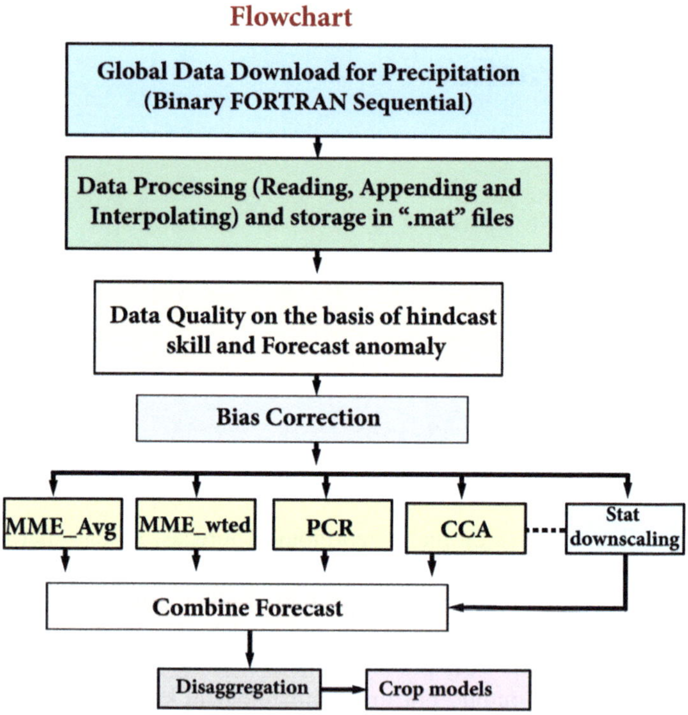

Fig. 11.1 The schematic diagram of the generation of the seasonal forecast procedure. Storage in ".mat" files in the diagram is an example to store the data in MATLAB format, however, the user may choose different format such netCDF, wgrib, wgrib2, ASCII, binary, or in any format as per their requirements/choice

Next, it is necessary to disaggregate monthly/seasonal scale forecasts into the daily scale to provide inputs to crop models. The details of the disaggregation methods and usability of the suitable technique have been discussed in Chap. 7. The user can use MATLAB, python, or other programming languages to develop and carry out the hands-on practice; however, the user needs to write their own programming code based on the next description to follow the processes. The flow chart of the generation of a single forecast and the use of disaggregation is provided in Fig. 11.1.

11.2 Components in Generating Forecast

In order to generate forecasts using multi-GCMs and various statistical approaches, several steps are involved, which are provided below.

11.2.1 Obtain Observed Data

The observed (weather/climate variables/parameters) data for a longer period (such as 30 years or more) is needed for understanding, developing a statistical model and model verification, model bias corrections, etc. The observed data may be station data, analysis data, or reanalysis data. The observed data may be available at different time intervals, i.e., hourly, daily, monthly, or so on. After collecting it, 'quality control' of the observed data is recommended. The errors in the observed data may be due to observational, and /or instrumental, and/or transmission. A number of quality control checks are used, and some of the widely used are (i) format code, (ii) internal consistency within one observation (like radiosonde observations of different layers), (iii) temporal consistency on observations from the same source, (iv) closeness to climatology (gross-check), and (v) consistency with an observation from surrounding observatories (buddy check). The brief details of the procedure of quality control checks can be found in Chap. 4, and for more details in https://library.wmo.int/docnum.php?explnumid=9869 and https://citeseerx.ist.psu.edu/viewdoc/download?doi=10.1.1.620.6457&rep=rep1&type=pdf. If the observed data are already available after quality control, then the user may ignore following the quality control of the observed data. The user then needs to convert the quality checked observed data at monthly/seasonal average (precipitation or temperature) or monthly/seasonal total (precipitation). The prepared observed data at a monthly/seasonal scale can be used as reference data (observed data, verification data, predictand field, or targeted data).

11.2.2 Downloading CMs Data

As discussed in Chap. 5, the user needs to download GCM data from the available source. The GCMs data are available in different formats such as grib, netCDF, ASCII, binary, etc. It is necessary to download each model variable separately and store each model data individually. It may be noted that the user needs to have the model and observed data for a similar period and over the same location, except for the forecasted data of the model. For example, suppose one has downloaded observed data for the period of 65 years from 1951 to 2015. Next, if the model data are available for the period of 41 years say from 1981 to 2022 (where the forecasted year is 2022), then the user needs to consider the common period (1981–2015) of the observed and model data to use for the training and test period.

11.2.3 Pre-processing of GCM Data

The GCMs data are mostly available for the global or continental region. In addition, the spatial scale or horizontal scale varies model-to-model. Thus, the user needs to translate the model data so that it will coincide with the observed horizontal resolutions/station points. For example, the user may use a suitable interpolation technique, such as bilinear interpolation, nearest neighbour, weighted interpolation techniques, etc., to interpolate the model data to the observed grid points. The user is advised to verify the interpolated data with raw data to understand whether the generated data have been produced/translated appropriately. The following points can be checked immediately after the interpolation:

(i) The spatial pattern of climatology, standard deviation, coefficient of variation, etc., of the raw and interpolated data, are consistent.
(ii) The time series (hourly/daily/monthly/seasonal) of the raw and interpolated data (station level/random grid points and area average) are consistent. The user may consider checking the standardised anomaly index of both data sets to confirm the check.
(iii) It will be better if the user further uses some advanced techniques to compare the interpolated and raw data.
(iv) After successfully reproducing the model data at observed locations, the user needs to extract the model data for the domain exactly the same as the observed data domain/station points.

Next, the interpolated data should be converted at the same time interval as presented in the observed/reference data. For example, if the reference data is prepared at a seasonal time scale (average over a season), the interpolated model data should also be converted at a seasonal scale. The interpolated data will then be considered and will be referred to as a model, and the model raw data will not be considered for the practices/generation of the monthly/seasonal forecast. It is strongly advised to the users/readers to follow the procedure discussed in Chap. 6 for evaluating the performance of GCMs, which will instantly give a glimpse of the efficacy of the model. Foremost, examining the raw products of each GCM output is suggested. For this, knowledge of the targeted region is useful for the climatological (Observed and model climatology and standard deviation). Several methods can be used to verify the model's pre-processed data (refer to Chap. 5). However, as stated earlier, spatial patterns, standard deviation, temporal skills, etc., can be evaluated to understand the performance of the model data.

11.2.4 Applying Bias Correction Methods

It is necessary to understand a few steps before entering into the bias correction methods. First, please check the common period of observed and model data. If the data length is sufficiently large, then the user can split the whole common period data into two periods, one for the training period and the other for the test period. For example, the user may consider doing the exercise for the Indian summer monsoon rainfall. The summer monsoon spans from June to September

(JJAS), thus, JJAS seasonal mean/total rainfall data from the model and observation are needed and to be obtained for the historical period (common period). Now, if the common period data length is around 60 years or so (i.e., seasonal mean/total rainfall data at each grid point/station level are available for the past 60 years or so), then this data length is quite good to split into two periods having length each period 30 years for training and test. Let us consider that the model and observed summer monsoon rainfall data are available for 60 years period (1961–2020); thus, the data length is 60 (each grid point/station will have 60 data), and one can consider the first 30 years (i.e., 1961–1990) for the training of the bias correction, and the method can be tested for the next 30 years (1991–2020). However, many global model generated products are available from 1981 or its nearby years, because of the improvements in the observations owing to the incorporation of satellite data (satellite era). Thus, for those models, the data are available since 1981, and the past data are available say up to 2020; thus, for training and testing using statistical techniques, a common period for observed and model data is only up to 2020, i.e., 40 years. This 40-year's data length is insufficient to split into two periods since each period will have 20 years of data and is inadequate to represent the climatological features and train the statistical model. In this case, one can consider the leave-out cross-validation method. In this method, the test year should be removed from the whole common period of the observed and model data and train the model and generate training coefficients which will be used for the prediction of the test year. The below example may provide a glimpse of the cross-validation method. Here the example is provided for the leave-one-out cross-validation method.

Let's consider both the observed and model data are present for the period 1981–2020. Remove the 1981 year data from both the observations and model data. Thus, after removing 1981, observed and model data will contain the data for the period 1982–2020. Using the time series 1982–2020 of both the data, one can estimate the coefficient (C_{1981}) to remove the bias. Then this coefficient should be used to generate a bias-corrected prediction for the year 1981. Similarly, for the bias-corrected prediction of the year 1982, remove the 1982 data from the observations and model time series, then generate C_{1982}, and finally generate the bias-corrected prediction using C_{1982}. In this way, one can get bias-corrected model data from 1981 to 2020.

The user needs to analyse the bias-corrected data (test data or obtained through cross-validation mode) and compare it with the observations to understand efficiency of the bias correction method employed to remove the bias. It may be noted that the user can compare qualitatively and quantitatively (such as spatial plot climatology, standard deviations, standardised anomaly index, coefficient of variations); in addition, the user can also use several skill scores such as root mean square error, mean bias, correlation coefficient, Index of agreement, etc. presented in Chap. 5 to understand the robustness of the bias correction methodology.

Although the leave-one-out cross-validation method is widely used for short data length, one can use different leave windows (such as leave-three-out or leave-five-out) instead of leave-one-window. In leave-three-out, the user can consider leaving the predicted/test year, previous and after the predicted/test years, which will leave three years. In many practices, randomly, two years, along with the predicted/test years, are left instead of before and after the predicted/test years in the training period (Sinha et al. 2013). It is not recommended to leave more years from the training period since a lot of information may be missing in the training period, and the efficiency of the statistical model will be reduced. On the other hand, it is strongly recommended to leave the predicted year from the training series; else the training period data will have the observed information, which will tend to be closed with the observations after bias correction; hence, observed bias will be introduced in the training period, and hence in the coefficient.

In the bias correction methods, a number of techniques such as mean bias removal technique, multiplicative shift technique, standardised re-constructions technique (without transformations, WOTF), regression technique, and quantile–quantile mapping. Principal Component Regression (with transformations, WTF), etc. can be used and detailed descriptions with mathematical formulations of these techniques are provided in Chap. 6. It is necessary to apply each of the bias correction techniques on each GCM product, and then raw, and bias-corrected GCM products should be extensively compared with the reference data for identifying one suitable bias correction method.

For example, all the bias correction techniques are employed on the CFSv2 global model to remove the model's systematic bias and improve summer monsoon rainfall prediction. In this example, the Indian summer monsoon rainfall from June to September (JJAS) for the period 1982–2018 has been considered, and the raw and bias-corrected CFSv2 is compared with the IMD rainfall analysis. Before bias correction, the CFSv2 model data were re-gridded to the IMD grid points and used as raw data. After implementing the bias correction techniques, it is necessary to estimate the skill or quality of these techniques, i.e., which method reproduces data closer to the observations. In this example, all the bias correction techniques stated above are applied to the summer monsoon rainfall averaged over all-India. Model raw outputs (no bias correction; NBC) and the bias-corrected outcomes are evaluated with the observations. The quality of bias-corrected rainfall on a monthly scale is evaluated using the different skill-score tests viz. Climatological mean, standard deviation, root mean square error, index of agreement, correlation coefficient.

The Table 11.1 shows the skill scores of CFSv2 for JJAS all India rainfall for the period 1982–2018 without a bias-corrected model (NBC) and all bias correction techniques: mean bias remove technique (MBRT), multiplicative shift technique (MST), standardized reconstruction technique (SZT), regression technique (RT), quantile–quantile mapping method (QMM) and principal component regression (PCR) along with IMD observations.

The observed climatology for the study period (1982–2008) is 7.63 mm day^{-1}, whereas the climatology of NBC is 5.68 mm day^{-1}, which is much lesser than

Table 11.1 Skill score of raw and bias-corrected CFSv2 model JJAS rainfall over India using several bias correction techniques

Statistics	Obs	NBC	MBRT	MST	SZT	RT	QMM	PCR
Mean (mm day^{-1})	7.63	5.68	7.63	7.63	7.63	7.63	7.61	7.61
SD (mm day^{-1})	0.75	0.25	0.25	0.34	0.78	0.34	0.76	0.72
RMSE (mm day^{-1})		2.06	0.69	0.44	0.83	0.72	0.80	0.83
Index of agreement		0.73	0.45	0.58	0.65	0.48	0.66	0.60

the observation. All of the bias correction methods are able to capture the long-term mean or climatology. All the methods can adjust the long-term mean biases. Therefore, there is a significant improvement in removing the bias in climatology by each method. It is observed that the standard deviation of NBC is only 0.25 mm day^{-1} while the observed standard deviation is 0.75 mm day^{-1}. This reflects that the interannual variability of the model is much lesser than the observations. This example, therefore, reveals that although the climatology of all bias-corrected methods is the same as observation, their standard deviations (SD) differ from each other. The SD of the Mean Bias-remove Technique (MBRT) is the same as the raw (NBC) model, and the SD of the multiplicative shift technique (MST) is also much lesser than that observed. Among the WOTF technique, only the standardised technique (SZT) method simulates standard deviation (SD) correctly. It may be noted that only this method projects the observations. This may be because the SD is affected by the change of scale due to the transformation in the case of the regression method. Therefore, out of six methods, three methods, the standardised technique (SZT), quantile–quantile mapping method (QMM), and principal component regression (PCR), simulate the observed inter-annual variability well.

Root mean square error (RMSE) is used to calculate each method's prevailing accuracy explicitly with NBC. It is seen that the RMSE of the NBC is 2.06 mm/day. Among all the techniques, the multiplicative method has the least RMSE (0.44 mm day^{-1}). It is seen that the RMSE of the standardised reconstruction technique is higher than other techniques. It should be noted that the RMSE only estimates the average error, and there is no scope for measuring the relative size of the average difference between the model and observation. In the present example, the Index of Agreement skill metric is calculated for each bias correction method along with NBC. It is seen that NBC has a minimum score of 0.37. Among the WOTF methods, SZT has the highest skill (0.65), and QMM has the highest skill (0.66) among the WTF methods. Therefore, in view of the index of agreement, both SZT and QMM methods are the most skillful among all six bias correction methods. The user can use other comparison methods using the other statistical skill scores, time series, and spatial patterns to understand which method provides better results compared to the observations.

11.2.5 Employing Different Statistical Techniques Using Several GCMs

The details of several statistical techniques are presented in Chap. 6. Here a brief description of a few techniques is provided for the reader's quick reference.

- **SVD Based Regression**: This technique is based on the method of Superensemble. This is a multilinear regression method (using singular value decomposition) on models, and the predictand (dependent variables) is the observed Precipitation or Temperature.
- **Supervised Principal Component Regression**: The PCR approach can forecast precipitation and mean temperature over the targeted region. It is an EOF-based regression technique in which the predictors (independent variables) are different general circulation models' precipitation or temperature products while the predictand (dependent variables) is the observed precipitation or temperature. The model products are supervised based on their correlation in a hind cast.
- **Canonical Correlation Analyses**: In this approach set of basis vectors is found so that the correlation between projections of two multivariate data sets will be maximised. One can use the model-predicted precipitation/temperature data sets as the independent variable and observed precipitation/temperature as the dependent variable.
- **Probabilistic Forecast**: All these above-stated forecasts can be combined using some statistical techniques and delivered.
  The same forecast has been converted into a probabilistic forecast to see the chance of occurrence of a particular event. A probabilistic forecast is made in three categories viz. below normal, near normal, and above normal. These categories are based on the past observed data set. The model and observed data sets were normalized to have zero mean and one standard deviation before subjecting to any analysis. As the mean was shifted to zero and the standard deviation changed to unity, the bounds for the categories became as follows.
  Below Normal category: ∞ to -0.43.
  Near Normal category: -0.43 to 0.43.
  Above normal category: 0.43 to ∞.
- **Other Methods**: In addition to the above, the user has the flexibility to use several other advanced statistical approaches, such as artificial neural networks (ANN), fuzzy clustering-based self-organizing maps (SOMs), etc., for downscaling the GCMs information into the targeted regions. During the last few decades, technological advancement has significantly contributed in improving forecast skills. Recently, the evolution of artificial intelligence (AI)/Machine Learning process and its application on the weather/climate forecast has made a breakthrough in forecasting systems. The user can use any of the methods based on AI/machine learning using GCM products. However, we will not describe those processes here.

11.3 Procedure

(i) **Choice of predictor fields**

(a) The user may consider the targeted variable/predictor field (from GCMs) the same as the observation/predictand field over the targeted domain. For example, for precipitation prediction, GCMs precipitation data can be used as predictor fields.
(b) The user may also consider other fields/auxiliary variables of GCMs as predictor fields; however, it is noted that these predictor fields should have a physical relationship with the predictand field, and play a dominating role in the occurrence of that predictand field.
(c) The domain of predictor fields (if those are large-scale fields) may differ from the domain of the predictand field; however, the reason for considering that predictor field should satisfy the above point (point b).
(d) The identification of the predictor fields and associated domain can be found using a teleconnection map between the predictand predictor fields, keeping in mind that point (b) should be valid.

(ii) **Inputs**

(a) Reference data for the historical period (predictand field).
(b) Global model data for the same historical period and forecast year.
(c) Quality control and preprocessing of model data.

(iii) **Bias correction**

(a) Apply each bias correction method to individual model products.
(b) Compare the bias-corrected results with observations to find a suitable bias-corrected method.
(c) Employ each statistical technique described in Sect. 11.2.5 (SVD, PCA, CCA, etc.) on the outputs of bias-corrected (obtained through the suitable bias correction method).
(d) Use several advanced statistical skill scores to understand each statistical technique's robustness and uncertainty.
(e) Do the above exercise for all the models.

11.4 Generating Final Consensus Forecast

It is very clear from the above procedure that a user will have multiple GCMs raw products (GCM raw; quality controlled GCM products interpolated on the observed grid points/station level), bias-corrected GCMs products (GCM_BC), and products after applying several statistical techniques (e.g. three techniques MLR_SVD, CCA, and SPCR as described in Chap. 6) on the GCM_BC. It is already established that bias corrected GCM has a higher skill than the GCM raw. It is indeed interesting to combine the products obtained from the several statistical techniques to have a single prediction that will be robust and consistently provide a reliable and skillful prediction for the targeted region. This single forecast can be termed a Final Consensus Prediction (FCP). Several methods are used in practice for generating single forecast. Note that the individual forecast is used as input for generating a single forecast using the following methods. Some widely used are:

(i) **Simple average of all the available forecast**: In this method, the weight of each prediction is same. For example, if F_1, F_2, $\ldots F_n$ are the predictions using different techniques/models, then

$$F_{final} = \frac{1}{n}\sum_{i=1}^{n} F_i$$

The combination of the products from different models is usually termed as Multi-model ensemble (MME).

(ii) **Weighted average MME (MME_wted)**: Here, the weight of a model is higher than the other one if the skill of that model is higher in predicting the predictand variable in the historical period. Mathematically,

$$F_{final} = \sum_{i=1}^{n} \omega_i F_i$$

$$\sum_{i=1}^{n} \omega_i = 1$$

where, F_i and ω_i are the forecast of *i*th forecast and weight of that forecast.

In general, the weight can be estimated using different algorithms. One such is to examine the correlation between the model and observations; a higher correlation will have a higher weight. For example, consider F_1, F_2, F_3, F_4, F_5 forecast products have correlations values C_1, C_2, C_3, C_4, C_5 respectively with the observations for the past 30 year period, then

$$\omega_1 = C_1/(C_1 + C_2 + C_3 + C_4 + C_5)$$

$$\omega_1 + \omega_2 + \omega_3 + \omega_4 + \omega_5 = 1$$

Like ω_1, the other weight, i.e., ω_2, ω_2, ω_4, ω_5 can be calculated. One can consider the other statistical scores to set the weight for the forecasted products.

(i) Now the user can use MLR_SVD, CCA, and SPCR techniques (detailed in Chap. 6) on the processed GCM products and generate three different forecasts. These three forecasts (see Sect. 6.4.5.1) can be combined to produce one single Final Consensus Prediction through the weighted average, and the weight can be determined using the skill of the hindcast predictions of these products.

An example where the final consensus forecast has been generated over India (in $0.25° \times 0.25°$ latitude longitude grid points), then dissemination of the forecast has been made by averaging all grid points within a met-subdivision as represented in Fig. 11.2.

In Fig. 11.2, the maximum temperature anomalies have been found in the positive direction in most of the subdivisions except Coastal A.P., Rayalseema, Telangana, SIK, NIK, Konkan & Goa, Madhya Maharashtra, entire Gujarat, Punjab, Odisha, Jharkhand, Bihar, Gangetic WB, SHWB, Assam & Meghalaya and NMMT regions where the negative anomalies are expected. Moreover, strong positive anomalies are in the entire J&K region. The anomaly for the country as a whole is expected to be climatology (0 °C).

In Fig. 11.3, the probabilistic forecast reveals the possibility of above-normal maximum temperatures in Arunachal Pradesh and West M.P regions, while the normal category is in the CHD region. However, no clear signals are available for the remaining parts of the country. For the country as a whole, the chances are high for the normal category.

In Fig. 11.4, it is seen that during the summer monsoon period (JJAS), the entire country is expected to receive normal rains with about 96% of long-period average (LPA) for all India while deficit rains are expected in West Rajasthan, Himachal Pradesh and Uttarakhand regions. However, most of the subdivisions are going to experience the below climatological rainfall. In the case of the probabilistic

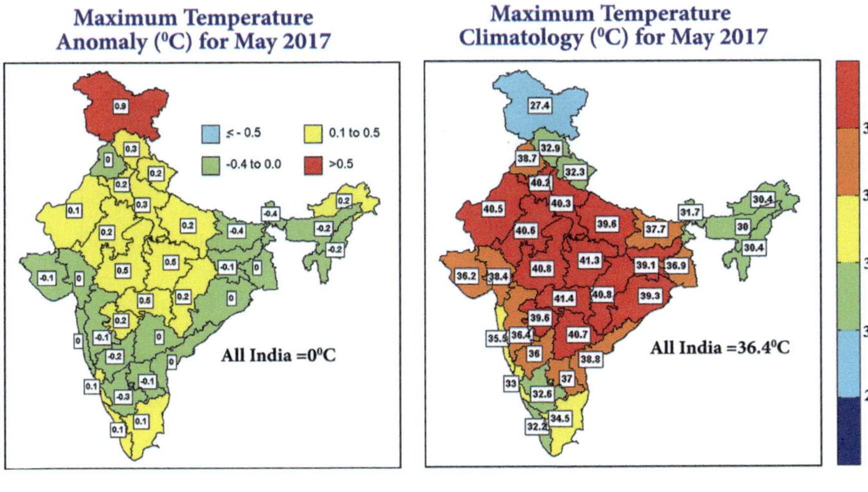

Fig. 11.2 Deterministic maximum temperature forecast for May 2017

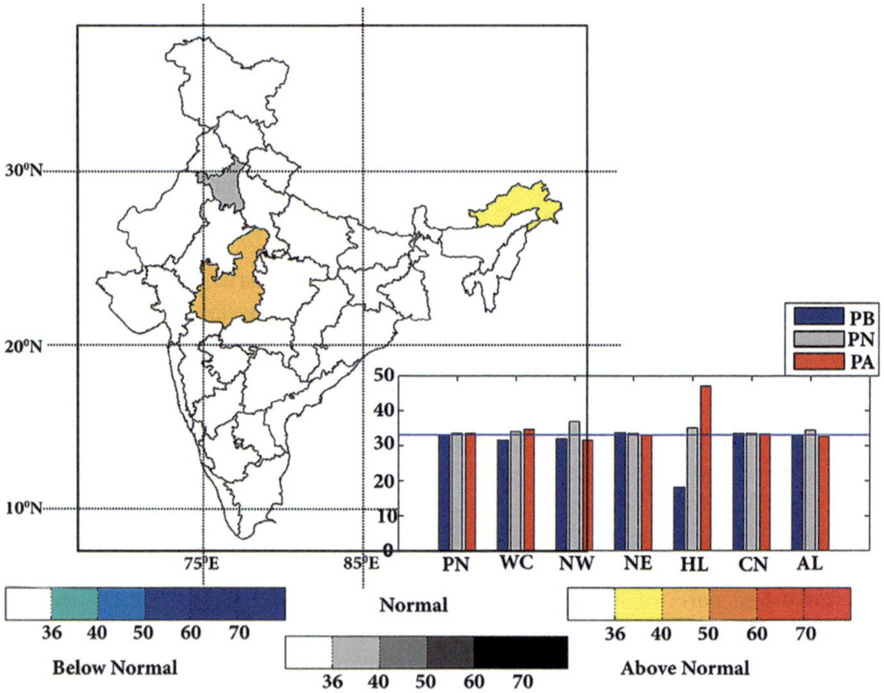

Fig. 11.3 Probabilistic maximum temperature forecast for May 2017

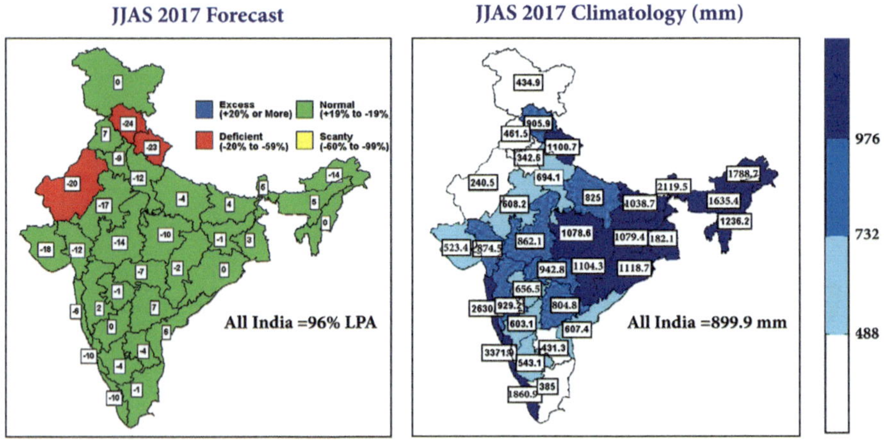

Fig. 11.4 Deterministic precipitation forecast (percent departure) for JJAS (lead 2)-2017

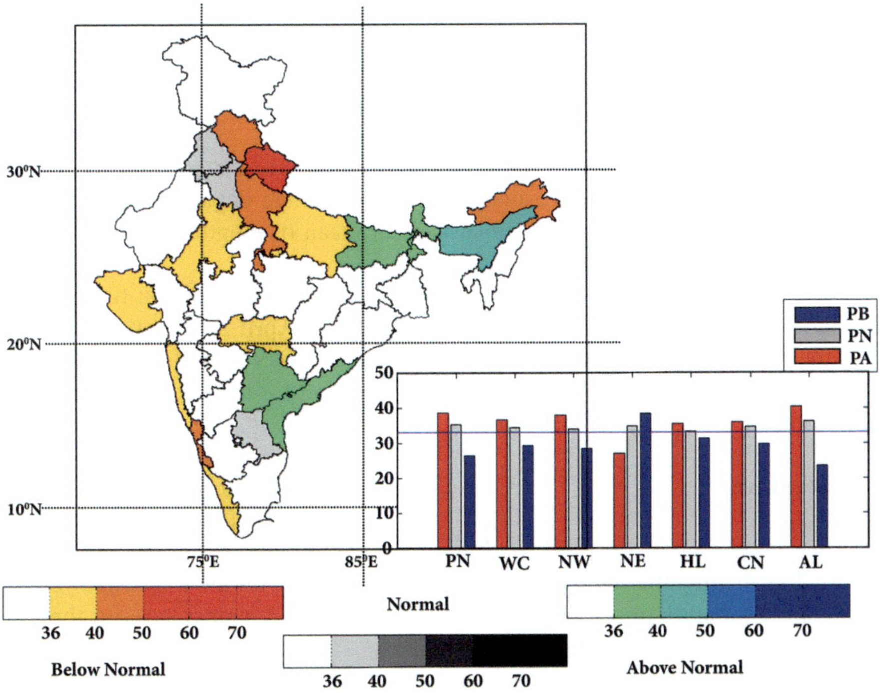

Fig. 11.5 Probabilistic precipitation forecast for JJAS (lead 2)-2017

forecast (Fig. 11.5), signals of above normal rains are accredited in Coastal A.P., Telangana, Bihar, SHWB and Assam & Meghalaya regions while below normal rains will be occurring in Kerala, Coastal Karnataka, Konkan & Goa, Vidarbha, Saurashtra & Kutch, East Rajasthan, entire U.P., Himachal Pradesh, Uttarakhand and Arunachal Pradesh regions, normal rains in Rayalaseema, Punjab and CHD regions. For the remaining part of the country, any clear signal is absent. For the country as a whole, the chance is high for the below-normal category rainfall.

11.5 Conversion of Monthly/Seasonal Scale Anomaly Forecasts for Using in Agriculture Models

For the integration of monthly/seasonal scale real-time anomaly/percentage departure forecast with crop models, the forecast of monthly/seasonal rainfall and temperatures with percentage departure and anomalies for the forecasted year are respectively needed to convert as a magnitude by using respective climatological mean (Details are described in Chap. 5 in Sect. 5). The following are conversion formulas for rainfall and temperature to magnitude from percentage departure and anomalies, respectively.

Rainfall

$$R_f = \left(\frac{100 + RPD}{100}\right) \times \overline{Y}$$

where, $R_f =$ s the forecasted year monthly or seasonal total rainfall, RPD is the forecasted year monthly or seasonal rainfall percentage departure from climatological mean, $\overline{Y}$ is the observed climatological mean of monthly or seasonal total rainfall for the hindcast period.

For example, the Southwest summer monsoon (JJAS) rainfall real-time forecast at Jharkhand is 3% for the forecasted year 2017 with April 2017 initial conditions and the observed climatological mean of seasonal JJAS rainfall (1982–2017) is 1079.4 mm. The real-time seasonal total JJAS rainfall for the forecasted year 2017 at Jharkhand $= \left(\frac{100+3}{100}\right) \times 1079.4$ mm $= 1111.782$ mm.

Temperature

$$T_f = T_a + \overline{T}$$

where T_f is the monthly or seasonal average temperature for the forecasted year.

T_a is the monthly or seasonal anomaly of the average temperature for the forecasted year.

$\overline{T}$is the observed climatological mean of monthly or seasonal average temperature for hind cast period.

For example, the maximum temperature (T_{max}) for the real-time Monthly anomaly forecast of May at Jammu & Kashmir is 0.9 °C for the forecasted year 2017 with April 2017 initial conditions and the observed climatological mean of May T_{max} (1982–2017) is 27.4 °C. The real-time Monthly T_{max} of May for the forecasted year 2017 at Jammu & Kashmir 27.4+0.9=28.3 °C.

11.6 Disaggregation of Seasonal Forecast and Use in the Crop Models

The real-time forecasts at monthly and seasonal scales for rainfall and temperatures are used to generate for the targeted domain. These forecasts are deterministic and probabilistic in nature and given in single value for each month or season one month in advance (Lead 1). Hence, to use this information in the crop model, it is necessary to perform a temporal downscaling of the forecasts daily. Several approaches for the temporal downscaling of monthly and seasonal climate forecasts are available (details are in Chap. 7). The user can use any of the disaggregated model/weather generators (WG) to generate daily sequences of the monthly/seasonal data. The following steps can be considered to generate daily sequences

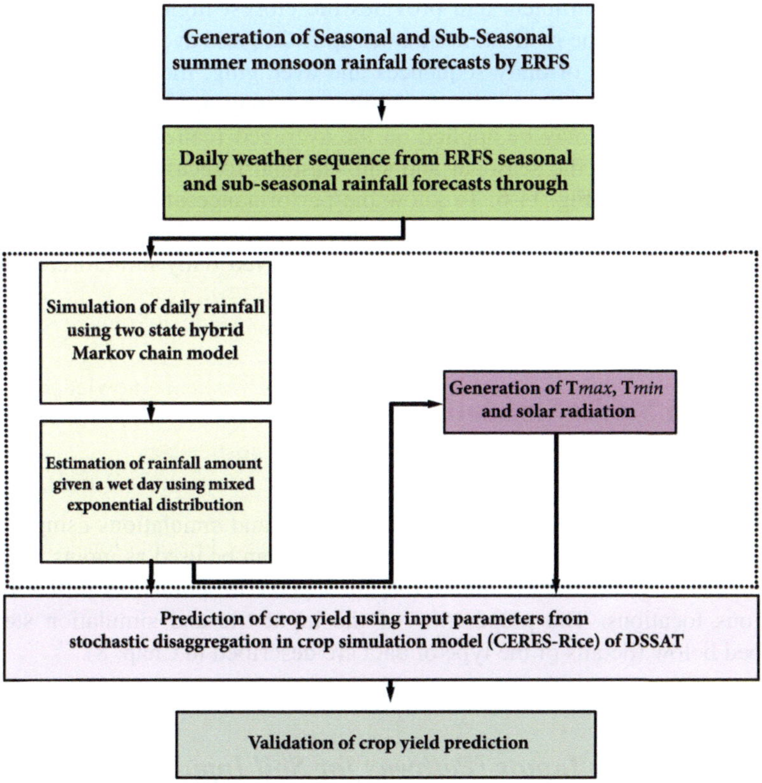

Fig. 11.6 Flow of methodology of current work of linkage for disaggregated monthly and seasonal rainfall forecast to crop simulation model (CERES-Rice) for prediction of rice yield at Kharagpur

using WG. Consider daily observed data (say rainfall, past data) for a station/point. Make a seasonal mean/total using daily data. Use the WG and disaggregate that seasonal data into a daily scale. Evaluate the performance of the disaggregated daily data with the observed data. Repeat the disaggregation processes, and generate different realisations of the daily sequence. Average the realisations and check the performance of the averaged sequence (say, disaggregation twice will produce two realisations and then average the two realisations).

Continue the process, and make sure that the performance of the averaged realization (e.g. temporal correlation coefficient) is increasing up to a certain number of realizations and no further improvements are seen. Consider generating a number of realisations up to that number or a bit more where saturation in skill has been reached. For example, in a case if the 190 or nearby (<200) number out of

1000 realisations is sufficient and provides the closest highest correlation value, then the user can fix the realization number up to 200 and average it. Keep in mind that in the generation of daily sequences and averaging, the number of dry days should be closer to the observations and the user may need to fix a threshold as the minimum, which may be applied on the averaged realization data. Details of the disaggregation of the seasonal and sub-seasonal forecast and linkage to crop model is depicted in Fig. 11.6. To know the performance of the stochastic disaggregation, correlations of disaggregated daily sequences from monthly observed rainfall and forecasted monthly rainfall with observed daily rainfall are obtained for rainfall intensity and total rainfall.

11.7 Crop Model Simulations and Generation of Climate Risk Matrices

This section describes the steps of crop modelling and simulations using monthly and seasonal forecasts of weather variables which can be used as inputs in DSSAT crop models to generate the climate risk management options for different crops in various locations. The preparation of the input file and simulation steps are described below (details of the type of data are described in Chap. 8).

11.7.1 Soil Data Inputs (Pathway for Soil Input)

(i) Open **DSSAT** → Select **Soil data** on the tools → **SBuild** screen will appear → Select **New** on the **Profile → fill the required details like** (country, site name, institute code, latitude and longitude, soil data source, soil series name, soil classification, colour, drainage, % slope and fertility factor) → Click Next.

(ii) Click Add layer (as per your requirement) → **fill in the required details like** (depth of the soil layer, master horizon, percentage of clay, silt, sand, total nitrogen and organic carbon, pH in water and cation exchange capacity for the respective layer) → Click **Next**.

(iii) Click **Calculate Missing Values** → check the values (lower and upper limit, saturation water content, bulk density, saturated hydraulic conductivity, root growth factor, runoff curve number, albedo, drainage rate) for your soil otherwise change it by clicking on the respective places → Click **Finish**.
(iv) Select **save** on the **Profile → Give 8 characters to save** the file → Click ok.
(v) Select save on the **File** → Select **Close** on the **Profile** → Select **Close** on the File.
(vi) Click **Soil folder** on the **Data** on **Selector** in the **DSSAT main window** → Click Refresh (you can see your soil folder and file).

The following images are the pictorial representation of the above steps.

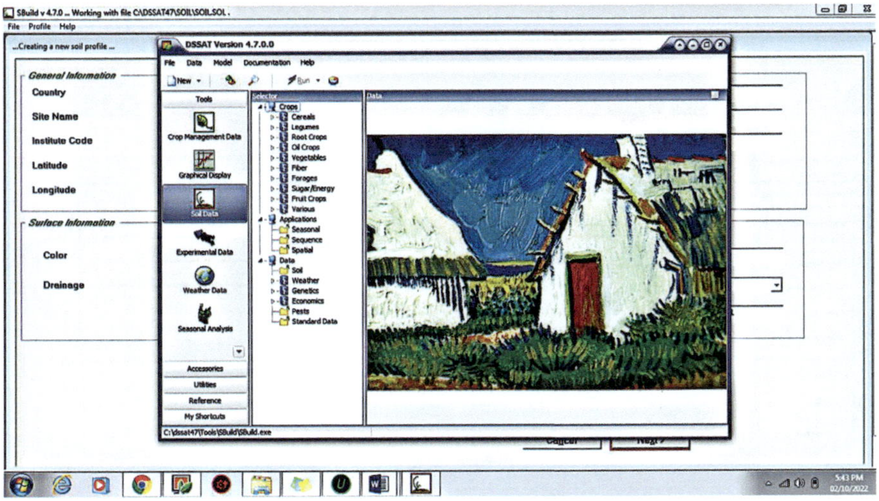

SBuild v 4.7.0 ... Working with file C:\DSSAT47\SOIL\SOIL.SOL .
File
Profile
Help
New
Edit
Delete
Close
Save As...
Save
Preview
Print
SBuild
DSSAT v 4.7.0.0

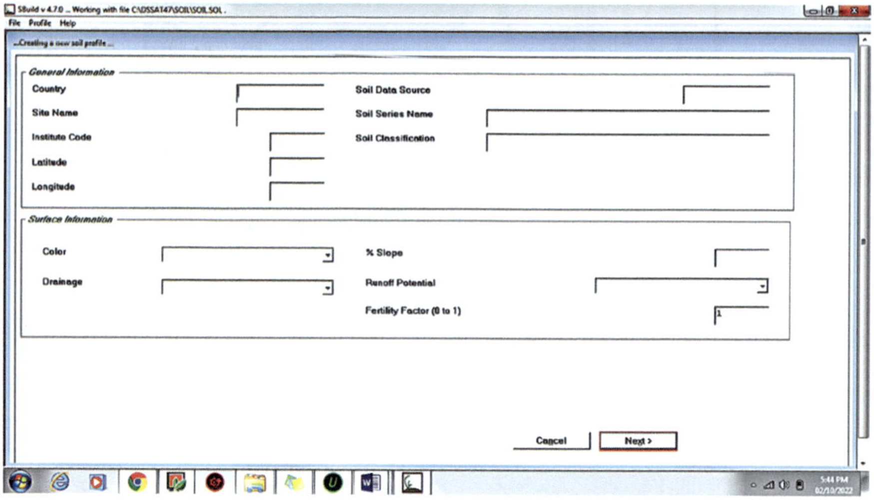
SBuild v 4.7.0 ... Working with file C:\DSSAT47\SOIL\SOIL.SOL .
File
Profile
Help
General Information
Country
Site Name
Institute Code
Latitude
Longitude
Soil Data Source
Soil Series Name
Soil Classification
Surface Information
Color
Drainage
% Slope
Runoff Potential
Fertility Factor (0 to 1)
Cancel
Next >

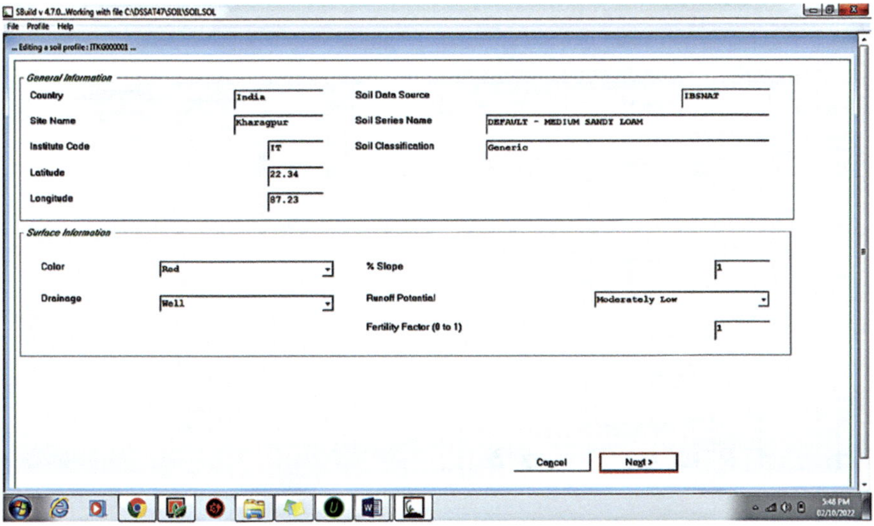
File Profile Help
General Information
Country
India
Site Name
Kharagpur
Institute Code
IT
Latitude
22.34
Longitude
87.23
Soil Data Source
IBSNAT
Soil Series Name
DEFAULT - MEDIUM SANDY LOAM
Soil Classification
Generic
Surface Information
Color
Red
Drainage
Well
% Slope
1
Runoff Potential
Moderately Low
Fertility Factor (0 to 1)
1
Cancel
Next >

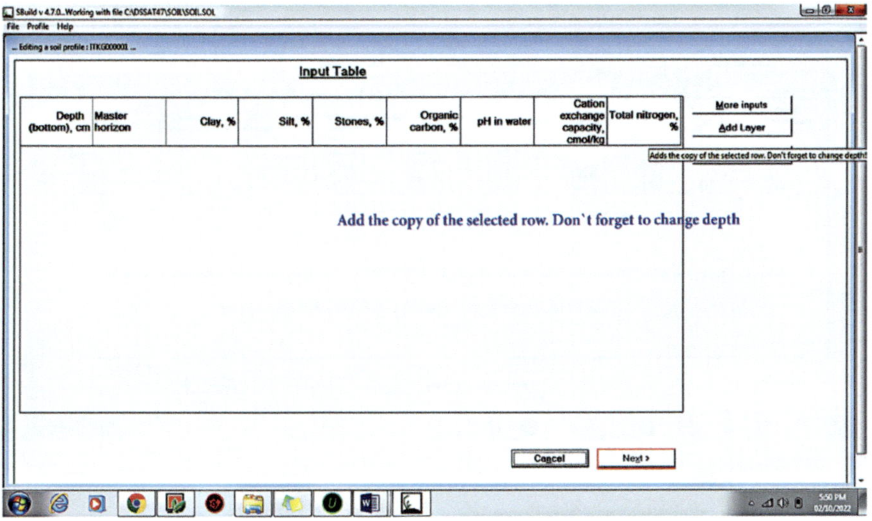
File Profile Help
Input Table
Depth (bottom), cm
Master horizon
Clay, %
Silt, %
Stones, %
Organic carbon, %
pH in water
Cation exchange capacity, cmol/kg
Total nitrogen, %
More inputs
Add Layer
Add the copy of the selected row. Don't forget to change depth
Cancel
Next >

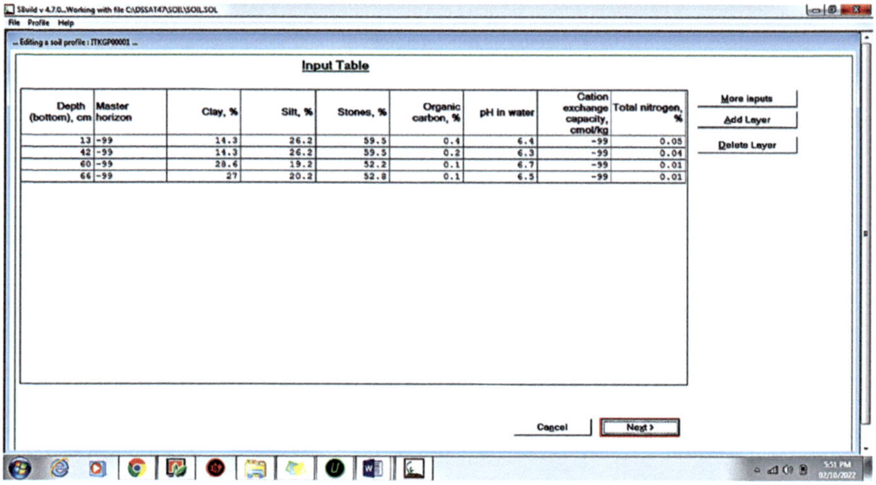
SBuild v 4.7.0...Working with file C:\DSSAT47\SOIL\SOIL.SOL
File Profile Help
... Editing a soil profile : ITKGP00001 ...
Input Table
Depth (bottom), cm
Master horizon
Clay, %
Silt, %
Stones, %
Organic carbon, %
pH in water
Cation exchange capacity, cmol/kg
Total nitrogen, %
13 -99 14.3 26.2 59.5 0.4 6.4 -99 0.05
42 -99 14.3 26.2 59.5 0.2 6.3 -99 0.04
60 -99 28.6 19.2 52.2 0.1 6.7 -99 0.01
66 -99 27 20.2 52.8 0.1 6.5 -99 0.01
More inputs
Add Layer
Delete Layer
Cancel
Next >

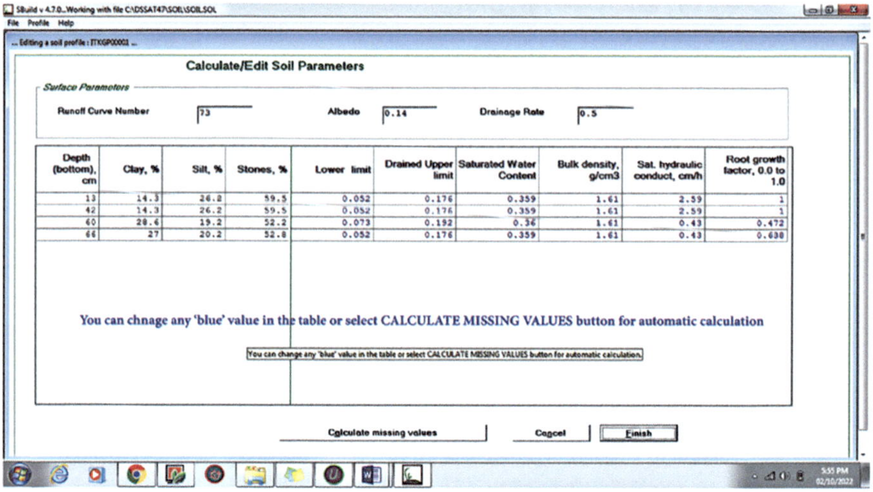
SBuild v 4.7.0...Working with file C:\DSSAT47\SOIL\SOIL.SOL
File Profile Help
... Editing a soil profile : ITKGP00001 ...
Calculate/Edit Soil Parameters
Surface Parameters
Runoff Curve Number 73
Albedo 0.14
Drainage Rate 0.5
Depth (bottom), cm
Clay, %
Silt, %
Stones, %
Lower limit
Drained Upper limit
Saturated Water Content
Bulk density, g/cm3
Sat. hydraulic conduct, cm/h
Root growth factor, 0.0 to 1.0
13 14.3 26.2 59.5 0.052 0.176 0.359 1.61 2.59 1
42 14.3 26.2 59.5 0.052 0.176 0.359 1.61 2.59 1
60 28.6 19.2 52.2 0.073 0.192 0.36 1.61 0.43 0.472
66 27 20.2 52.8 0.052 0.176 0.359 1.61 0.43 0.638
You can chnage any 'blue' value in the table or select CALCULATE MISSING VALUES button for automatic calculation
You can change any 'blue' value in the table or select CALCULATE MISSING VALUES button for automatic calculation.
Calculate missing values
Cancel
Finish

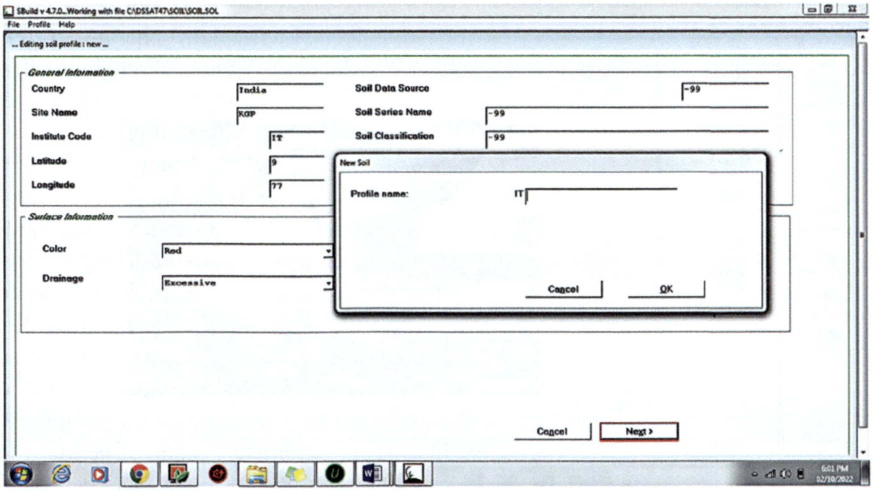
File Profile Help
General Information
Country
India
Site Name
KGP
Institute Code
IT
Latitude
9
Longitude
77
Soil Data Source
-99
Soil Series Name
-99
Soil Classification
-99
New Soil
Profile name:
IT
Cancel
OK
Surface Information
Color
Red
Drainage
Excessive
Cancel
Next >

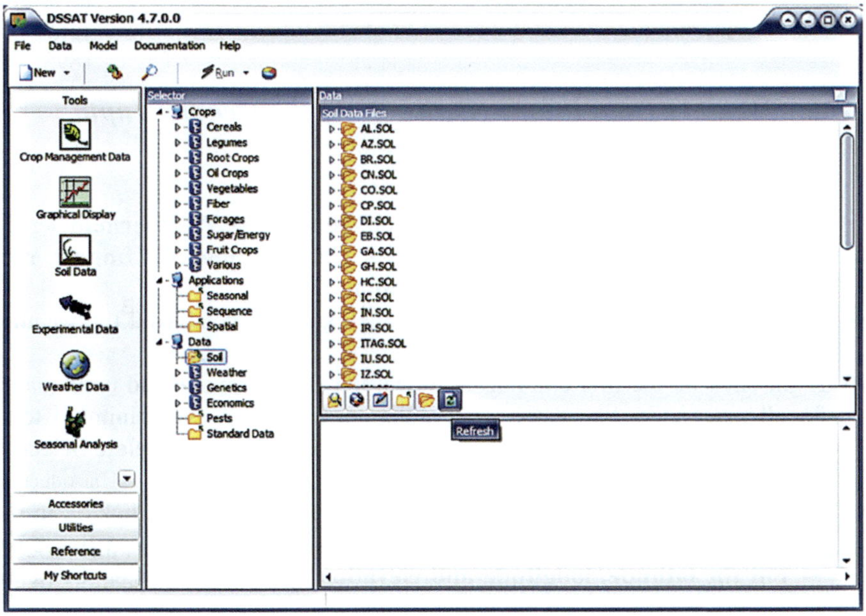
DSSAT Version 4.7.0.0
File Data Model Documentation Help
New
Run
Tools
Crop Management Data
Graphical Display
Soil Data
Experimental Data
Weather Data
Seasonal Analysis
Accessories
Utilities
Reference
My Shortcuts
Selector
Crops
Cereals
Legumes
Root Crops
Oil Crops
Vegetables
Fiber
Forages
Sugar/Energy
Fruit Crops
Various
Applications
Seasonal
Sequence
Spatial
Data
Soil
Weather
Genetics
Economics
Pests
Standard Data
Data
Soil Data Files
AL.SOL
AZ.SOL
BR.SOL
CN.SOL
CO.SOL
CP.SOL
DI.SOL
EB.SOL
GA.SOL
GH.SOL
HC.SOL
IC.SOL
IN.SOL
IR.SOL
ITAG.SOL
IU.SOL
IZ.SOL
Refresh

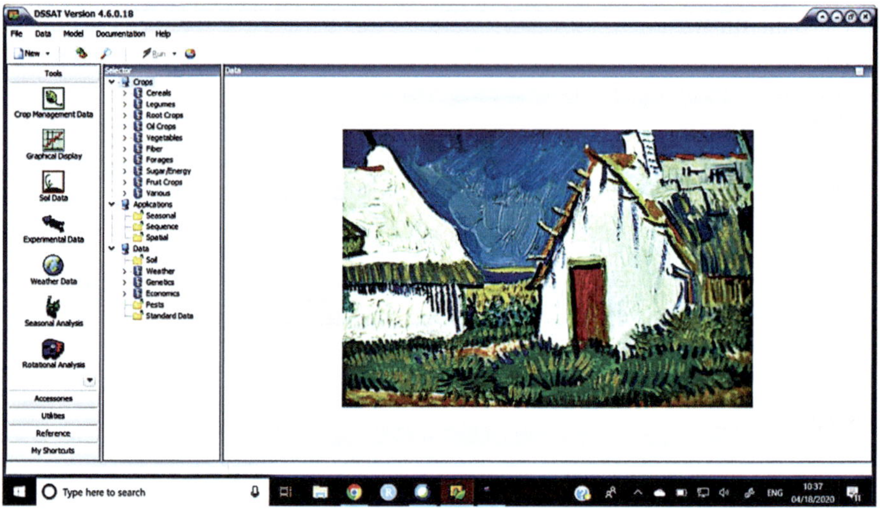

11.7.2 Weather Data Inputs (Pathway for the Weather Input in DSSAT)

(i) Open **DSSAT** → Select **weather data** on the tools → **Weatherman** screen will appear → Select **New Station** → Choose **Input or Import raw weather data and then save as new station**.

(ii) Select **Open** from file menu → Select the **excel file** you saved the weather data → Click **Open**.

(iii) Right click on the first column → Select the variable name and units (same for all parameters like date, solar radiation, maximum and minimum temperature and precipitation) → Delete the first using the icon **Delete Selected row**.

(iv) Select **Import data into weatherman** from the file menu → Click **ok** in the file option window → Give station name (4 Alphabets only) → Click **Save → Fill the latitude, longitude and elevation of your station** → Click **ok** → Click **Write files**.

(v) Select **Tools** → Select **Estimate parameter** → Select **Observed Data (Perform range check on data)** → Select **Analyze** → Select **Graph** → Select **Observed Data** → Click the **first icon to select the parameters** (solar radiation, maximum and minimum temperature and precipitation) → click **Ok (Graph will appear)**.

(vi) Click **Weather folder** on the **Data** on **Selector** in the **DSSAT main window** → Click **Daily folder** → **Click Refresh** (you can see you weather folder and respective year files).

The following images are the pictorial representation of the above steps.

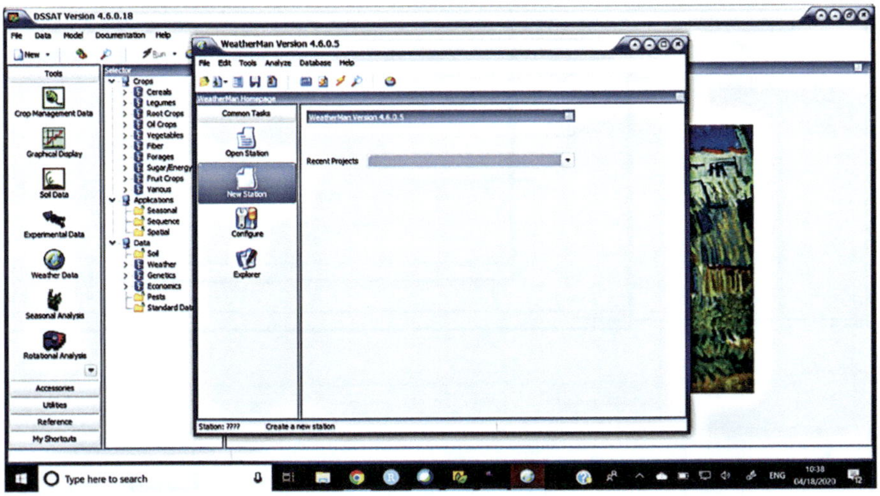

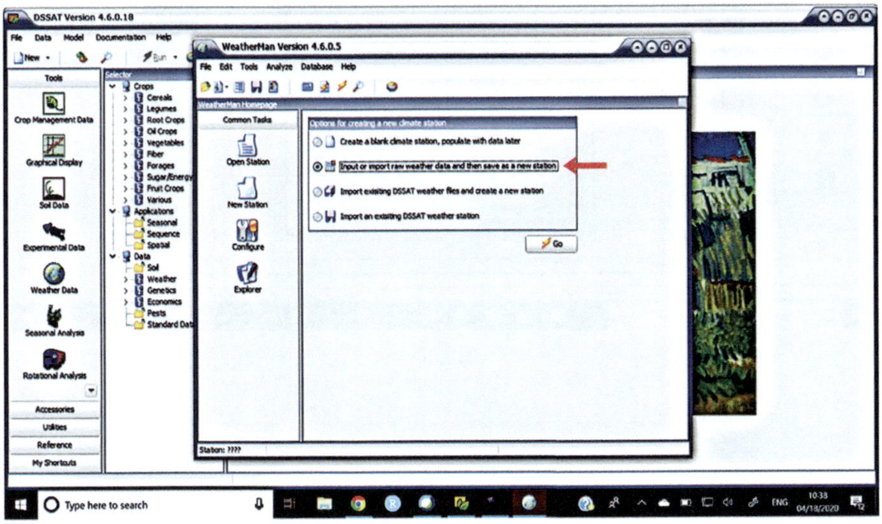

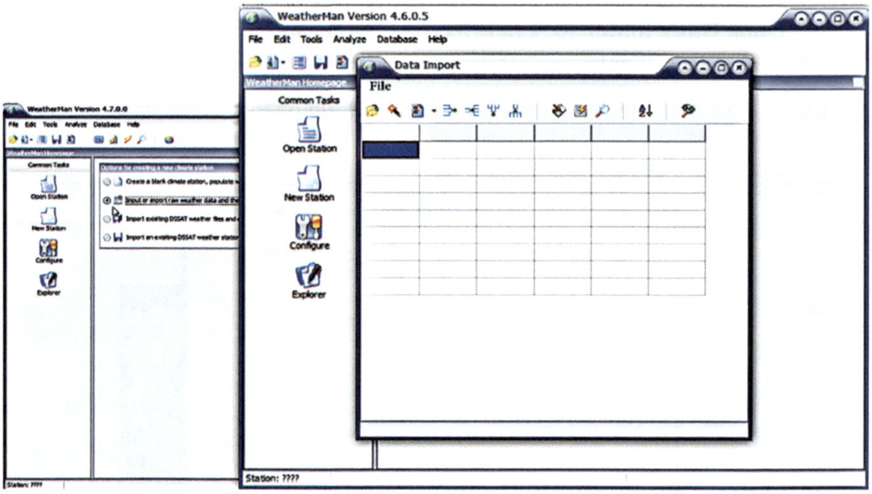
WeatherMan Version 4.7.0.0
WeatherMan Version 4.6.0.5
File Edit Tools Analyze Database Help
Data Import
File
Common Tasks
Open Station
New Station
Configure
Explorer
Station: ????

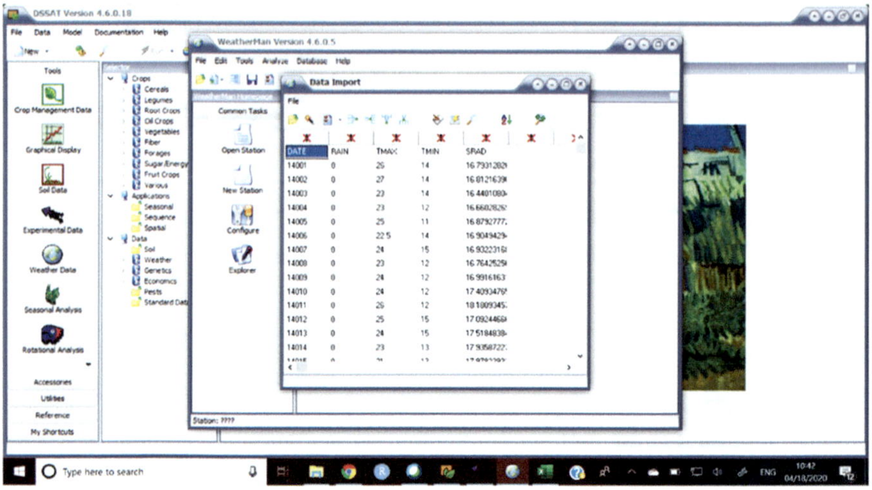
DSSAT Version 4.6.0.18
File Data Model Documentation Help
Tools
Crop Management Data
Graphical Display
Soil Data
Experimental Data
Weather Data
Seasonal Analysis
Rotational Analysis
Accessories
Utilities
Reference
My Shortcuts
WeatherMan Version 4.6.0.5
Data Import
DATE RAIN TMAX TMIN SRAD
Common Tasks
Open Station
New Station
Configure
Explorer
Station: ????
Type here to search

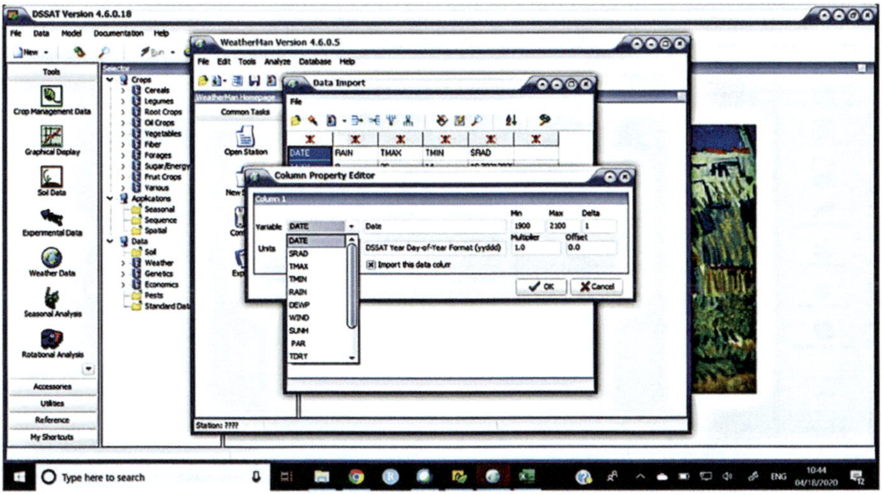
DSSAT Version 4.6.0.18
File Data Model Documentation Help
Tools
Crop Management Data
Graphical Display
Soil Data
Experimental Data
Weather Data
Seasonal Analysis
Rotational Analysis
Accessories
Utilities
Reference
My Shortcuts
Crops
Cereals
Legumes
Root Crops
Oil Crops
Vegetables
Fiber
Forages
Sugar/Energy
Fruit Crops
Various
Applications
Seasonal
Sequence
Spatial
Data
Soil
Weather
Genetics
Economics
Pests
Standard Data
WeatherMan Version 4.6.0.5
File Edit Tools Analyze Database Help
Common Tasks
Open Station
Station: ????
Data Import
File
DATE
RAIN
TMAX
TMIN
SRAD
Column Property Editor
Column 1
Variable
DATE
Date
Units
Min
Max
Delta
1900
2100
1
Multiplier
Offset
1.0
0.0
DSSAT Year Day-of-Year Format (yyddd)
Import this data column
DATE
SRAD
TMAX
TMIN
RAIN
DEWP
WIND
SUNH
PAR
TDRY
OK
Cancel
Type here to search
ENG
10:44
04/18/2020

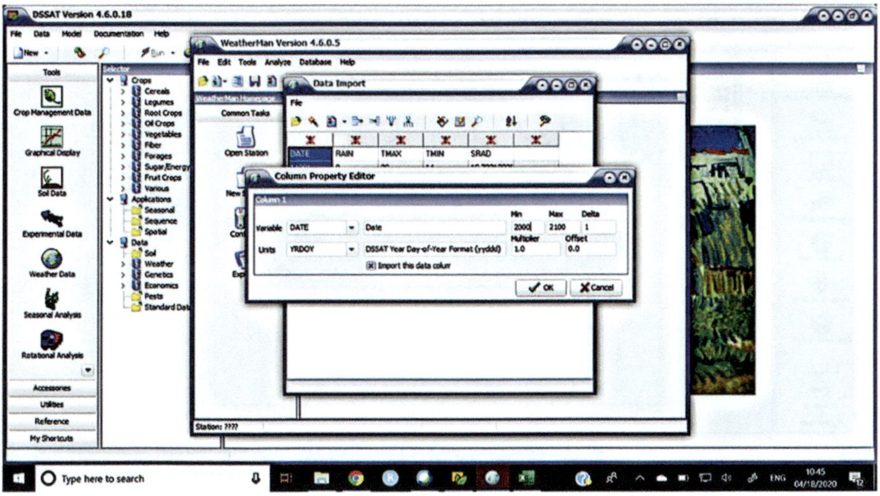
DSSAT Version 4.6.0.18
File Data Model Documentation Help
Tools
Crop Management Data
Graphical Display
Soil Data
Experimental Data
Weather Data
Seasonal Analysis
Rotational Analysis
Accessories
Utilities
Reference
My Shortcuts
Crops
Cereals
Legumes
Root Crops
Oil Crops
Vegetables
Fiber
Forages
Sugar/Energy
Fruit Crops
Various
Applications
Seasonal
Sequence
Spatial
Data
Soil
Weather
Genetics
Economics
Pests
Standard Data
WeatherMan Version 4.6.0.5
File Edit Tools Analyze Database Help
Common Tasks
Open Station
Station: ????
Data Import
File
DATE
RAIN
TMAX
TMIN
SRAD
Column Property Editor
Column 1
Variable
DATE
Date
Units
YRDOY
Min
Max
Delta
2000
2100
1
Multiplier
Offset
1.0
0.0
DSSAT Year Day-of-Year Format (yyddd)
Import this data column
OK
Cancel
Type here to search
ENG
10:45
04/18/2020

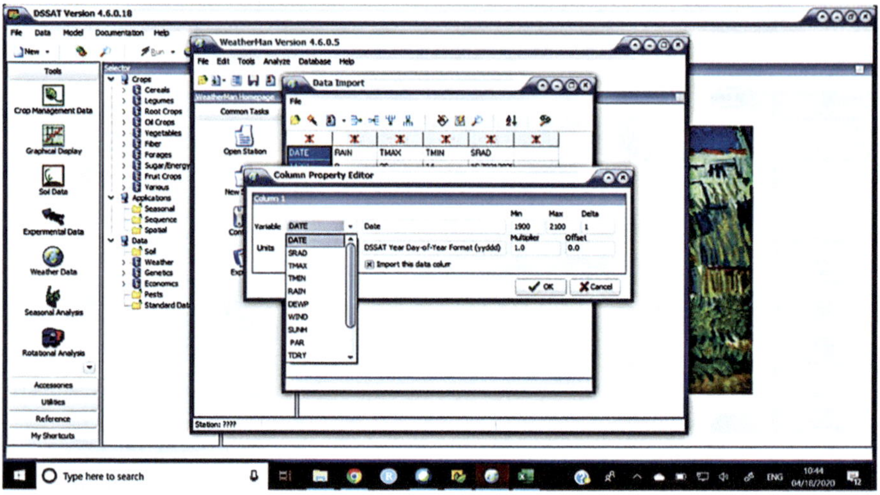
DSSAT Version 4.6.0.18
WeatherMan Version 4.6.0.5
Data Import
Column Property Editor
Column 1
Variable DATE
Units
Date
Min 1900
Max 2100
Delta 1
Multiplier 1.0
Offset 0.0
DSSAT Year Day-of-Year Format (yyddd)
Import this data column
DATE
SRAD
TMAX
TMIN
RAIN
DEWP
WIND
SUNH
PAR
TDRY
OK
Cancel
Station: ????
Type here to search

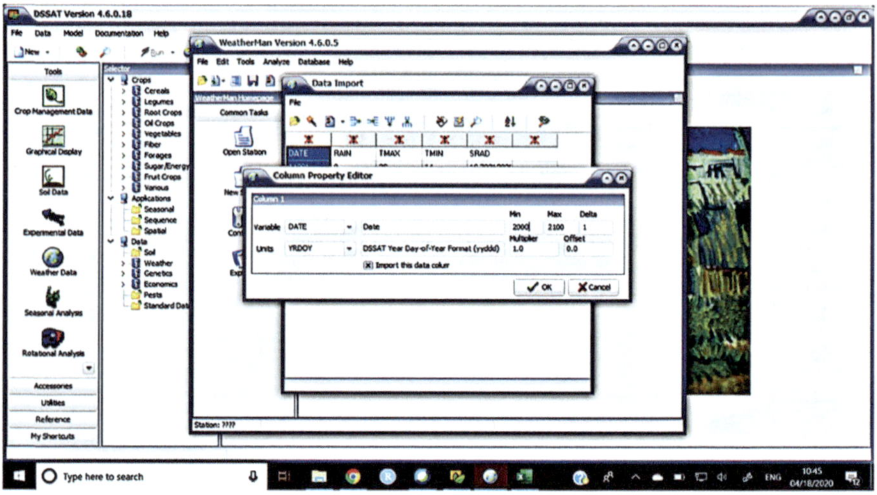
DSSAT Version 4.6.0.18
WeatherMan Version 4.6.0.5
Data Import
Column Property Editor
Column 1
Variable DATE
Units YRDOY
Date
Min 2000
Max 2100
Delta 1
Multiplier 1.0
Offset 0.0
DSSAT Year Day-of-Year Format (yyddd)
Import this data column
OK
Cancel
Station: ????
Type here to search

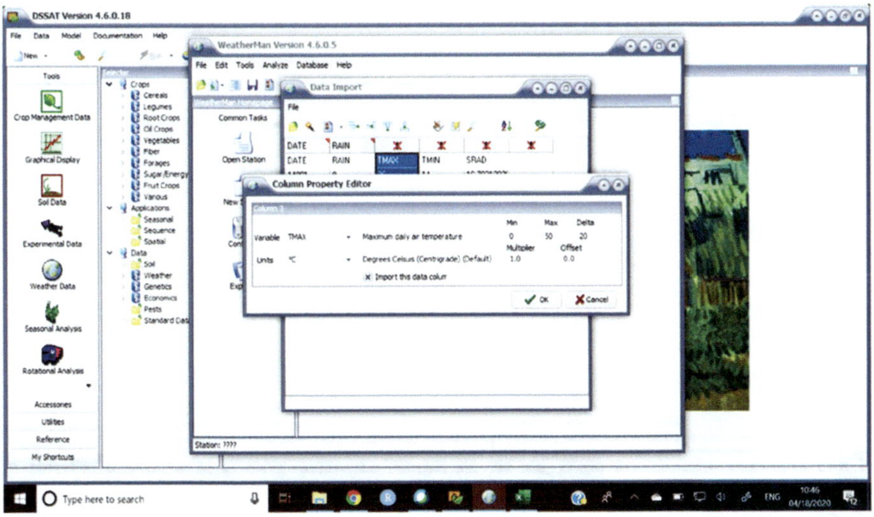
DSSAT Version 4.6.0.18
WeatherMan Version 4.6.0.5
Data Import
Column Property Editor
Variable TMAX
Maximum daily air temperature
Min 0
Max 50
Delta 20
Units °C
Degrees Celsius (Centigrade) (Default)
Multiplier 1.0
Offset 0.0
Import this data column
OK
Cancel
DATE RAIN TMAX TMIN SRAD
Station: ????

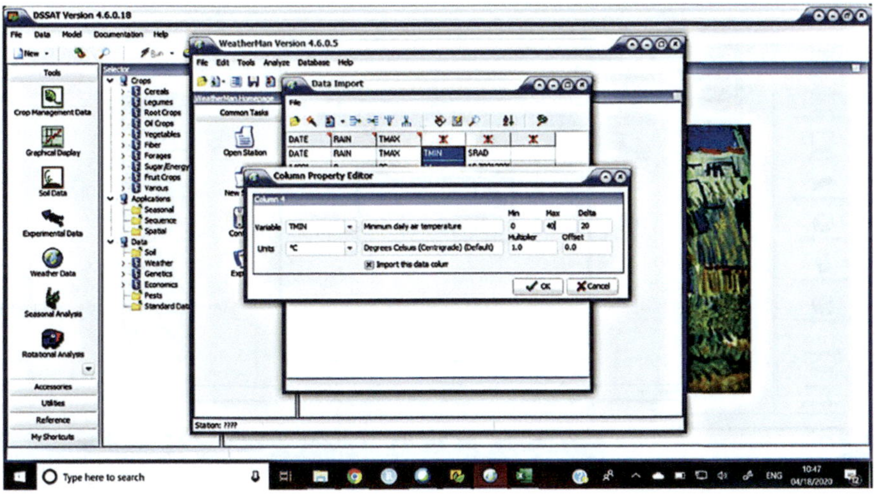
DSSAT Version 4.6.0.18
WeatherMan Version 4.6.0.5
Data Import
Column Property Editor
Column 4
Variable TMIN
Minimum daily air temperature
Min 0
Max 40
Delta 20
Units °C
Degrees Celsius (Centigrade) (Default)
Multiplier 1.0
Offset 0.0
Import this data column
OK
Cancel
DATE RAIN TMAX TMIN SRAD
Station: ????

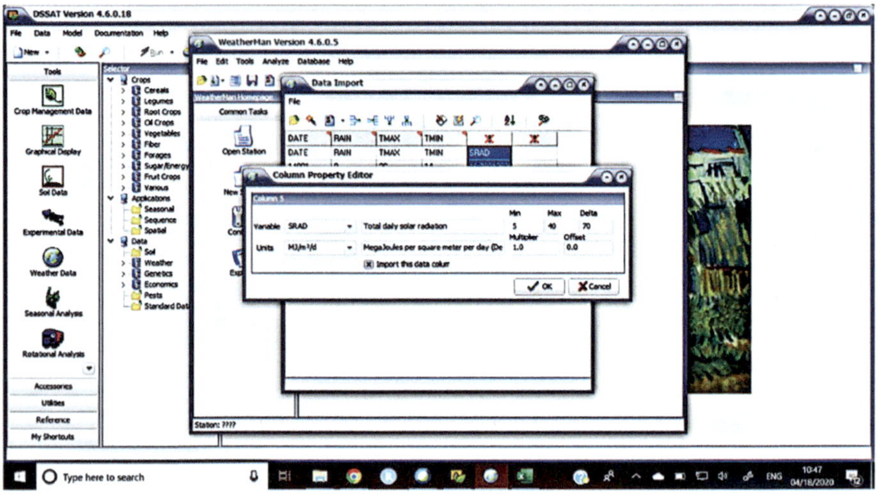
DSSAT Version 4.6.0.18
WeatherMan Version 4.6.0.5
Data Import
Column Property Editor
Variable SRAD
Total daily solar radiation
Units MJ/m²/d
MegaJoules per square meter per day
Import this data column
OK
Cancel
Type here to search

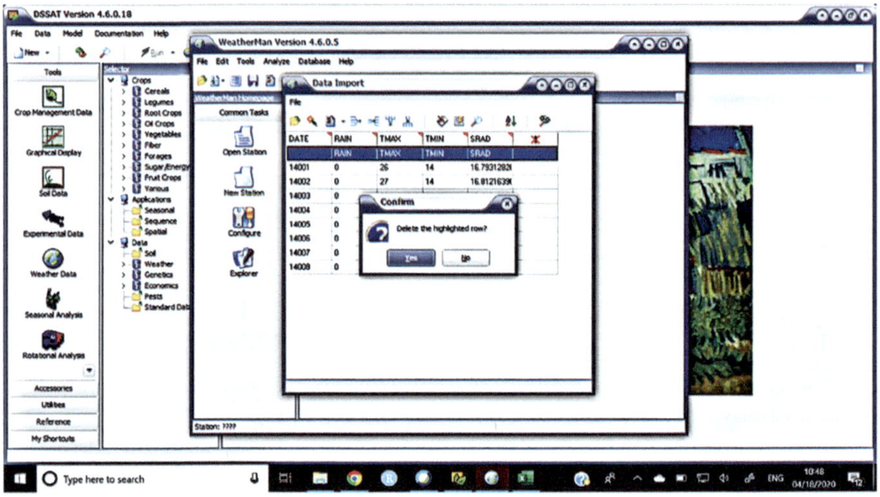
DSSAT Version 4.6.0.18
WeatherMan Version 4.6.0.5
Data Import
DATE
RAIN
TMAX
TMIN
SRAD
Confirm
Delete the highlighted row?
Yes
No
Type here to search

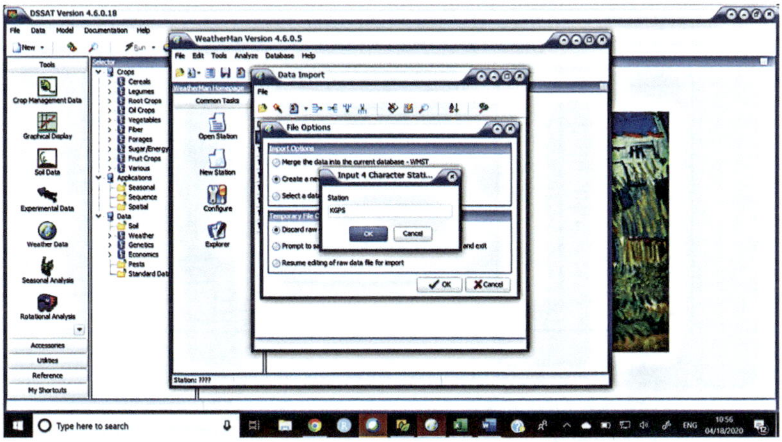
DSSAT Version 4.6.0.18
WeatherMan Version 4.6.0.5
Data Import
File Options
Import Options
Merge the data into the current database - WMST
Input 4 Character Stati...
Station
KGPS
OK
Cancel
Resume editing of raw data file for import
Station: ????
Type here to search

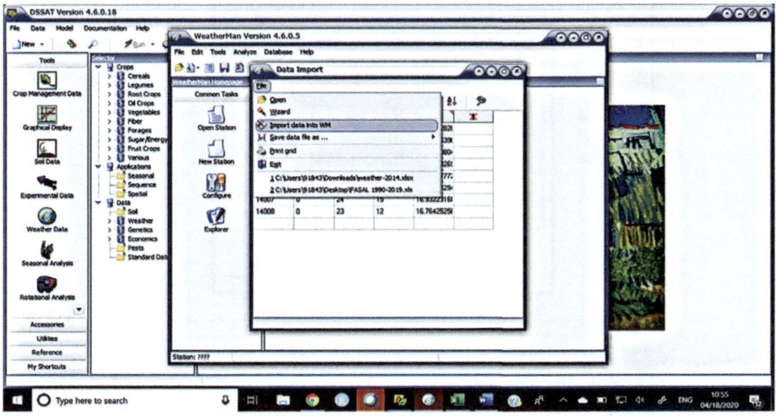
DSSAT Version 4.6.0.18
WeatherMan Version 4.6.0.5
Data Import
Open
Wizard
Import data into WM
Save data file as ...
Print grid
Exit
Station: ????
Type here to search

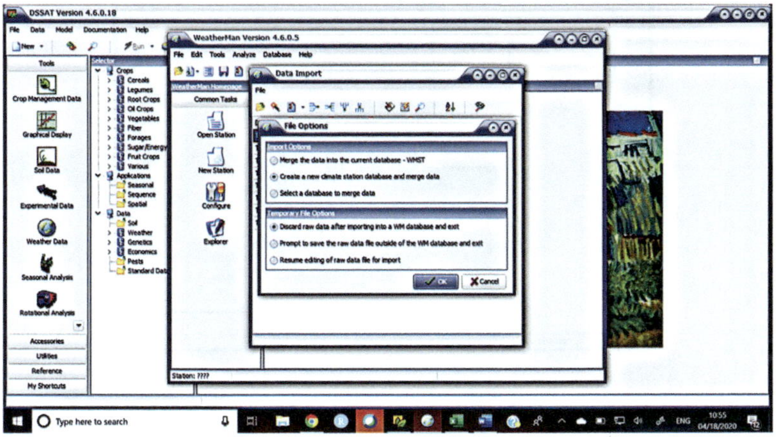
DSSAT Version 4.6.0.18
WeatherMan Version 4.6.0.5
Data Import
File Options
Import Options
Merge the data into the current database - WMST
Create a new climate station database and merge data
Select a database to merge data
Temporary File Options
Discard raw data after importing into a WM database and exit
Prompt to save the raw data file outside of the WM database and exit
Resume editing of raw data file for import
OK
Cancel
Station: ????
Type here to search

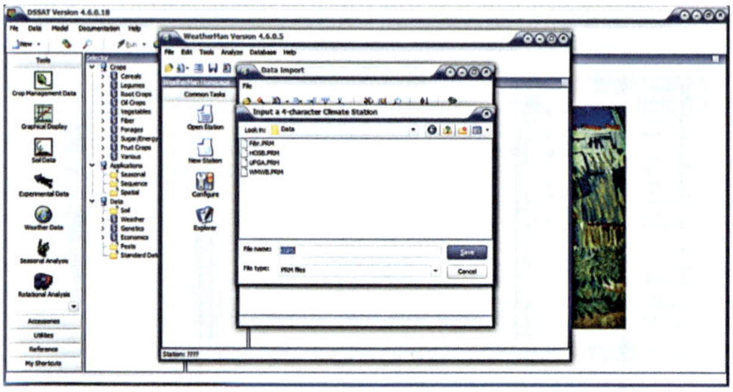

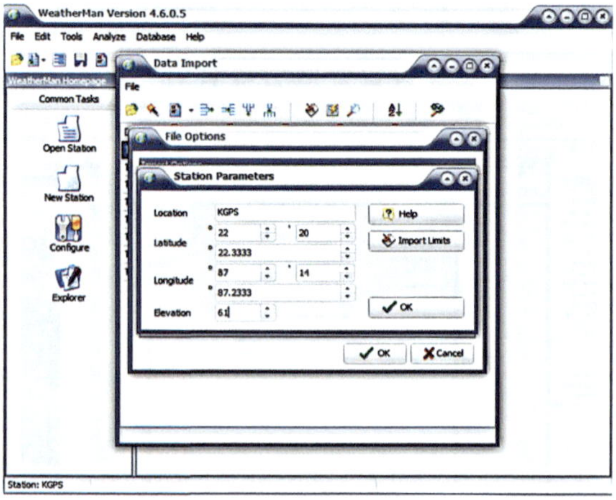
WeatherMan Version 4.6.0.5
File Edit Tools Analyze Database Help
Data Import
Common Tasks
Open Station
New Station
Configure
Explorer
File Options
Station Parameters
Location KGPS
Latitude 22 20 22.3333
Longitude 87 14 87.2333
Elevation 61
Help
Import Limits
OK
OK
Cancel
Station: KGPS

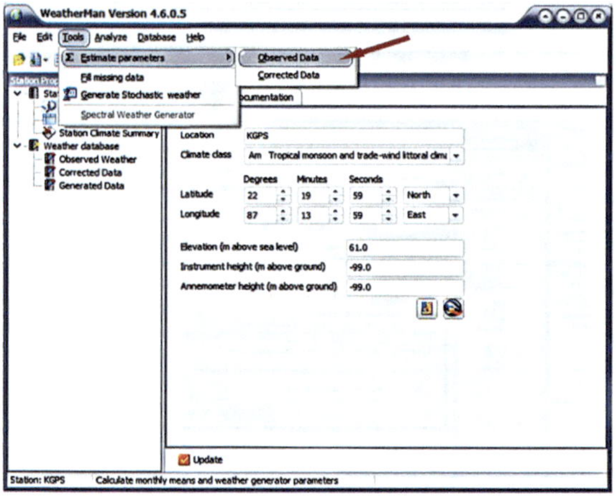
WeatherMan Version 4.6.0.5
File Edit Tools Analyze Database Help
Estimate parameters
Observed Data
Corrected Data
Fill missing data
Generate Stochastic weather
Spectral Weather Generator
Station Climate Summary
Weather database
Observed Weather
Corrected Data
Generated Data
Location KGPS
Climate class Am Tropical monsoon and trade-wind littoral clim
Degrees Minutes Seconds
Latitude 22 19 59 North
Longitude 87 13 59 East
Elevation (m above sea level) 61.0
Instrument height (m above ground) -99.0
Annemometer height (m above ground) -99.0
Update
Station: KGPS Calculate monthly means and weather generator parameters

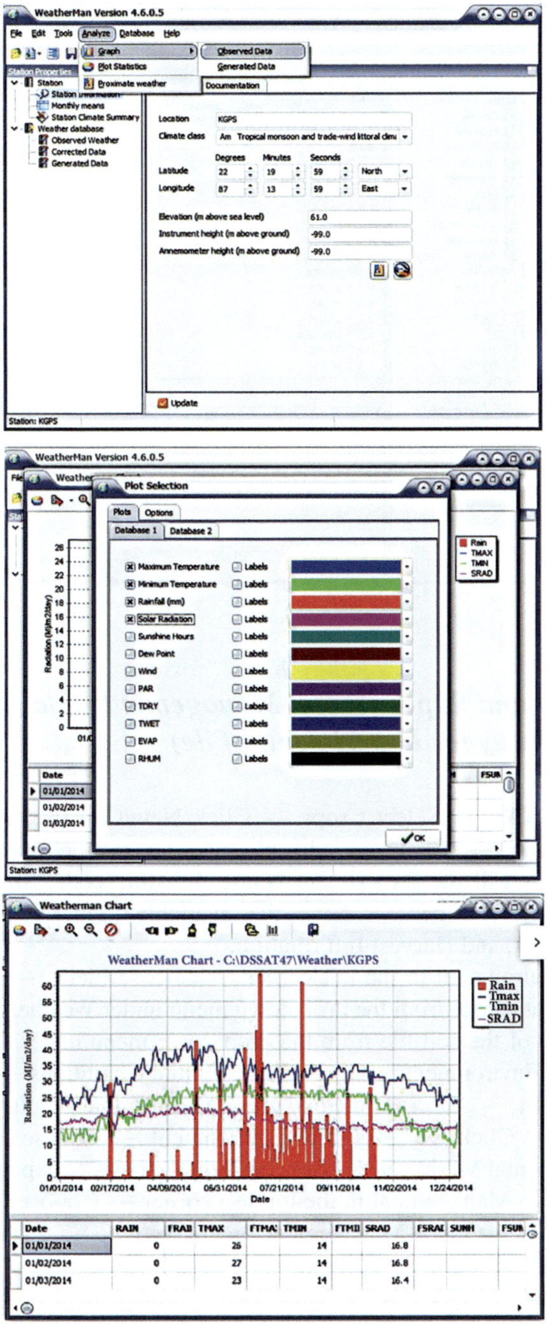

WeatherMan Version 4.6.0.5
File Edit Tools Analyze Database Help
Graph
Plot Statistics
Proximate weather
Observed Data
Generated Data
Documentation
Station
Monthly means
Station Climate Summary
Weather database
Observed Weather
Corrected Data
Generated Data
Location KGPS
Climate class Am Tropical monsoon and trade-wind littoral clim
Degrees Minutes Seconds
Latitude 22 19 59 North
Longitude 87 13 59 East
Elevation (m above sea level) 61.0
Instrument height (m above ground) -99.0
Anemometer height (m above ground) -99.0
Update
Station: KGPS
Plot Selection
Plots Options
Database 1 Database 2
Maximum Temperature
Minimum Temperature
Rainfall (mm)
Solar Radiation
Sunshine Hours
Dew Point
Wind
PAR
TDRY
TWET
EVAP
RHUM
Labels
OK
Rain
TMAX
TMIN
SRAD
Weatherman Chart
WeatherMan Chart - C:\DSSAT47\Weather\KGPS
Radiation (MJ/m2/day)
Date
Rain
Tmax
Tmin
SRAD
01/01/2014 02/17/2014 04/08/2014 05/31/2014 07/21/2014 09/11/2014 11/02/2014 12/24/2014
Date RAIN FRAI TMAX FTMA TMIN FTMI SRAD FSRAI SUNH FSUN
01/01/2014 0 26 14 16.8
01/02/2014 0 27 14 16.8
01/03/2014 0 23 14 16.4

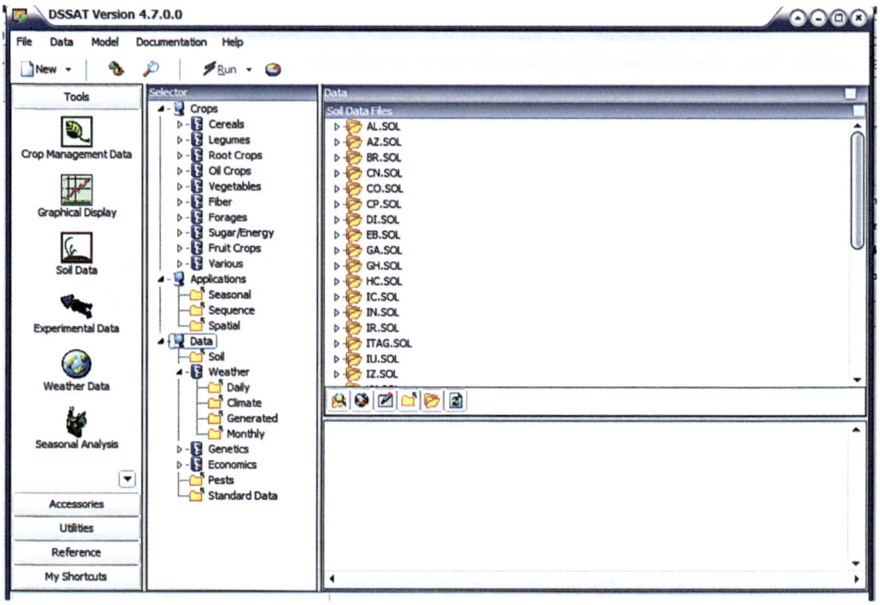

11.7.3 Crop and Experimental Management Data Inputs (Pathway for Experimental File)

(i) Open DSSAT → Select Crops → Click New/Choose any existing experimental file → Click on edit file just below the existing experimental files → Fill the Experiment Name, Institute code, Site code, Year and Experiment Number → Give the available General Information, Plot Information, and Harvest Information.

(ii) Go to Environment in the left corner → Choose Fields → Select the Name of the weather file from the drop down menu under Weather Station → Select the Name of the soil file from the drop down menu under Soil → Click OK → Go to Environment again → Choose Initial Conditions → Fill the Residue information → Select Profile and check the information on Initial Conditions for soil → Click OK → Go to Environment → Choose Soil Analysis and Environmental Modifications (if needed on the basis of experiment).

(iii) Then Go to Management in the top left corner → Choose Cultivar → Click OK → Then Go to Management again → Choose Planting → Fill the information under Sowing window → Go to Transplant → Fill the available details → Click OK → Again Go to Management in the top left corner → Choose Irrigation → Select the Management from the drop down under Irrigation Applications → Fill the irrigation details → Click OK → Go to Management in the top left corner → Add levels of fertilizer → Click Add application to add fertiliser information → Click OK → Choose Organic amendments/Tillage/Harvest/Chemical Applications under Management and similar fill in details.

(iv) Go to Simulation Options in the top left → Fill the Simulation Start Date (must be before planting or on planting date) → Go to the methods and fill in the details → Go to Management → choose details irrigation and water management, Nitrogen, Organic amendments and Harvest from the drop down → Click OK.

(v) Then Go to Treatments in the top left → Add levels → Select the description and details of the levels from the drop-down menu → Click OK.

(vi) Go to Refresh → OK → Save and Refresh.

(vii) Go to DSSAT main window → Click on Refresh just below the existing experimental files → Select the experimental file → Click on Run in the top left corner to run the file → Run model → OK → Analysis.

11.7.4 Calibration/Estimation of Crop Genotype Coefficients (Pathway for the Calibration of Genotype Coefficients)

(i) Open **DSSAT** → Click **Experimental Data** → if you see A on the window → fill the required details like treatment number, anthesis day, emergence day, etc.

(ii) Select **Save as** in the file menu → Select the respective crop folder → Select the file type as **A File** → Give the name as same as experiment name for the respective crop → Click **Save**.

(iii) Click **A** it will turn into **T** file → fill the required details like treatment number, grain weight, leaf weight, etc.

(iv) Repeat step (ii) for the T file.

(v) Open **DSSAT** → Click **Crops** on the Selector → Click rice folder on the Cereals folder (for example). Click as per your crop type → Click **Refresh** (you can find your experimental file).

(vi) Click on the respective file → Click **Data** → You can see the A and T files.

(vii) Click on the respective file → Click **Run** → Select the treatments (require) → Click **Run model** → Click **Ok**.

(viii) Select Analyze → Select Overview → Click View. A notepad file will display all the details like crop, treatments involved and yield details for both measure and simulated.

(ix) Both the measured and simulated values should match. If not we need to change the genotype coefficients.

(x) Select Genetics in the selector → Click Cereals and select the respective crop → Right click on the .CUL file and add the respective coefficients by giving the name (e.g. BT07).

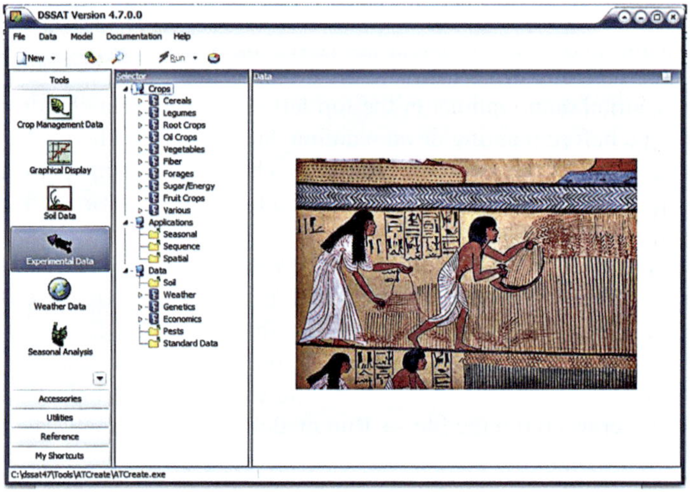

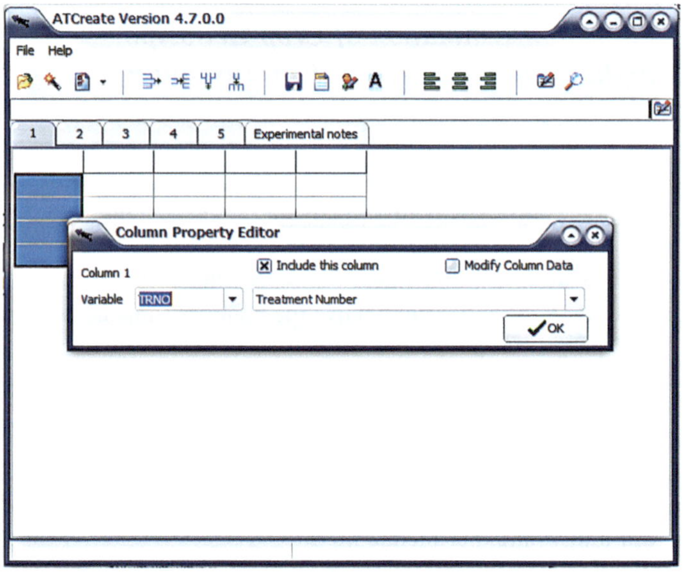

(xi) Select **Genotype Coefficient Calculator** in the tools → Click **Start GenSelect** → Click **Ok**.

(xii) Select a Crop and select a cultivar to calibrate → Click **Go** → Select Treatments to include → ClickGo → Type the filename (BT 07) and click Save → Click **Run GenCalc**.

(xiii) Click Files → Model Outputs → Overview → Check/Final Run. It will display the simulated values matched with the observed i.e. with less RMSE values.

(xiv) Repeat step (x) to see the modified Coefficients with the same name.

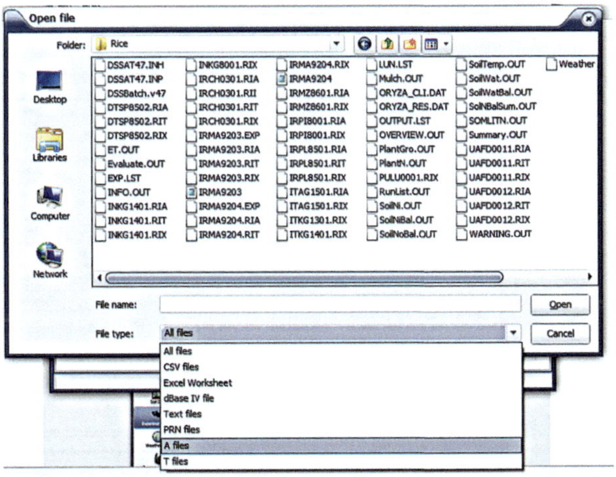
Open file
Folder: Rice
Desktop
Libraries
Computer
Network
DSSAT47.INH
DSSAT47.INP
DSSBatch.v47
DTSP8502.RIA
DTSP8502.RIT
DTSP8502.RIX
ET.OUT
Evaluate.OUT
EXP.LST
INFO.OUT
INKG1401.RIA
INKG1401.RIT
INKG1401.RIX
INKG8001.RIX
IRCH0301.RIA
IRCH0301.RII
IRCH0301.RIT
IRCH0301.RIX
IRMA9203.EXP
IRMA9203.RIA
IRMA9203.RIT
IRMA9203.RIX
IRMA9203
IRMA9204.EXP
IRMA9204.RIA
IRMA9204.RIT
IRMA9204.RIX
IRMA9204
IRMZ8601.RIA
IRMZ8601.RIX
IRPI8001.RIA
IRPI8001.RIX
IRPL8501.RIA
IRPL8501.RIT
IRPL8501.RIX
ITAG1501.RIA
ITAG1501.RIX
ITKG1301.RIX
ITKG1401.RIX
LUN.LST
Mulch.OUT
ORYZA_CLI.DAT
ORYZA_RES.DAT
OUTPUT.LST
OVERVIEW.OUT
PlantGro.OUT
PlantN.OUT
PULU0001.RIX
RunList.OUT
SoilN.OUT
SoilNiBal.OUT
SoilNoBal.OUT
SoilTemp.OUT
SoilWat.OUT
SoilWatBal.OUT
SoilNBalSum.OUT
SOMLITN.OUT
Summary.OUT
UAFD0011.RIA
UAFD0011.RIT
UAFD0011.RIX
UAFD0012.RIA
UAFD0012.RIT
UAFD0012.RIX
WARNING.OUT
Weather
File name:
File type: All files
Open
Cancel
All files
CSV files
Excel Worksheet
dBase IV file
Text files
PRN files
A files
T files

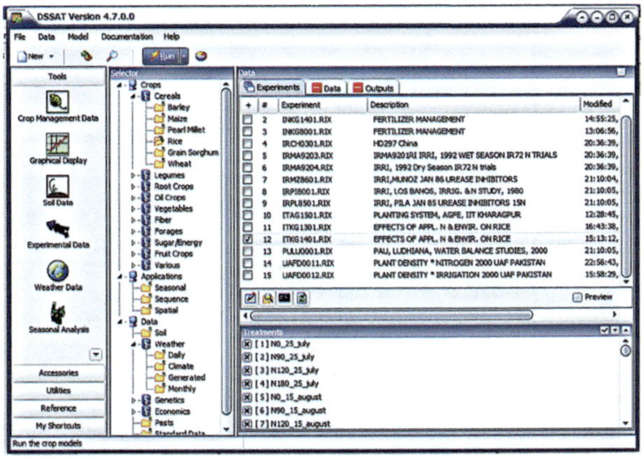
DSSAT Version 4.7.0.0
File Data Model Documentation Help
New
Run
Tools
Crop Management Data
Graphical Display
Soil Data
Experimental Data
Weather Data
Seasonal Analysis
Accessories
Utilities
Reference
My Shortcuts
Selector
Crops
Cereals
Barley
Maize
Pearl Millet
Rice
Grain Sorghum
Wheat
Legumes
Root Crops
Oil Crops
Vegetables
Fiber
Forages
Sugar/Energy
Fruit Crops
Various
Applications
Seasonal
Sequence
Spatial
Data
Soil
Weather
Daily
Climate
Generated
Monthly
Genetics
Economics
Pests
Data
Experiments
Data
Outputs
Experiment
Description
Modified
2 INKG1401.RIX FERTILIZER MANAGEMENT 14:55:25,
3 INKG8001.RIX FERTILIZER MANAGEMENT 13:06:56,
4 IRCH0301.RIX HD297 China 20:36:39,
5 IRMA9203.RIX IRMA9201RI IRRI, 1992 WET SEASON IR72 N TRIALS 20:36:39,
6 IRMA9204.RIX IRRI, 1992 Dry Season IR72 N trials 20:36:39,
7 IRMZ8601.RIX IRRI,MUNOZ JAN 86 UREASE INHIBITORS 21:10:04,
8 IRPI8001.RIX IRRI, LOS BANOS, IRRIG. &N STUDY, 1980 21:10:05,
9 IRPL8501.RIX IRRI, PILA JAN 85 UREASE INHIBITORS 15N 21:10:05,
10 ITAG1501.RIX PLANTING SYSTEM, AGFE, IIT KHARAGPUR 12:28:45,
11 ITKG1301.RIX EFFECTS OF APPL. N &ENVIR. ON RICE 16:43:38,
12 ITKG1401.RIX EFFECTS OF APPL. N &ENVIR. ON RICE 15:13:12,
13 PULU0001.RIX PAU, LUDHIANA, WATER BALANCE STUDIES, 2000 21:10:05,
14 UAFD0011.RIX PLANT DENSITY * NITROGEN 2000 UAF PAKISTAN 22:56:43,
15 UAFD0012.RIX PLANT DENSITY * IRRIGATION 2000 UAF PAKISTAN 15:58:29,
Preview
Treatments
[1] N0_25_july
[2] N90_25_july
[3] N120_25_july
[4] N180_25_july
[5] N0_15_august
[6] N90_15_august
[7] N120_15_august
Run the crop models

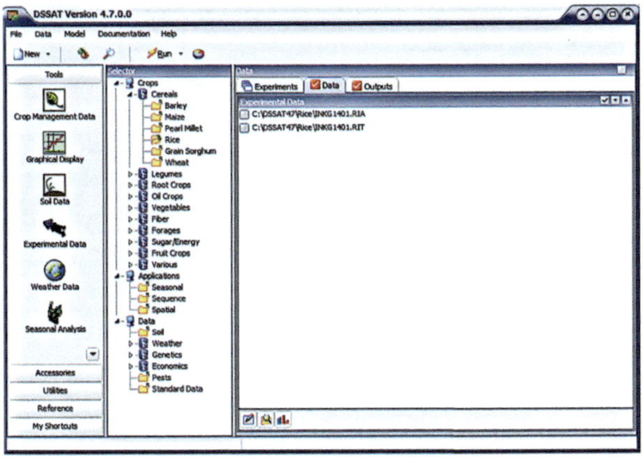
DSSAT Version 4.7.0.0
File Data Model Documentation Help
New
Run
Tools
Crop Management Data
Graphical Display
Soil Data
Experimental Data
Weather Data
Seasonal Analysis
Accessories
Utilities
Reference
My Shortcuts
Selector
Crops
Cereals
Barley
Maize
Pearl Millet
Rice
Grain Sorghum
Wheat
Legumes
Root Crops
Oil Crops
Vegetables
Fiber
Forages
Sugar/Energy
Fruit Crops
Various
Applications
Seasonal
Sequence
Spatial
Data
Soil
Weather
Genetics
Economics
Pests
Standard Data
Data
Experiments
Data
Outputs
Experimental Data
C:\DSSAT47\Rice\INKG1401.RIA
C:\DSSAT47\Rice\INKG1401.RIT

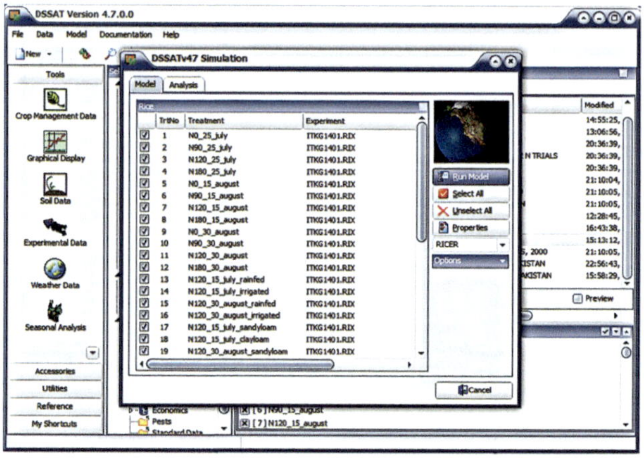
DSSAT Version 4.7.0.0
File Data Model Documentation Help
New
Tools
Crop Management Data
Graphical Display
Soil Data
Experimental Data
Weather Data
Seasonal Analysis
Accessories
Utilities
Reference
My Shortcuts
DSSATv47 Simulation
Model Analysis
Rice
TrtNo Treatment Experiment
1 N0_25_July ITKG1401.RIX
2 N90_25_July ITKG1401.RIX
3 N120_25_July ITKG1401.RIX
4 N180_25_July ITKG1401.RIX
5 N0_15_august ITKG1401.RIX
6 N90_15_august ITKG1401.RIX
7 N120_15_august ITKG1401.RIX
8 N180_15_august ITKG1401.RIX
9 N0_30_august ITKG1401.RIX
10 N90_30_august ITKG1401.RIX
11 N120_30_august ITKG1401.RIX
12 N180_30_august ITKG1401.RIX
13 N120_15_July_rainfed ITKG1401.RIX
14 N120_15_July_irrigated ITKG1401.RIX
15 N120_30_august_rainfed ITKG1401.RIX
16 N120_30_august_irrigated ITKG1401.RIX
17 N120_15_July_sandyloam ITKG1401.RIX
18 N120_15_July_clayloam ITKG1401.RIX
19 N120_30_august_sandyloam ITKG1401.RIX
Run Model
Select All
Unselect All
Properties
RICER
Options
Cancel
Modified
Preview

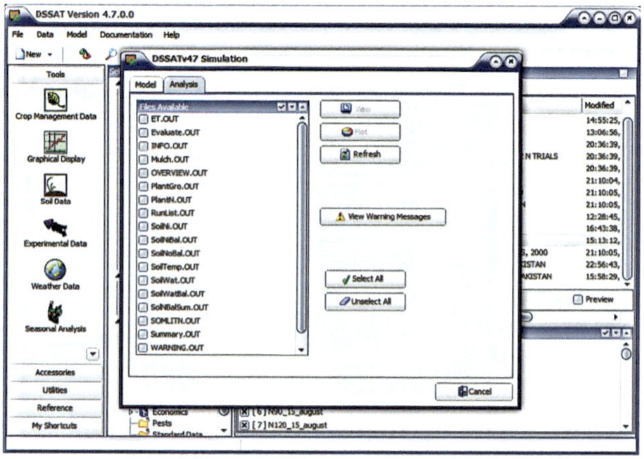
DSSAT Version 4.7.0.0
File Data Model Documentation Help
DSSATv47 Simulation
Model Analysis
Files Available
ET.OUT
Evaluate.OUT
INFO.OUT
Mulch.OUT
OVERVIEW.OUT
PlantGro.OUT
PlantN.OUT
RunList.OUT
SoilNi.OUT
SoilNBal.OUT
SoilNoBal.OUT
SoilTemp.OUT
SoilWat.OUT
SoilWatBal.OUT
SoilNBalSum.OUT
SOMLITN.OUT
Summary.OUT
WARNING.OUT
View
Plot
Refresh
View Warning Messages
Select All
Unselect All
Cancel

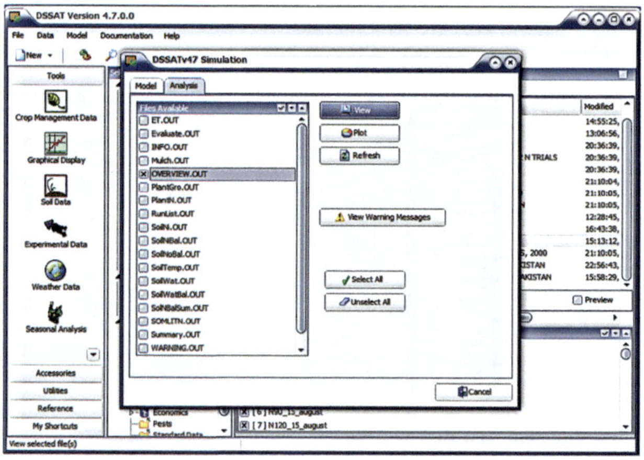
DSSAT Version 4.7.0.0
File Data Model Documentation Help
DSSATv47 Simulation
Model Analysis
Files Available
ET.OUT
Evaluate.OUT
INFO.OUT
Mulch.OUT
OVERVIEW.OUT
PlantGro.OUT
PlantN.OUT
RunList.OUT
SoilNi.OUT
SoilNBal.OUT
SoilNoBal.OUT
SoilTemp.OUT
SoilWat.OUT
SoilWatBal.OUT
SoilNBalSum.OUT
SOMLITN.OUT
Summary.OUT
WARNING.OUT
View
Plot
Refresh
View Warning Messages
Select All
Unselect All
Cancel
View selected file(s)

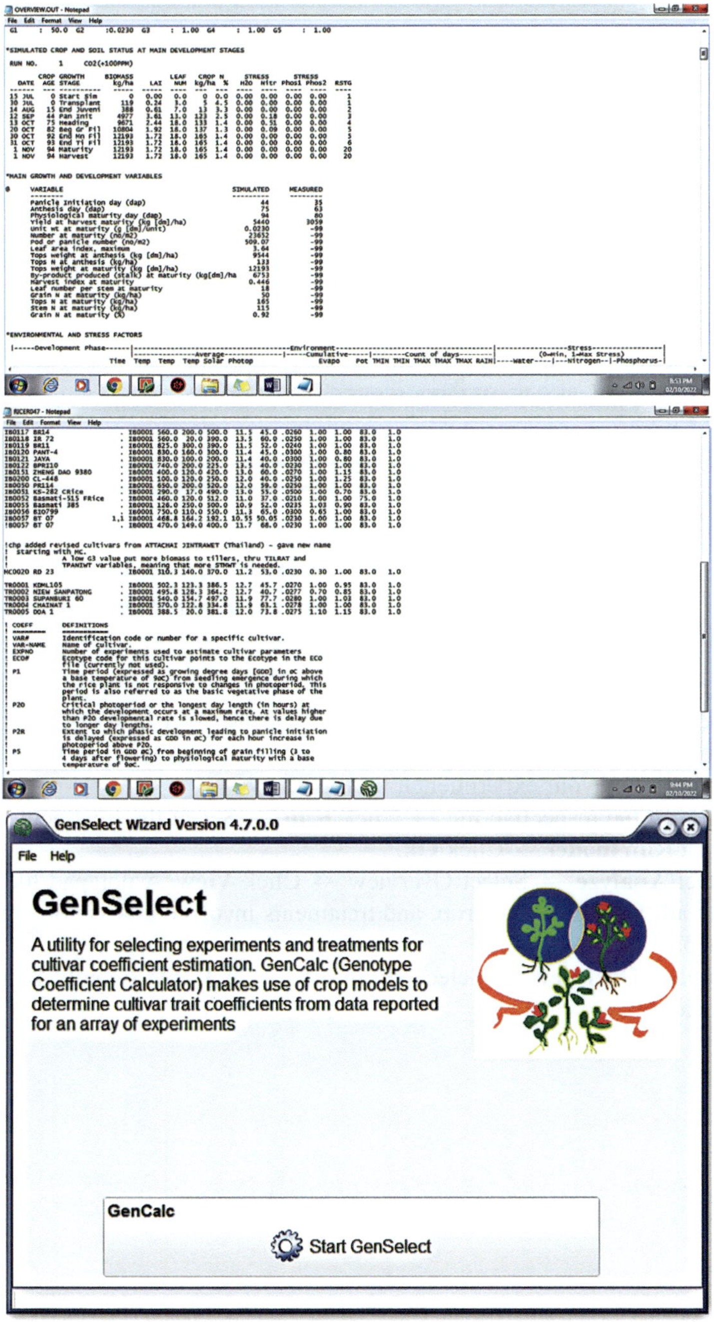
GenSelect Wizard Version 4.7.0.0
File Help
GenSelect
A utility for selecting experiments and treatments for cultivar coefficient estimation. GenCalc (Genotype Coefficient Calculator) makes use of crop models to determine cultivar trait coefficients from data reported for an array of experiments
GenCalc
Start GenSelect

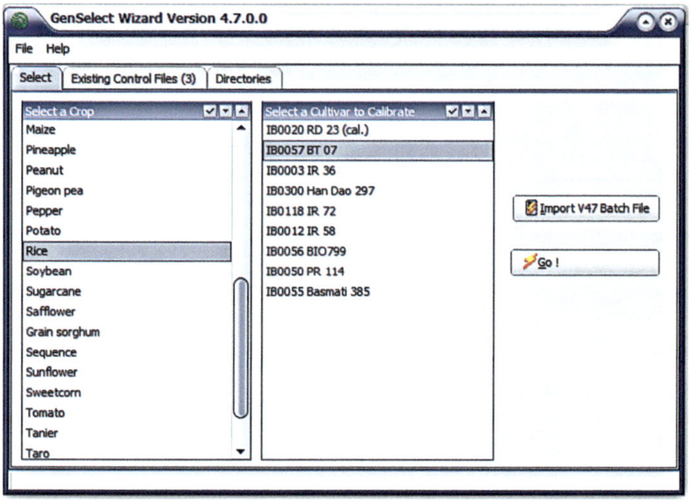

11.7.5 General Crop Yield Simulations (Data View and Graphical View) Pathway for Crop Simulations (Individual Years)

(i) Open **DSSAT** → Click **Crops** on the Selector → Click rice folder on the Cereal folder (for example). Click as per your crop type → Click **Refresh** (you can find your experimental file).

(ii) Click on the respective file → Click **Run** → Select the treatments (require) → Cl **Run model** → Click **Ok**.

(iii) Select **Analyze** → Select Overview → Click **View**. A notepad file will display all the details like crop, and treatments involved and yield details in the notepad file.

(iv) Unselect **Overview** → Select **PlantGro** → Click **Plot** (Gbuild Window will appear).

(v) Select the **variable** (like Grain number, Grain weight, etc.) and select the **treatments** on the **Run** → Click **Next** (Graph will appear) → You can print or export the data into text or excel file by clicking the respective options → Click Back (generate new graph with different combination).

Pathway for Seasonal Analysis

(xv) **Open** DSSAT → **Click** Crop Management on Tools (**Xbuild window will appear**) → **Click** file type as Seasonal **on the drop-down menu**.
(xvi) Fill the remaining as it is explained in the Pathway for the Experimental File and save it.
(xvii) Click Applications on the Selector → Click Seasonal folder → Click Refresh (you can find your seasonal file) ix) Follow steps (ii) to (v) for the **results**.

The above images are the pictorial representation of the above steps of both pathways.

11.7.6 Disaggregated Weather Information and Crop Yield Simulation (Using Matlab for Disaggregation)

(i) Open DSSAT → Program → Select Weatherman → Select New Station → Choose input or import weather data → save as a new station → go → Open folder (sign) → select file → open it → Unit selection of each weather parameter (like date, solar radiation, Tmax, Tmin and Rainfall) → Assign unit and range of values for the weather parameters (solar radiation, Tmax, Tmin and Rainfall) → delete the upper row → Import data into weatherman → give station name → save station → station name → fill Lat. Long. Elevation of your station → writes files.

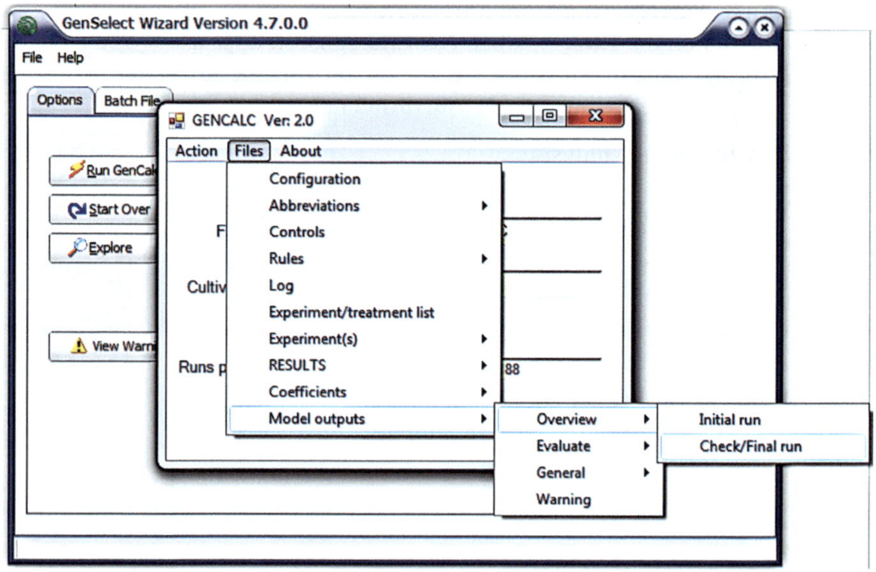
GenSelect Wizard Version 4.7.0.0
File Help
Options
Batch File
Run GenCal
Start Over
Explore
View Warn
GENCALC Ver: 2.0
Action Files About
Configuration
Abbreviations
Controls
Rules
Log
Experiment/treatment list
Experiment(s)
RESULTS
Coefficients
Model outputs
Overview
Evaluate
General
Warning
Initial run
Check/Final run
Cultiv
Runs p
88

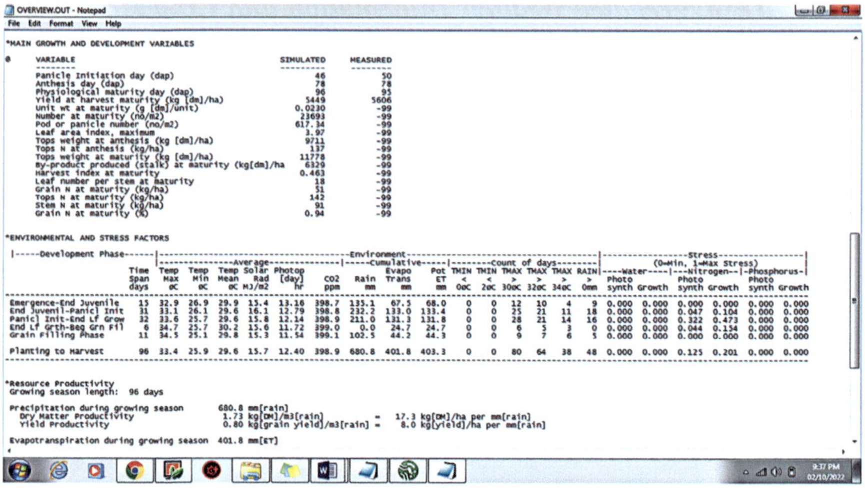
OVERVIEW.OUT - Notepad
File Edit Format View Help
*MAIN GROWTH AND DEVELOPMENT VARIABLES
@ VARIABLE SIMULATED MEASURED
Panicle Initiation day (dap) 46 50
Anthesis day (dap) 78 78
Physiological maturity day (dap) 96 95
Yield at harvest maturity (kg [dm]/ha) 5449 5606
Unit wt at maturity (g [dm]/unit) 0.0230 -99
Number at maturity (no/m2) 23693 -99
Pod or panicle number (no/m2) 617.34 -99
Leaf area index, maximum 3.97 -99
Tops weight at anthesis (kg [dm]/ha) 9711 -99
Tops N at anthesis (kg/ha) 137 -99
Tops weight at maturity (kg [dm]/ha) 11778 -99
By-product produced (stalk) at maturity (kg[dm]/ha 6329 -99
Harvest index at maturity 0.463 -99
Leaf number per stem at maturity 18 -99
Grain N at maturity (kg/ha) 51 -99
Tops N at maturity (kg/ha) 142 -99
Stem N at maturity (kg/ha) 91 -99
Grain N at maturity (%) 0.94 -99
*ENVIRONMENTAL AND STRESS FACTORS
|-----Development Phase------|----Environment----|----Stress----|
Average Cumulative Count of days (0=Min, 1=Max Stress)
Time Span days | Temp Max ºC | Temp Min ºC | Temp Mean ºC | Solar Rad MJ/m2 | Photop [day] hr | CO2 ppm | Rain mm | Evapo Trans mm | Pot ET mm | TMIN < 0ºC | TMIN < 2ºC | TMAX > 30ºC | TMAX > 32ºC | TMAX > 34ºC | RAIN > 0mm | Water Photo synth | Water Growth | Nitrogen Photo synth | Nitrogen Growth | Phosphorus Photo synth | Phosphorus Growth
Emergence-End Juvenile 15 32.9 26.9 29.9 15.4 13.16 398.7 135.1 67.5 68.0 0 0 12 10 4 9 0.000 0.000 0.000 0.000 0.000 0.000
End Juvenil-Panicl Init 31 33.1 26.1 29.6 16.1 12.79 398.8 232.2 133.0 133.4 0 0 25 21 11 18 0.000 0.000 0.047 0.104 0.000 0.000
Panicl Init-End Lf Grow 32 33.6 25.7 29.6 15.8 12.14 398.9 211.0 131.3 131.8 0 0 28 21 14 16 0.000 0.000 0.322 0.473 0.000 0.000
End Lf Grth-Beg Grn Fil 6 34.7 25.7 30.2 15.6 11.72 399.0 0.0 24.7 24.7 0 0 6 5 3 0 0.000 0.000 0.044 0.154 0.000 0.000
Grain Filling Phase 11 34.5 25.1 29.8 15.3 11.54 399.1 102.5 44.2 44.3 0 0 9 7 6 5 0.000 0.000 0.000 0.000 0.000 0.000
Planting to Harvest 96 33.4 25.9 29.6 15.7 12.40 398.9 680.8 401.8 403.3 0 0 80 64 38 48 0.000 0.000 0.125 0.201 0.000 0.000
*Resource Productivity
Growing season length: 96 days
Precipitation during growing season 680.8 mm[rain]
Dry Matter Productivity 1.73 kg[DM]/m3[rain] = 17.3 kg[DM]/ha per mm[rain]
Yield Productivity 0.80 kg[grain yield]/m3[rain] = 8.0 kg[yield]/ha per mm[rain]
Evapotranspiration during growing season 401.8 mm[ET]

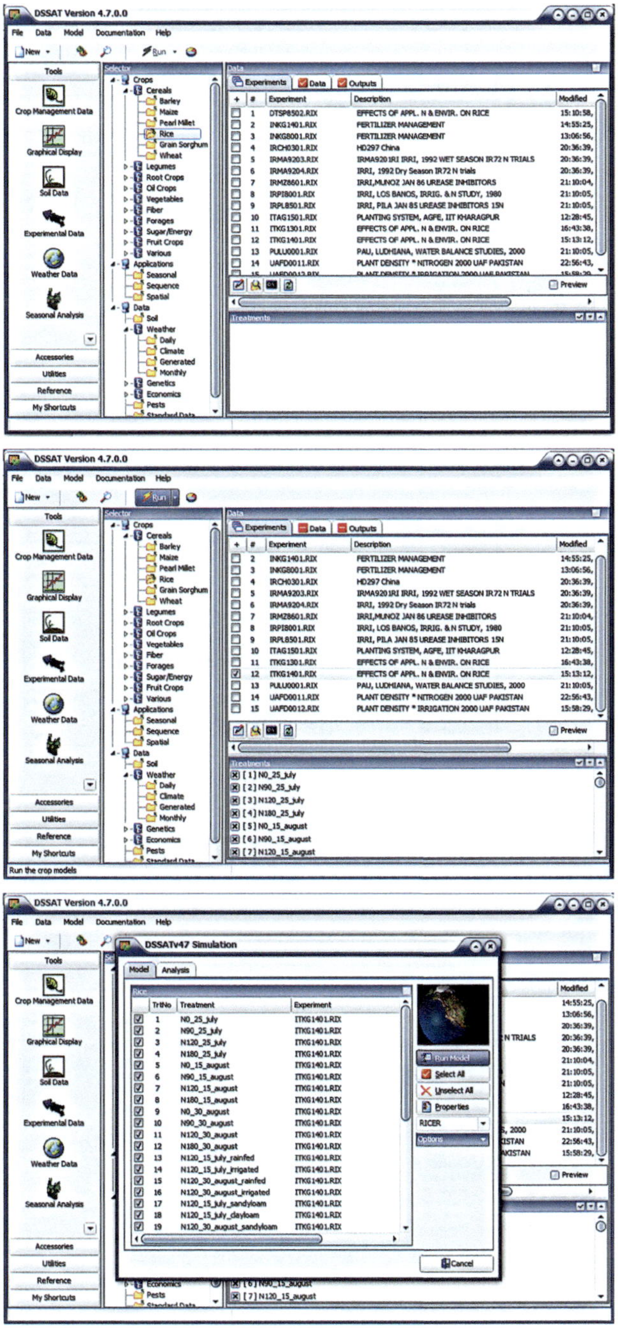
DSSAT Version 4.7.0.0
File Data Model Documentation Help
Tools
Crop Management Data
Graphical Display
Soil Data
Experimental Data
Weather Data
Seasonal Analysis
Accessories
Utilities
Reference
My Shortcuts
Experiments
Data
Outputs
Treatments
Preview
Run the crop models
DSSATv47 Simulation
Model
Analysis
Run Model
Select All
Unselect All
Properties
Cancel

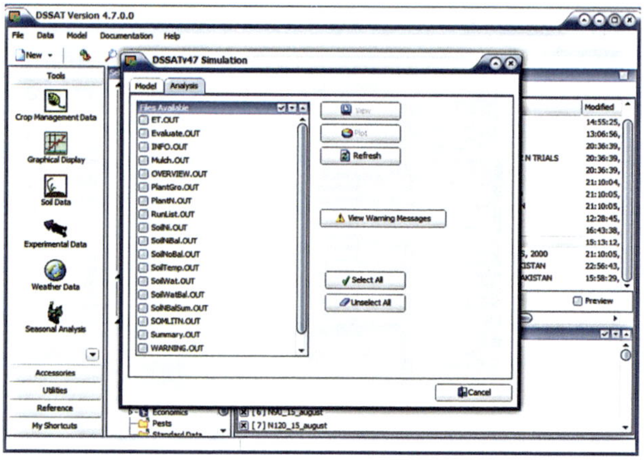
DSSAT Version 4.7.0.0
File Data Model Documentation Help
New
Tools
Crop Management Data
Graphical Display
Soil Data
Experimental Data
Weather Data
Seasonal Analysis
Accessories
Utilities
Reference
My Shortcuts
DSSATv47 Simulation
Model
Analysis
Files Available
ET.OUT
Evaluate.OUT
INFO.OUT
Mulch.OUT
OVERVIEW.OUT
PlantGro.OUT
PlantN.OUT
RunList.OUT
SoilNi.OUT
SoilNBal.OUT
SoilNoBal.OUT
SoilTemp.OUT
SoilWat.OUT
SoilWatBal.OUT
SoilNBalSum.OUT
SOMLITN.OUT
Summary.OUT
WARNING.OUT
View
Plot
Refresh
View Warning Messages
Select All
Unselect All
Cancel
Modified
Preview
Pests
[7] N120_15_august

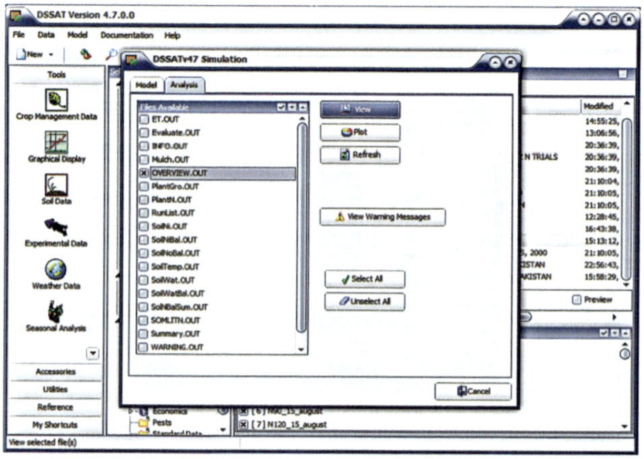
DSSAT Version 4.7.0.0
File Data Model Documentation Help
DSSATv47 Simulation
Model
Analysis
Files Available
ET.OUT
Evaluate.OUT
INFO.OUT
Mulch.OUT
OVERVIEW.OUT
PlantGro.OUT
PlantN.OUT
RunList.OUT
SoilNi.OUT
SoilNBal.OUT
SoilNoBal.OUT
SoilTemp.OUT
SoilWat.OUT
SoilWatBal.OUT
SoilNBalSum.OUT
SOMLITN.OUT
Summary.OUT
WARNING.OUT
View
Plot
Refresh
View Warning Messages
Select All
Unselect All
Cancel
View selected file(s)

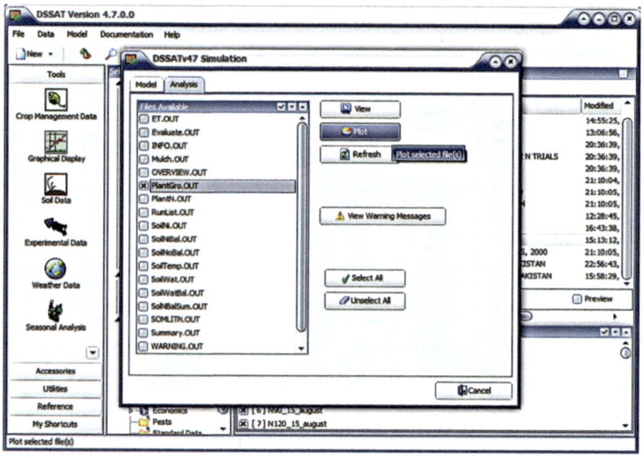
DSSAT Version 4.7.0.0
File Data Model Documentation Help
DSSATv47 Simulation
Model
Analysis
Files Available
ET.OUT
Evaluate.OUT
INFO.OUT
Mulch.OUT
OVERVIEW.OUT
PlantGro.OUT
PlantN.OUT
RunList.OUT
SoilNi.OUT
SoilNBal.OUT
SoilNoBal.OUT
SoilTemp.OUT
SoilWat.OUT
SoilWatBal.OUT
SoilNBalSum.OUT
SOMLITN.OUT
Summary.OUT
WARNING.OUT
View
Plot
Refresh
Plot selected file(s)
View Warning Messages
Select All
Unselect All
Cancel
Plot selected file(s)

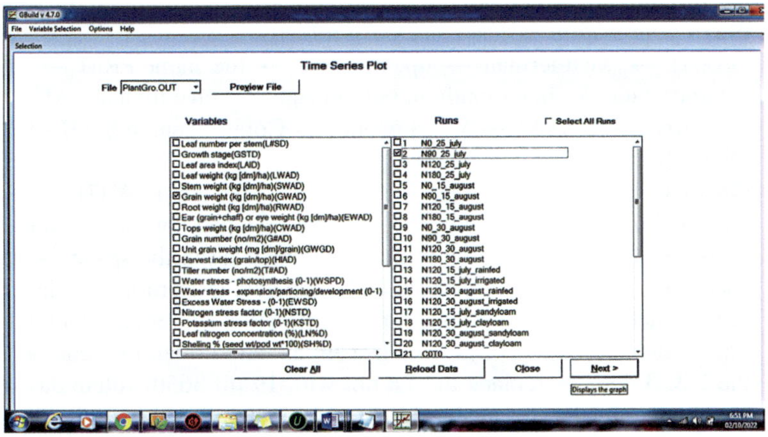
GBuild v 4.7.0
File Variable Selection Options Help
Selection
Time Series Plot
File PlantGro.OUT
Preview File
Variables
Runs
Select All Runs
Leaf number per stem(L#SD)
Growth stage(GSTD)
Leaf area index(LAID)
Leaf weight (kg [dm]/ha)(LWAD)
Stem weight (kg [dm]/ha)(SWAD)
Grain weight (kg [dm]/ha)(GWAD)
Root weight (kg [dm]/ha)(RWAD)
Ear (grain+chaff) or eye weight (kg [dm]/ha)(EWAD)
Tops weight (kg [dm]/ha)(CWAD)
Grain number (no/m2)(G#AD)
Unit grain weight (mg [dm]/grain)(GWGD)
Harvest index (grain/top)(HIAD)
Tiller number (no/m2)(T#AD)
Water stress - photosynthesis (0-1)(WSPD)
Water stress - expansion/partioning/development (0-1)
Excess Water Stress - (0-1)(EWSD)
Nitrogen stress factor (0-1)(NSTD)
Potassium stress factor (0-1)(KSTD)
Leaf nitrogen concentration (%)(LN%D)
Shelling % (seed wt/pod wt*100)(SH%D)
1 N0_25_july
2 N90_25_july
3 N120_25_july
4 N180_25_july
5 N0_15_august
6 N90_15_august
7 N120_15_august
8 N180_15_august
9 N0_30_august
10 N90_30_august
11 N120_30_august
12 N180_30_august
13 N120_15_july_rainfed
14 N120_15_july_irrigated
15 N120_30_august_rainfed
16 N120_30_august_irrigated
17 N120_15_july_sandyloam
18 N120_15_july_clayloam
19 N120_30_august_sandyloam
20 N120_30_august_clayloam
21 COTO
Clear All
Reload Data
Close
Next >
Displays the graph
6:51 PM
02/10/2022

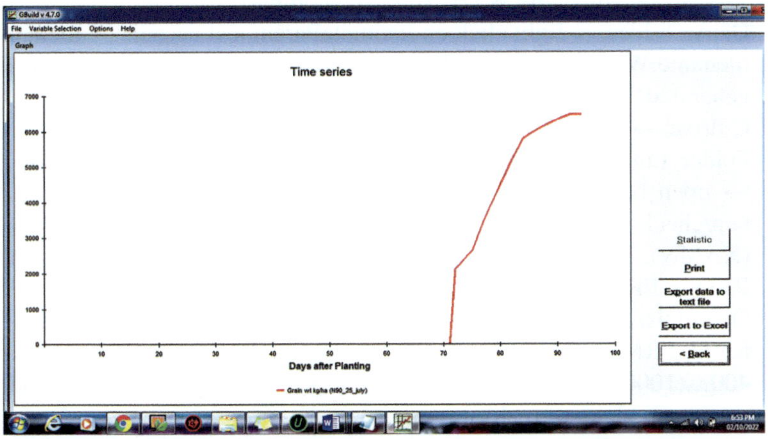
GBuild v 4.7.0
File Variable Selection Options Help
Graph
Time series
7000
6000
5000
4000
3000
2000
1000
0
10
20
30
40
50
60
70
80
90
100
Days after Planting
Grain wt kg/ha (N90_25_july)
Statistic
Print
Export data to text file
Export to Excel
< Back
6:53 PM
02/10/2022

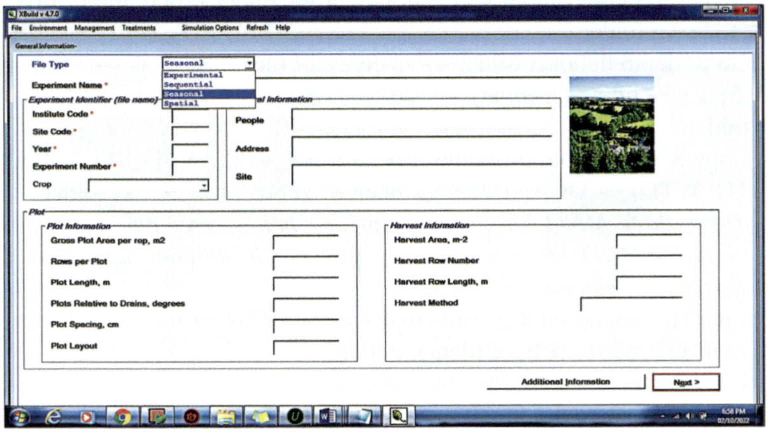
XBuild v 4.7.0
File Environment Management Treatments Simulation Options Refresh Help
General Information
File Type
Seasonal
Experimental
Sequential
Seasonal
Spatial
Experiment Name *
Experiment Identifier (file name)
Institute Code *
Site Code *
Year *
Experiment Number *
Crop
People
Address
Site
Plot
Plot Information
Gross Plot Area per rep, m2
Rows per Plot
Plot Length, m
Plots Relative to Drains, degrees
Plot Spacing, cm
Plot Layout
Harvest Information
Harvest Area, m-2
Harvest Row Number
Harvest Row Length, m
Harvest Method
Additional Information
Next >
6:58 PM
02/10/2022

(ii) Go to C drive → DSSAT → Weather → Climate → copy file name.CLI file of your station → paste in DisAg folder.

(iii) DSSAT → weatherman → open station → file name.PRM → observed weather data → click month on bottom right → save filename.MTH → Go to C drive → DSSAT → W → Month → Copy filename.MTH → Paste in DisAg folder.

(iv) Go to C drive → DSSAT → Weather → file name.WTH → Paste in DisAg. Folder → open the same file in Text pad format → delete 1st 3 lines → press indent option to maintain/check/align the space → give two spaces (..) before date → delete all parameters after rain → select configure → block select mode → go to search option → replace 1st (.) dot with 4 (...) dots after Julian days → select Julian days coloum → cut → paste in the 2 & 3 dots → replace the 1st dot with 19 till 365th Julian day of 1999 → replace the 1st dot again with 20 from 1st Julian day of 2000 → 2014 → save → "filename.WTH" (inverted comma).

(v) DisAg folder → right click on EstimateParam test → edit → filename.CLI, filename.WTD click (run) EstimateParam test → filename .PRM will be generated.

(vi) C drive → DSSAT → weather → copy filename .MTH → Paste DisAg. Folder (another. MTH file named as ERFS.MTH in the DisAg. Folder → open both .MTH files in Text Pad → press indent ¶ option to maintain/check/align the space → Copy the values of SRad (SAMN), Tmax (XAMN), Tmin (NAMN), and Rainfall (RAMN) from filename .MTH → Paste in ERFS.MTH (check the space/alignment).

(vii) Open DisAg folder → right click on DisAggregate HOSH → edit → filename .PRM filename .MTH (and the required no. of realizations say 200, 400…,1000) → Run DisAggregate HOSH → program will run and generate specified no. of realizations.

(viii) Copy the entire output realization data → folder (1000_realization_WTD) in DisAg folder.

(ix) Go to demo data set folder → open excel file → save as → formatted text (Space delimited format) → save → copy → filename .PRN → DisAg folder.

(x) Copy 3 Matlab tool programs (corr_for_kg_work, WTD_TO_WTH, WTH_TO_WTD) → DisAg folder → open → (corr for kg work) folder.
These three MATLAB code is provided below, user can copy those program for MATLAB or write their own code in different languages such as Fortran, python etc.
Line 16: specify path of 1000 realisations .WTD ensure "‘’"
Line 17: specify path of filename .PRN
Line 20: specify no. of years, first year
Line 28: change the file name
Line 33: change the no. of years
"Run the model"
Four windows with plotted graphs will be generated.

(xi) Open → demo data set → filename .WHX → open → text pad → open your experimental file also → select fields (e.g. HOSB change it to your name weather file name) → soil ID → replace it with your soil profile file ID. This will be the 1st line of your treatment, from 2nd line onwards copy from 2 → 201 treatments from filename .WHX → paste it below.

(xii) Name the treatments (2 → 201) same as above → don't disturb the alignments → make simulation options as seasonal → close → save → paste → c drive → DSSAT → specified crop. Generating WTD → WTH → open → Dis.Ag folder → open → WTD TO WTH.m Matlab file →.

Line 02: specify path of 1000 realisations. WTD ensure"'.

Line 03: Specify station name.

Line 04: rem file name (e.g. starting year and no. Years).

Line 05: No. of realisations = 200 → Save → run.

(xiii) Open 1000_realisation folder → a new folder created name WTH →.

(xiv) DSSAT → crop → cereals (respective crops) → run the model → experimental file.

(xv) Training folder → HOSB.BAT file → copy → paste → open with text pad → minimize → Go to C drive → DSSAT → Crop → Cereals (respective crop) → open → DSSBatch.

V → Copy command line path → paste in → HOSB.BAT file → save → renames.

→ filename .BAT Copy and Paste it in C Drive → DSSAT in specified crop.

(xvi) Open DisAg → 1000 realization WTD → open → WTH → copy → 200 realizations → paste → C Drive DSSAT → weather.

(xvii) DisAg → copy → weather file .WTH (e.g. HOSB8728) → paste → 1000 realisation.WTD folder → again paste → rename as filename starting year 01 (e.g. HOSB8701) → copy this file → c drive → DSSAT → weather.

(xviii) Run → HOSB.BAT file which is in C Drive → DSSAT respective crop_ folder RUN the BAT file → go to → DSSAT → respective crop → copy summary OUT or OSU file → go to training folder → D is-Ag → open folder yield with dly Wx sequences 1 folder → paste → summary → output file → open goodness of fits plots simulated rice → path of summary output file in MATLAB → give name of output file → mention years everywhere → mention yield column no. (4 places) → then run Plot graphs.

(xix) Go to command window in MATLAB → Work space → select → rice yield c → select all the contents → copy → paste → new excel sheet → 1 column (observed yield) then 1–200 (simulated yield) → correlate with observed → take averages of 2–200 realizations → copy observed and average of 200 realisation in sheet two → plot line graph → mention years.

(xx) For using ERFS data copy the previously created filename .MTH file → Dis.Ag folder → copy filename .MTH file → paste in the same folder → open .MTH file → Text Pad → del all the years by retaining.

MATLAB Code: WTD_TO_WTH.M and WTH_TO WTD.M

```
%%%%%%%%%%%%%%%%%%%%%%%%%%%%%%%%%%%%%%%%%%%%%%%%%%
%%%%%%%%                                  %%%%%%%%
%%%%%%%%       WTD_TO_WTH.m               %%%%%%%%
%%%%%%%%                                  %%%%%%%%
%%%%%%%%%%%%%%%%%%%%%%%%%%%%%%%%%%%%%%%%%%%%%%%%%%
clc; clear all; close all;
%% Eiditing Section
wtddir='C:\Users\admin\....\CRM WORK\Scenario\'; % Location Directory of WTD file
stn_name='STNM';
rem_file_name='8728';
no_of_realization=200;
%% Do Not Touching Section
wthdir=strcat(wtddir,'WTH\');
mkdir(wthdir);
for i=1:no_of_realization
  flnmin=strcat(wtddir,stn_name,rem_file_name,'.WTH');
  if(i<10)
    fextn=strcat('000',num2str(i));
  elseif(i<=10||i<100)
    fextn=strcat('00',num2str(i));
  elseif(i<=100||i<1000)
    fextn=strcat('0',num2str(i));
  else
    fextn=strcat(num2str(i));
  end
  flnmo=strcat(wtddir,stn_name,fextn,'.WTD');
  flnmout=strcat(wthdir,fextn,rem_file_name(1:2),'01.WTH');
  fpi=fopen(flnmin,'r');
  fpo=fopen(flnmo,'r');
  fpout=fopen(flnmout,'wt');
  l1=fgetl(fpi);
  l2=fgetl(fpi);
  l3=fgetl(fpi);
  l4=fgetl(fpi);
  l5=fgetl(fpi);
  fprintf(fpout,l1);fprintf(fpout,'\n');
  fprintf(fpout,l2);fprintf(fpout,'\n');
  fprintf(fpout,l3);fprintf(fpout,'\n');
  fprintf(fpout,l4);fprintf(fpout,'\n');
  fprintf(fpout,l5);fprintf(fpout,'\n');
  ll=fgetl(fpo);
  while(feof(fpo)==0)
    line=fgetl(fpo);
    ll=line(3:32);
    fprintf(fpout,ll);fprintf(fpout,'\n');
    clear line;
  end
```

```
%%%%%%%%%%%%%%%%%%%%%%%%%%%%%%%%%%%%%%%%%%%%%%%%%%%%%%%%
%%%%%%%%                                        %%%%%%%%
%%%%%%%%        WTH_TO_WTD.m                    %%%%%%%%
%%%%%%%%                                        %%%%%%%%
%%%%%%%%%%%%%%%%%%%%%%%%%%%%%%%%%%%%%%%%%%%%%%%%%%%%%%%%
clc; clear all; close all;
%% Eiditing Section
dir='C:\Users\admin\...\CRM WORK\'; % Location Directory of WTH file
filename='STNM8728.WTH';
%% Do Not Touching Section
flnmin=strcat(dir,filename);
flnmo=strcat(dir,filename(1:4),'.WTD');
fpi=fopen(flnmin,'r');
fpo=fopen(flnmo,'wt');
l=fgetl(fpi);
l=fgetl(fpi);
l=fgetl(fpi);
l=fgetl(fpi);
line=fgetl(fpi);
fprintf(fpo,line(1));fprintf(fpo,' ');fprintf(fpo,line(2:30));fprintf(fpo,'\n');
while(feof(fpi)==0)
  line=fgetl(fpi);
  if(str2num(line(1:2))<str2num(filename(5:6)))
    fprintf(fpo,'20');fprintf(fpo,line(1:29));fprintf(fpo,'\n');
  else
    fprintf(fpo,'19');fprintf(fpo,line(1:29));fprintf(fpo,'\n');
  end
  clear line;
end
clear l*;
fclose all;
```

current year (In case of *rabi* crops keep previous as well as current year) → replace values of with-99 of Tmax, Tmin, SR → change year in the first line → paste ERFS rainfall values in the file by using ERFS probability for every month → if ERFS data is not available → take mean value of RNUM from .CLI file → divided by respective months → copy → paste in .MTH file → save → Open Dis.Ag HOSH.BAT file from Dis. Ag folder → double click → 1000 WTD realization will be again created.

(xxi) Then go to ERFS hindcast JJAS precip.xlsx (for *kharif* crops) or ERFS hindcast DJ (for *rabi* crops) → open the excel file → select your state met. Division data → copy your station data → paste and transpose it in sheet 2 → calculate monthly rainfall based on ERFS forecast → again will transpose monthly data in sheet 3 of the same file → now save as excel file in formatted text format (filename .PRN) → now open the PRN file and monthly file → copy monthly rainfall mean from PRN file → paste → monthly file → save → run di-seg HOSH.BAT file and 1000 realisation will be generated → continue the steps from serial nos. 8–19.

11.8 Guidelines for Utilisation of CRM in Practice

The CRM tools assist policy makers/engineers/farmers in timely decision making in agricultural operations to minimise the adverse impact of variable weather and climate on crop production. These tools can be generated through several techniques using remote sensing, crop modelling, statistical modelling, AI/ML techniques, etc. Crop modelling is one of the decision-making tools to generate CRM options for different crops, seasons and specific locations. The use of crop models to answer the *what if* scenarios in generating various managements options i.e. planting time or sowing window, tillage, seedbed techniques, seeding depth, irrigation water management, intercultural operations, fertiliser management, pesticide applications and harvesting time to avoid the loss of crop yield due to in-seasonal weather variability. Based upon the simulation outputs (crop yield and economics), the best suited agricultural management options can be formulated as CRM options, which can be communicated to farmers in the form of weather based agro-advisory for implementation at the farm level.

References

Abeysingha, N., Singh, M., Islam, A. and Sehgal, V. (2016). Climate change impacts on irrigated rice and wheat production in Gomti river basin of India: A case study. *SpringerPlus,* **5(1):** 1–20.

Abrol, Y.P., Gadgil, S. (1999). Rice in a variable climate. APC Publications, New Delhi.

Achaya, K.T. (1994). Indian food: A historical companion. Oxford University Press, Delhi, xvi, 322 pages.

Acharya, N., Chattopadhyay, S., Mohanty, U., Dash, S. and Sahoo, L. (2013). On the bias correction of general circulation model output for Indian summer monsoon. *Meteorological Applications,* **20(3):** 349–356.

Acock, B., Charles-Edwards, D., Fitter, D., Hand, D., Ludwig, L., Warren Wilson, J. and Withers, A. (1978). The contribution of leaves from different levels within a tomato crop to canopy net photosynthesis: An experimental examination of two canopy models. *Journal of Experimental Botany,* **29(4):** 815–827.

Aggarwal, P.K., Kalra, N., Chander, S. and Pathak, H. (2006). Infocrop: A dynamic simulation model for the assessment of crop yields, losses due to pests, and environmental impact of agro-ecosystems in tropical environments. i. model description. *Agricultural Systems,* **89(1):** 1–25.

Ahmad, I., Ahmad, B., Boote, K. and Hoogenboom, G. (2020). Adaptation strategies for maize production under climate change for semi-arid environments. *European Journal of Agronomy,* **115:** 126040.

Allen, L., Pan, D., Boote, K., Pickering, N. and Jones, J. (2003). Carbon dioxide and temperature effects on evapotranspiration and water use efficiency of soybean. *Agronomy Journal,* **95(4):** 1071–1081.

Anandaraja, N., Rathakrishnan, T., Ramasubramanian, M., Saravanan, P. and Suganthi, N. (2008). Indigenous weather and forecast practices of Coimbatore district farmers.

Anthes, R.A. (1977). A cumulus parameterization scheme utilizing a one dimensional cloud model. *Monthly Weather Review,* **105(3):** 270–286.

Artikov, I., Hoffman, S.J., Lynne, G.D., Zillig, L.M.P., Hu, Q., Tomkins, A.J., Hubbard, K.G., Hayes, M.J. and Waltman, W. (2006). Understanding the influence of climate forecasts on farmer decisions as planned behavior. *Journal of Applied Meteorology and Climatology,* **45(9):** 1202–1214.

Attri, S. and Rathore, L. (2003). Simulation of impact of projected climate change on wheat in India. *International Journal of Climatology: A Journal of the Royal Meteorological Society,* **23(6):** 693–705.

Ayana, G.F., Megento, T.L. & Kussa, F.G. (2022). The extent of livelihood diversification on the determinants of livelihood diversification in Assosa Wereda, Western Ethiopia. *Geo Journal,* **87:** 2525–2549. https://doi.org/10.1007/s10708-021-10379-5

U. Mohanty et al., *Climate Risk Management in Agriculture*,
https://doi.org/10.1007/978-3-031-51862-1

Baffaut, C., Nearing, M.A. and Nicks, A.D. (1996). Impact of cligen parameters on wepp-predicted average annual soil loss. Transactions of the ASAE, **39(2):** 447–457.

Bal, S.K. and Minhas, P.S. (2017). Atmospheric stressors: Challenges and coping strategies. *In:* Abiotic stress management for resilient agriculture, pages 9–50. Springer

Balan Sarojini, B., Stott, P.A., Black, E. and Polson, D. (2012). Fingerprints of changes in annual and seasonal precipitation from cmip5 models overland and ocean. Geophysical Research Letters, 39(21).

Banerjee, S.N. (1956). On the incidence of paddy pests in West Bengal. *In:* Proc. Zool. Soc, **9:** 65–83.

Bannayan, M., Crout, N. and Hoogenboom, G. (2003). Application of the ceres-wheat model for within-season prediction of winter wheat yield in the United Kingdom. *Agronomy Journal,* **95(1):** 114–125.

Barnes, S.L. (1964). A technique for maximizing details in numerical weather map analysis. *Journal of Applied Meteorology,* **3(4):** 396–409.

Barnett, T. and Preisendorfer, R. (1987). Origins and levels of monthly and seasonal forecast skill for united states surface air temperatures determined by canonical correlation analysis. *Monthly Weather Review,* **115(9):** 1825–1850.

Basu, S. (1953). Weather lore in India. *MAUSAM,* **4(1)**: 3–12.

Batchelor, W., Jones, J., Boote, K. and Pinnschmidt, H. (1993). Extending the use of crop models to study pest damage. Transactions of the ASAE, **36(2):** 551–558.

Baumgard, L.H. and Rhoads Jr, R.P. (2013). Effects of heat stress on post absorptive metabolism and energetics. *Annu. Rev. Anim. Biosci.,* **1(1):** 311–337.

Bhatt, D., Sonkar, G. and Mall, R. (2019). Impact of climate variability on the rice yield in Uttar Pradesh: An agro-climatic zone based study. Environmental Processes, **6(1):** 135–153.

Bindi, M., Fibbi, L., Gozzini, B., Orlandini, S. and Miglietta, F. (1996). Modelling the impact of future climate scenarios on yield and yield variability of grapevine. *Climate Research,* **7(3):** 213–224.

Birthal, P.S., Khan, T., Negi, D.S. and Agarwal, S. (2014). Impact of climate change on yields of major food crops in India: Implications for food security. *Agricultural Economics Research Review,* **27(347-2016-17126):** 145–155.

Boote, K., Jones, J., Mishoe, J. and Berger, R. (1983). Coupling pests to crop growth simulators to predict yield reductions [mathematical models]. Phytopathology (USA).

Boote, K., Jones, J., Mishoe, J. and Wilkerson, G. (1986). Modeling growth and yield of groundnut. *In: International Symposium: Agrometeorology of Groundnut, Niamey (Niger)*, 21–26 Aug 1985. ICRISAT.

Boote, K., Batchelor, W., Jones, J., Pinnschmidt, H. and Bourgeois, G. (1993). Pest damage relations at the field level. *In:* Systems approaches for agricultural development, pages 277–296. Springer.

Boote, K. and Pickering, N. (1994). Modeling photosynthesis of row crop canopies. *Hort Science,* **29(12):** 1423–1434.

Boote, K., Pickering, N. and Allen Jr, L. (1997). Plant Modeling: Advances and gaps in our capability to predict future crop growth and yield in response to global climate change. Advances in carbon dioxide effects research, pages 179–228.

Boote, K., Jones, J., Hoogenboom, G. and Pickering, N. (1998). The crop gromodel for grain legumes. *In: Understanding Options for Agricultural Production*, pages 99–128. Springer.

Boote, K.J., Mínguez, M.I. and Sau, F. (2002). Adapting the CROPGRO legume model to simulate growth of faba bean. *Agronomy Journal,* **94(4):** 743–756.

Bourke, P. and Robertson, G.W. (1955). Forecasting from weather data of potato blight and other plant diseases and pests.

Brocker, J. and Smith, L.A. (2007). Scoring probabilistic forecasts: The importance of being proper. *Weather and Forecasting,* **22(2):** 382–388.

Brown, M.E. and Kshirsagar, V. (2015). Weather and international price shocks on food prices in the developing world. *Global Environmental Change,* **35:** 31–40.

Brown, L. and Cocheme, J. (1973). Study of the agroclimatology of the high-lands of eastern Africa.

Buishand, T. (1978). Some remarks on the use of daily rainfall models. *Journal of Hydrology,* **36(3-4):** 295–308.

Cannon, A.J. and McKendry, I.G. (1999). Forecasting all-India summer monsoon rainfall using regional circulation principal components: A comparison between neural network and multiple regression models. *International Journal of Climatology: A Journal of the Royal Meteorological Society,***19(14):** 1561–1578.

Chakrabarty, M. (2016). Climate change and food security in India. Observer Research Foundation (ORF): New Delhi, India.

Chattopadhyay, N., Rao, K., Sahai, A., Balasubramanian, R., Pai, D., Pattanaik, D., Chandras, S. and Khedikar, S. (2018). Usability of extended range and seasonal weather forecast in Indian agriculture. *Mausam,* **69(1):** 29–44.

Chaturvedi, R.K., Joshi, J., Jayaraman, M., Bala, G. and Ravindranath, N. (2012). Multi-model climate change projections for India under representative concentration pathways. *Current Science,* pages 791–802.

Chen, J., Zhang, X.-c., Liu, W.-z. and Li, Z. (2009). Evaluating and extending cligen precipitation generation for the loess plateau of China 1. *JAWRA Journal of the American Water Resources Association,* **45(2):** 378–396.

Chen, J., Brissette, F.P. and Leconte, R. (2010). A daily stochastic weather generator for preserving low-frequency of climate variability. *Journal of Hydrology,* **388(3-4):** 480–490.

Chen, J., Brissette, F., Leconte, R. and Caron, A. (2012a). A versatile weather generator for daily precipitation and temperature. Transactions of the ASABE, **55(3):** 895–906.

Chen, J., Brissette, F.P. and Leconte, R. (2012b). Downscaling of weather generator parameters to quantify hydrological impacts of climate change. *Climate Research,* **51(3):** 185–200.

Chen, J. and Brissette, F.P. (2014). Stochastic generation of daily precipitation amounts: Review and evaluation of different models. Climate Research, **59(3):** 189–206.

Clark, M., Gangopadhyay, S., Hay, L., Rajagopalan, B. and Wilby, R. (2004a). The schaake shuffle: A method for reconstructing space–time variability inforecasted precipitation and temperature fields. *Journal of Hydrometeorology,* **5(1):** 243–262.

Clark, M.P., Gangopadhyay, S., Brandon, D., Werner, K., Hay, L., Rajagopalan, B. and Yates, D. (2004b). A resampling procedure for generating conditioned daily weather sequences. *Water Resources Research,* 40(4).

Cocheme, J. and Franquin, P. (1967). An agroclimatology survey of a semi arid area in Africa south of the sahara. tech. Technical report, Note.

Coiffier, J. (2011). Fundamentals of numerical weather prediction. Cambridge University Press.

Cressman, G.P. (1959). An operational objective analysis system. Monthly Weather Review, **87(10):** 367–374.

Cruz, R., Lasco, R., Pulhin, J., Pulhin, F. and Garcia, K. (2006). Climate change impact on water resources in Pantabangan watershed, Philippines. *In:* AIACC Final Technical Report, pages 9–107.

Das, S., Ashrit, R., Iyengar, G.R., Mohandas, S., Das Gupta, M., George, J.P., Rajagopal, E. and Dutta, S.K. (2008). Skills of different mesoscale models over Indian region during monsoon season: Forecast errors. *Journal of Earth System Science,* **117(5):** 603–620.

DeWitt, D.G. (2005). Retrospective forecasts of interannual sea surface temperature anomalies from 1982 to present using a directly coupled atmosphere–ocean general circulation model. *Monthly Weather Review,* **133(10):** 2972–2995.

Deser, C., Knutti, R., Solomon, S. and Phillips, A.S. (2012). Communication of the role of natural variability in future North American climate. *Nature Climate Change,* **2(11):** 775–779.

Doblas-Reyes, F.J., Déqué M. and Piedelievre, J.-P. (2000). Multi-model spread and probabilistic seasonal forecasts in provost. *Quarterly Journal of the Royal Meteorological Society,* **126(567):** 2069–2087.

Doorenbos, J. (1977). Guidelines for predicting crop water requirements. FAO irrigation and drainage paper, **24:** 1–179.

Draper, N.R. and Smith, H. (1981). Applied regression analysis. 2nd. Edition. John Wiley and Sons, New York, New York.

Economic Survey 2020–21 (2021). Ministry of Finance of the Government of India. Volume II. Agriculture & Food Management. In: Ministry of Finance, Chapter 7, New Delhi. https://www.indiabudget.gov.in/budget2021-22/economicsurvey/doc/vol2chapter/echap07_vol2.pdf

Eliassen, A. (1949). The quasi-static equations of motion with pressure as independent variable. *Geofys. Publ.*, **17:** 3.

Emanuel, K.A. (1991). A scheme for representing cumulus convection in large-scale models. *Journal of the Atmospheric Sciences,* **48(21):** 2313–2329.

Emanuel, K.A. and Zivković-Rothman, M. (1999). Development and evaluation of a convection scheme for use in climate models. *Journal of the Atmospheric Sciences,* **56(11):** 1766–1782

Eubank, R. and Webster, J. (1985). The singular-value decomposition as a tool for solving estimability problems. *The American Statistician,* **39(1):** 64–66.

Eum, H.-I., Simonovic, S.P. and Kim, Y.-O. (2010). Climate change impact assessment using k-nearest neighbor weather generator: Case study of the Nakdong river basin in Korea. *Journal of Hydrologic Engineering,* **15(10):** 772–785.

Fan, Y. and van den Dool, H. (2011). Bias correction and forecast skill of ncepgfs ensemble week-1 and week-2 precipitation, 2-m surface air temperature, and soil moisture forecasts. *Weather and Forecasting,* **26(3):** 355–370.

Feudale, L. and Tompkins, A. (2011). A simple bias correction technique for modeled monsoon precipitation applied to west Africa. *Geophysical Research Letters,* 38(3).

Fiebrich, C.A. (2009). History of surface weather observations in the United States. *Earth-Science Reviews,* **93(3-4)**: 77–84.

Fisher, R.A. (1922). The goodness of fit of regression formulae, and the distribution of regression coefficients. *Journal of the Royal Statistical Society,* **85(4):** 597–612.

Fishman, R. (2016). More uneven distributions overturn benefits of higher precipitation for crop yields. *Environmental Research Letters,* **11(2):** 024004.

Forrester, Jay W. (1961). Industrial Dynamics. Portland, OR: Productivity Press. pp. 464.

Fowler, H., Kilsby, C., O'connell, P. and Burton, A. (2005). A weather-type conditioned multi-site stochastic rainfall model for the generation of scenarios of climatic variability and change. *Journal of Hydrology,* **308(1-4):** 50–66.

Fowler, H.J., Blenkinsop, S. and Tebaldi, C. (2007). Linking climate change modelling to impacts studies: Recent advances in downscaling techniques for hydrological modelling. *International Journal of Climatology: A Journal of the Royal Meteorological Society,* **27(12):** 1547–1578.

Frisvold, G.B. and Murugesan, A. (2013). Use of weather information for agricultural decision making. *Weather, Climate, and Society,* **5(1):** 55–69.

Fritsch, J. and Chappell, C. (1980). Numerical prediction of convectively driven mesoscale pressure systems. part i: Convective parameterization. *Journal of Atmospheric Sciences,* **37(8):** 1722–1733.

Fuss, S., Havlík, P., Szolgayová, J., Schmid, E., Reuter, W. H., Khabarov, N., Obersteiner, M., Ermoliev, Y., Ermolieva, T. and Kraxner, F. (2015). Global food security & adaptation under crop yield volatility. *Technological Forecasting and Social Change,* **98:** 223–233.

Gadgil, S., Rao, P.S. and Rao, K.N. (2002). Use of climate information for farm-level decision making: Rainfed groundnut in southern India. *Agricultural Systems,* **74(3)**: 431–457.

Gadgil, S. and Gadgil, S. (2006). The Indian monsoon, gdp and agriculture. Economic and political weekly, pages 4887–4895.

Gandhi, G.S., Chaudhary, J. and Sahu, K.K. (2018). Farmers feedback about the agromet advisory services (aas) at Mahasamund district of Chattisgarh. *Journal of Pharmacognancy and Phytochemistry,* **5:** 2522–2524.

Gandin, L.S. (1966). Objective analysis of meteorological field. *Quarterly Journal of the Royal Meteorological Society*, **92:** 393.

Garrity, D., Oldeman, L., Morris, R. and Lenka, D. (1986). Rainfed lowland rice ecosystems: Characterization and distribution. Progress in rainfed low-land rice, **3:** 23.

Gaughan, J. and Cawdell-Smith, A. (2015). Impact of climate change on livestock production and reproduction. *In:* Climate change Impact on livestock: adaptation and mitigation, pages 51–60. Springer.

Geng, S., de Vries, F.W.P. and Supit, I. (1986). A simple method for generating daily rainfall data. *Agricultural and Forest Meteorology,* **36(4):** 363–376.

Geng, S., Auburn, J., Brandstetter, E. and Li, B. (1988). A program to simulate meteorological variables: Documentation for simmeteo. *Agronomy Progress Rep*, 204.

Ghajarnia, N., Liaghat, A. and Arasteh, P.D. (2015). Comparison and evaluation of high resolution precipitation estimation products in urmia basin iran. *Atmospheric Research,* **158**: 50–65.

Ghosh, K., Singh, A., Mohanty, U., Acharya, N., Pal, R., Singh, K. and Pasupalak, S. (2015). Development of a rice yield prediction system over Bhubaneswar, India: Combination of extended range forecast and ceres-rice model. *Meteorological Applications,* **22(3):** 525–533.

Gijsman, A.J., Hoogenboom, G., Parton, W.J. and Kerridge, P.C. (2002). Modifying DSSAT crop models for low-input agricultural systems using a soil organic matter–residue module from century. *Agronomy Journal,* **94(3):** 462–474.

Gilchrist, B. and Cressman, G.P. (1954). An experiment in objective analysis. *Tellus,* **6(4):** 309–318.

Gill, J.C. and Malamud, B.D. (2014). Reviewing and visualizing the interactions of natural hazards. *Reviews of Geophysics,* **52(4):** 680–722.

Gitz, V., Meybeck, A., Lipper, L., Young, C.D. and Braatz, S. (2016). Climate change and food security: Risks and responses. *Food and Agriculture Organization of the United Nations (FAO) Report,* **110:** 2–4.

Glahn, H.R. (1968). Canonical correlation and its relationship to discriminant analysis and multiple regression. *Journal of Atmospheric Sciences*, **25(1):** 23–31.

Godwin, D. and Allan Jones, C. (1991). Nitrogen dynamics in soil-plant systems. Modeling plant and soil systems, **31:** 287–321.

Godwin, D. and Singh, U. (1998). Nitrogen balance and crop response to nitrogen in upland and lowland cropping systems. *In: Understanding Options for Agricultural Production*, pages 55–77. Springer.

Govindaswamy, C. (1964). Temperature and blast incidence. *Madras Agric.,* **51(6):** 255–257.

Grell, G.A. (1993). Prognostic evaluation of assumptions used by cumulus parameterizations. *Monthly Weather Review,* **121(3):** 764–787.

Grell, G.A., Dudhia, J. and Stauffer, D.R. (1994). A description of the fifth generation penn state/ncar mesoscale model (mm5). Technical report.

Griffin, T.S., Johnson, B.S. and Ritchie, J.T. (1993). A simulation model for potato growth and development: Substor-potato Version 2.0. Michigan State University, Department of Crop and Soil Sciences.

Grimm, S.S., Jones, J.W., Boote, K.J. and Herzog, D. (1994). Modeling the occurrence of reproductive stages after flowering for four soybean cultivars. *Agronomy Journal,* **86(1):** 31–38.

Grimm, S.S., Jones, J.W., Boote, K.J. and Hesketh, J.D. (1993). Parameter estimation for predicting flowering date of soybean cultivars. *Crop Science*, **33(1):** 137–144.

Guido, Z., Zimmer, A., Lopus, S., Hannah, C., Gower, D., Waldman, K., Krell,N., Sheffield, J., Caylor, K. and Evans, T. (2020). Farmer forecasts: Impacts of seasonal rainfall expectations on agricultural decision-making in sub Saharan Africa. *Climate Risk Management,* **30:** 100247.

Guiot, J., Berger, A. and Munaut, A. (1982). Response functions. Climate from tree rings. Cambridge Univ. Press.

Gunst, R.F. and Mason, R.L. (1980). Regression analysis and its application: A data-oriented approach, volume 34. CRC Press.

Guntukula, R. and Goyari, P. (2020). The impact of climate change on maize yields and its variability in Telangana, India: A panel approach study. *Journal of Public Affairs*, **20(3):** e2088.
Guo, Y. and Senior, M.J. (2006). Climate model simulation of point rainfall frequency characteristics. *Journal of Hydrologic Engineering,* **11(6):** 547–554.
Gupta, S., Rai, K., Singh, P., Ameta, V., Gupta, S. K., Jayalekha, A., Mahala, R., Pareek, S., Swami, M. and Verma, Y. (2015). Seed set variability under high temperatures during flowering period in pearl millet (Pennisetum glaucum l.(r.) br.). *Field Crops Research,* **171:** 41–53.
Haerter, J., Hagemann, S., Moseley, C. and Piani, C. (2011). Climate model bias correction and the role of timescales. *Hydrology and Earth System Sciences,* **15(3):** 1065–1079.
Hagedorn, R., Doblas-Reyes, F. and Palmer, T. (2005). The rationale behind the climate prediction using demeter forecast. *Tellus A,* **57:** 280–289.
Hale, J.D. (1992). Best of the Old Farmer's Almanac: The First 200 Years. Random House.
Hamlet, A.F., Huppert, D. and Lettenmaier, D.P. (2002). Economic value of long-lead streamflow forecasts for columbia river hydropower. *Journal of Water Resources Planning and Management,* **128(2):** 91–101.
Hansen, J.W. and Indeje, M. (2004). Linking dynamic seasonal climate forecasts with crop simulation for maize yield prediction in semi-arid Kenya. *Agricultural and Forest Meteorology,* **125(1-2):** 143–157.
Hansen, J.W. and Mavromatis, T. (2001). Correcting low-frequency variability bias in stochastic weather generators. *Agricultural and Forest Meteorology*, **109(4):** 297–310.
Hanson, C.L. (1998). Multiple applications: Its application in areas of Hydrology, Water Resources and Ecology in Headwaters, **248:** 27.
Harmel, R.D., Richardson, C.W., Hanson, C. and Johnson, G. (2002). Evaluating the adequacy of simulating maximum and minimum daily air temperature with the normal distribution. *Journal of Applied Meteorology,* **41(7):** 744–753.
Hashino, T., Bradley, A. and Schwartz, S. (2006). Evaluation of bias correction methods for ensemble streamflow volume forecasts. *Hydrology and Earth System Sciences Discussions,* **3(2):** 561–594.
Hazra, K.K. and Bohra, A. (2021). Increasing relevance of pulse crops to sustainable intensification of Indian agriculture. *National Academy Science Letters*, **44:** 1–3. https://doi.org/10.1007/s40009-020-00948-6
Hazra, S., Swain, D.K. and Bhadoria, P.B.S. (2019). Wheat grown under elevated CO_2 was more responsive to nitrogen fertilizer in eastern India. *European Journal of Agronomy,* **105:** 1–12.
He, Y. and Barnston, A.G. (1996). Long-lead forecasts of seasonal precipitation in the tropical pacific islands using cca. *Journal of Climate,* **9(9):** 2020–2035.
Heidke, P. (1926). Berechnung des erfolges und der güte der windstärkevorhersagen im sturmwarnungsdienst. *Geografiska Annaler*, **8(4):** 301–349.
Herdt, R. (1996). Summary conclusions and implication. Rice research in Asia: Progress and priorities. Wallingford, CAB International/IRRI, pages 17–33.
Holzworth, D.P., Huth, N.I., deVoil, P.G., Zurcher, E.J., Herrmann, N.I., McLean, G., Chenu, K., van Oosterom, E.J., Snow, V., Murphy, C., et al. (2014). Apsim–evolution towards a new generation of agricultural systems simulation. *Environmental Modelling & Software,* **62:** 327–350.
Hoogenboom, G., White, J.W., Jones, J.W. and Boote, K.J. (1994). Beangro: A process-oriented dry bean model with a versatile user interface. *Agronomy Journal,* **86(1):** 182–190.
Hoogenboom, G., Porter, C.H., Boote, K.J., Shelia, V., Wilkens, P.W., Singh, U., White, J.W., Asseng, S., Lizaso, J.I., Moreno, L.P., Pavan, W., Ogoshi, R., Hunt, L.A., Tsuji, G.Y. and Jones, J.W. (2019). The DSSAT crop modeling ecosystem. *In:* p.173–216 [K.J. Boote, editor] Advances in Crop Modeling for a Sustainable Agriculture. Burleigh Dodds Science Publishing, Cambridge, United Kingdom (https://doi.org/10.19103/AS.2019.0061.10).
Hoogenboom, G., Porter, C.H., Shelia, V., Boote, K.J., Singh, U., White, J.W., Pavan, W., Oliveira, F.A.A., Moreno-Cadena, L.P., Lizaso, J.I., Asseng, S., Pequeno, D.L.N., Kimball, B.A., Alderman, P.D., Thorp, K.R., Jones, M.R., Cuadra, S.V., Vianna, M.S., Villalobos, F.J.,

Ferreira, T.B., Batchelor, W.D., Koo, J., Hunt, L.A. and Jones, J.W. (2021). Decision Support System for Agrotechnology Transfer (DSSAT) Version 4.8 (HYPERLINK "https://dssat.net/" DSSAT.net). DSSAT Foundation, Gainesville, Florida, USA.

Hotelling, H. (1936). Relations between two sets of variates. *Biometrika,* **28(3-4):** 321–377.

Houghton, J.T., Ding, Y., Griggs, D.J., Noguer, M., van der Linden, P.J., Dai, X., Maskell, K. and Johnson, C. (2001). Climate Change 2001: The scientific basis: Contribution of Working Group I to the third assessment report of the Intergovernmental Panel on Climate Change. Cambridge University Press.

Hundal, S. S. and Prabhjyot-Kaur, (1996). Climate change and its impact on crop productivity. In Climate Variability and Agriculture (eds Abrol, Y. P. et al.), Narosa Publishing House, New Delhi, pp. 377–393.

Hunt, L., White, J. and Hoogenboom, G. (2001). Agronomic data: Advances in documentation and protocols for exchange and use. *Agricultural Systems,* **70(2-3):** 477–492.

Hwang, S.-O., Schemm, J.-K.E., Barnston, A.G. and Kwon, W.-T. (2001). Long-lead seasonal forecast skill in far eastern Asia using canonical correlation analysis. *Journal of Climate,* **14(13):** 3005–3016.

Idso, S.B., Kimball, B.A., Shaw, P.E., Widmer, W., Vanderslice, J.T., Higgs, D.J., Montanari, A. and Clark, W.D. (2002). The effect of elevated atmospheric CO_2 on the vitamin C concentration of (sour) orange juice. *Agriculture, Ecosystems & Environment,* **90(1):** 1–7.

Iizumi, T., Sakuma, H., Yokozawa, M., Luo, J.-J., Challinor, A.J., Brown, M. E., Sakurai, G. and Yamagata, T. (2013). Prediction of seasonal climate-induced variations in global food production. *Nature Climate Change,* **3(10):** 904–908.

Ines, A.V. and Hansen, J.W. (2006). Bias correction of daily gcm rainfall for crop simulation studies. *Agriculture and Forest Meteorology,* **138(1-4):** 44–53.

Islam, M.S., Matsui, T. and Yoshida, Y. (1996). Effect of carbon dioxide enrichment on physico-chemical and enzymatic changes in tomato fruits at various stages of maturity. *Scientia Horticulturae,* **65(2–3):** 137–149.

Jahn, M. (2015). Economics of extreme weather events: Terminology and regional impact models. *Weather and Climate Extremes,* **10:** 29–39.

Jain, S.K. and Kumar, V. (2012). Trend analysis of rainfall and temperature data for India. *Current Science,* pages 37–49.

Jakob Themeßl, M., Gobiet, A. and Leuprecht, A. (2011). Empirical-statistical downscaling and error correction of daily precipitation from regional climate models. *International Journal of Climatology,* **31(10):** 1530–1544.

Jame, Y. and Cutforth, H. (1996). Crop growth models for decision support systems. *Canadian Journal of Plant Science,* **76(1):** 9–19.

Jha, B. and Tripathi, A. (2011). Isn't climate change affecting wheat productivity in India? *Indian Journal of Agricultural Economics,* 66(902-2016-67321).

Johnson, C. and Bowler, N. (2009). On the reliability and calibration of ensemble forecasts. *Monthly Weather Review,* **137(5):** 1717–1720.

Johnson, G.L., Hanson, C.L., Hardegree, S.P. and Ballard, E.B. (1996). Stochastic weather simulation: Overview and analysis of two commonly used models. *Journal of Applied Meteorology and Climatology,* **35(10):** 1878–1896.

Jones, J.W., Hansen, J.W., Royce, F.S. and Messina, C.D. (2000). Potential benefits of climate forecasting to agriculture. *Agriculture, Ecosystems & Environment,* **82(1-3):** 169–184.

Jones, J.W., Hoogenboom, G., Porter, C.H., Boote, K.J., Batchelor, W.D., Hunt, L., Wilkens, P.W., Singh, U., Gijsman, A.J. and Ritchie, J.T. (2003). The dssat cropping system model. *European Journal of Agronomy,* **18(3-4):** 235–265.

Jones, C.A. (1986). CERES-Maize; A simulation model of maize growth and development. Number 04; SB91. M2, J6.

Jones, J. (1993). Decision support systems for agricultural development. *In:* Systems approaches for agricultural development, pages 459–471. Springer.

Jones, J., Boote, K., Hoogenboom, G., Jagtap, S. and Wilkerson, G. (1989). Soygro v5. 42, soybean crop growth simulation model. user's guide. *Florida Agricultural Experiment Station Journal,* **8304:** 83.

Jones, J., Tsuji, G., Hoogenboom, G., Hunt, L., Thornton, P., Wilkens, P., Imamura, D., Bowen, W. and Singh, U. (1998). Decision support system for agrotechnology transfer: Dssat v3. *In:* Understanding options for agricultural production, pages 157–177. Springer.

Jones, J.W., Hansen, J.W., Royce, F.S. and Messina, C.D. (2000). Potential benefits of climate forecasting to agriculture. *Agriculture, Ecosystems & Environment,* **82(1-3):** 169–184

Joshi, P.K. (2015). Has Indian agriculture become crowded and risky? Status, implications and the way forward. *Indian Journal of Agricultural Economics,* 70(902-2016-68367).

Kang, Y., Khan, S. and Ma, X. (2009). Climate change impacts on crop yield, crop water productivity and food security–a review. *Progress in Natural Science,* **19(12):** 1665–1674.

Kar, S.C., Acharya, N., Mohanty, U. and Kulkarni, M.A. (2012). Skill of monthly rainfall forecasts over India using multi-model ensemble schemes. *International Journal of Climatology,* **32(8):** 1271–1286.

Kashyap, D. and Agarwal, T. (2021). Temporal trends of climatic variables and water footprint of rice and wheat production in Punjab, India from 1986 to 2017. *Journal of Water and Climate Change,* **12(4):** 1203–1219.

Katz, R.W. (1977). Precipitation as a chain-dependent process. *Journal of Applied Meteorology* (1962-1982), pages 671–676.

Katz, R.W. and Parlange, M.B. (1993). Effects of an index of atmospheric circulation on stochastic properties of precipitation. *Water Resources Research,* **29(7):** 2335–2344.

Katz, R.W. and Parlange, M.B. (1998). Over dispersion phenomenon in stochastic modeling of precipitation. *Journal of Climate,* **11(4):** 591–601.

Kaur, P. and Hundal, S.S. (1999). Forecasting growth and yield of groundnut with a dynamic simulation model 'PNUTGRO' under Punjab conditions. *J. Agric. Sci.,* Cambridge **133:** 167-173.

Keating, B.A., Carberry, P.S., Hammer, G.L., Probert, M.E., Robertson, M.J., Holzworth, D., Huth, N.I., Hargreaves, J.N., Meinke, H., Hochman, Z., et al. (2003). An overview of apsim, a model designed for farming systems simulation. *European Journal of Agronomy,* **18(3-4):** 267–288.

Kharin, V.V. and Zwiers, F.W. (2002). Climate predictions with multi model ensembles. *Journal of Climate,* **15(7):** 793–799.

Kharin, V.V. and Zwiers, F.W. (2003). Improved seasonal probability forecasts. *Journal of Climate,* **16(11):** 1684–1701.

Kim, I.-W., Oh, J., Woo, S. and Kripalani, R. (2019). Evaluation of precipitation extremes over the asian domain: Observation and modelling studies. *Climate Dynamics,* **52(3):** 1317–1342.

Klemm, T. and McPherson, R.A. (2017). The development of seasonal climate forecasting for agricultural producers. *Agricultural and Forest Meteorology,* **232**: 384–399.

Kogbe, J. and Adediran, J. (2003). Influence of nitrogen, phosphorus and potassium application on the yield of maize in the savanna zone of Nigeria. *African Journal of Biotechnology,* **2(10):** 345–349.

Kothawale, D. and Rupa Kumar, K. (2005). On the recent changes in surface temperature trends over India. *Geophysical Research Letters,* **32(18)**.

Kottegoda, N., Natale, L. and Raiteri, E. (2003). A parsimonious approach to stochastic multisite modelling and disaggregation of daily rainfall. *Journal of Hydrology,* **274(1-4):** 47–61.

Krishna Kumar, K., Rupa Kumar, K., Ashrit, R., Deshpande, N., and Hansen, J.W. (2004). Climate impacts on Indian agriculture. *International Journal of Climatology: A Journal of the Royal Meteorological Society,* **24(11):** 1375–1393.

Krishnamurti, T., Kishtawal, C., Shin, D. and Williford, C.E. (2000a). Improving tropical precipitation forecasts from a multi analysis superensemble. *Journal of Climate,* **13(23):** 4217–4227.

Krishnamurti, T.N., Kishtawal, C., Zhang, Z., LaRow, T., Bachiochi, D., Williford, E., Gadgil, S. and Surendran, S. (2000b). Multimodel ensemble forecasts for weather and seasonal climate. *Journal of Climate,* **13(23):** 4196–4216.

Kulkarni, M.A., Acharya, N., Kar, S.C., Mohanty, U., Tippett, M.K., Robertson, A.W., Luo, J.-J. and Yamagata, T. (2012). Probabilistic prediction of Indian summer monsoon rainfall using global climate models. *Theoretical and Applied Climatology,* **107(3-4):** 441–450.

Kulkarni, M.A., Singh, A. and Mohanty, U. (2012b). Effect of spatial correlation on regional trends in rain events over India. *Theoretical and Applied Climatology,* **109(3):** 497–505.

Kumar, A., Dudhia, J., Rotunno, R., Niyogi, D. and Mohanty, U. (2008). Analysis of the 26 july 2005 heavy rain event over Mumbai, India using the weather research and forecasting (wrf) model. *Quarterly Journal of the Royal Meteorological Society,* **134(636):** 1897–1910.

Kumar, A. and Sharma, P. (2013). Impact of climate variation on agricultural productivity and food security in rural India. Technical report, Economics Discussion Papers.

Kumar, A. and Singh, A. (2014). Climate change and its impact on wheat production and mitigation through agroforestry technologies. *Int. J. Environ. Sci.* **5(1):** 73–90.

Kumar, A., Sharma, P. and Ambrammal, S.K. (2015). Effects of climatic factors on productivity of cash crops in India: Evidence from state-wise panel data. *Global Journal of Research in Social Sciences* **1(1)**.

Kumar, A., Sharma, P. and Ambrammal, S.K. (2015a). Climatic effects on sugarcane productivity in India: A stochastic production function application. *International Journal of Economics and Business Research,* **10(2):** 179–203.

Kumar, A., Sharma, P. and Ambrammal, S.K. (2015b). Effects of climatic factors on productivity of cash crops in India: Evidence from state-wise panel data. *Global Journal of Research in Social Sciences* **1(1)**.

Kumar, A., Sharma, P. and Joshi, S. (2016). Assessing the impacts of climate change on land productivity in Indian crop agriculture: An evidence from panel data analysis. *Journal of Agricultural Science and Technology,* **18(1):** 1–13.

Kumar, A., Singh, B. and Sharma, P. (2020). Assessing the climate change impact on rice and wheat production in Uttar Pradesh and Haryana states of India. *Climate Change,* **6(21):** 74–93.

Kuo, H.-L. (1974). Further studies of the parameterization of the influence of cumulus convection on large-scale flow. *Journal of Atmospheric Sciences,* **31(5):** 1232–1240.

Laing AG, Evans JL (2011) Introduction to tropical meteorology (2nd ed.). Retrieved from http://www.goes-r.gov/users/comet/tropical/textbook_2nd_edition/index.htm.

Lal, M., Singh, K., Rathore, L., Srinivasan, G. and Saseendran, S. (1998). Vulnerability of rice and wheat yields in NW India to future changes in climate. *Agricultural and Forest Meteorology,* **89(2):** 101–114.

Lee, D. and De Witt, D. (2009). New hybrid coupled forecast system utilizing the cfssst, 2009 forecasts http://portal.iri.Columbia.edu/portal/server.pt/gateway/ptargs0497257340018.NOAA-abstract2009.

Leng, G. and Hall, J. (2019). Crop yield sensitivity of global major agricultural countries to droughts and the projected changes in the future. *Science of the Total Environment,* **654:** 811–821.

Li, H., Sheffield, J. and Wood, E.F. (2010). Bias correction of monthly precipitation and temperature fields from intergovernmental panel on climate change ar4 models using equidistant quantile matching. *Journal of Geophysical Research: Atmospheres,* **115(D10)**.

Lobell, D.B. and Gourdji, S.M. (2012). The influence of climate change on global crop productivity. *Plant Physiology,* **160(4):** 1686–1697.

Lobo, C., Chattopadhyay, N. and Rao, K. (2017). Making small holder farming climate-smart. Economic and Political Weekly, LII, **52(1):** 53–58.

Loehlin, J.C. (1998). Latent variable models: An introduction to factor, path, and structural analysis. Lawrence Erlbaum Associates Publishers.

Luo, J.-J., Masson, S., Behera, S., Shingu, S. and Yamagata, T. (2005). Seasonal climate predictability in a coupled oagcm using a different approach for ensemble forecasts. *Journal of Climate,* **18(21):** 4474–4497.

Mahato, A. (2014). Climate change and its impact on agriculture. *International Journal of Scientific and Research Publications,* **4(4):** 1–6.

Maini, P. and Rathore, L. (2011). Economic impact assessment of the agrometeorological advisory service of India. *Current Science,* pages 1296–1310

Mall, R.K., Gupta, A., Singh, R., Singh, R.S. and Rathore, L. (2006). Water resources and climate change: An Indian perspective. *Current Science,* pages 1610–1626.

Mallya, G., Mishra, V., Niyogi, D., Tripathi, S. and Govindaraju, R.S. (2016). Trends and variability of droughts over the Indian monsoon region. *Weather and Climate Extremes,* **12:** 43–68.

Mandel, J. (1982). Use of the singular value decomposition in regression analysis. *The American Statistician,* **36(1):** 15–24.

Maraun, D., Wetterhall, F., Ireson, A., Chandler, R., Kendon, E., Widmann, M., Brienen, S., Rust, H., Sauter, T., Themeßl, M., et al. (2010). Precipitation downscaling under climate change: Recent developments to bridge the gap between dynamical models and the end user. *Reviews of Geophysics,* **48(3)**.

Marza, B., Angelescu, C. and Tindeche, C. (2015). Agricultural insurances and food security. The new climate change challenges. *Procedia Economics and Finance,* **27:** 594–599.

Matalas, N.C. (1967). Mathematical assessment of synthetic hydrology. *Water Resources Research,* **3(4):** 937–945.

Maurya, R., Sinha, P., Mohanty, M. and Mohanty, U. (2017). Coupling of community land model with regcm4 for Indian summer monsoon simulation. *Pure and Applied Geophysics,* **174(11):** 4251–4270.

Maurya, R., Sinha, P., Mohanty, M. and Mohanty, U. (2018). Regcm4 model sensitivity to horizontal resolution and domain size in simulating the Indian summer monsoon. *Atmospheric Research,* **210:** 15–33.

Maurya, R., Mohanty, M., Sinha, P. and Mohanty, U. (2020). Performance of hydrostatic and non-hydrostatic dynamical cores in regcm4. 6 for Indian summer monsoon simulation. *Meteorological Applications*, **27(3):** e1915.

Mavromatis, T. and Hansen, J.W. (2001). Interannual variability characteristics and simulated crop response of four stochastic weather generators. *Agricultural and Forest Meteorology,* **109(4):** 283–296.

McGuffie, K. and Henderson-Sellers, A. (2014). The climate modelling primer. John Wiley & Sons.

McKague, K., Rudra, R., Ogilvie, J., Ahmed, I. and Gharabaghi, B. (2005). Evaluation of weather generator climgen for southern Ontario. *Canadian Water Resources Journal,* **30(4):** 315–330.

Mearns, L., Giorgi, F., McDaniel, L. and Shields, C. (1995). Analysis of daily variability of precipitation in a nested regional climate model: Comparison with observations and doubled CO_2 results. *Global and Planetary Change,* **10(1-4):** 55–78.

Mearns, L.O., Rosenzweig, C. and Goldberg, R. (1996). The effect of changes in daily and interannual climatic variability on ceres-wheat: A sensitivity study. *Climatic Change,* **32(3):** 257–292.

Mearns, L., Giorgi, F., McDaniel, L. and Shields, C. (2003). Climate scenarios for the southeastern us based on gcm and regional model simulations. *In:* Issues in the impacts of climate variability and change on agriculture, pages 7–35. Springer.

Mehrotra, R. and Sharma, A. (2007). A semi-parametric model for stochastic generation of multi-site daily rainfall exhibiting low-frequency variability. *Journal of Hydrology,* **335(1-2):** 180–193.

Meinke, H. and Hammer, G. (1995). Climatic risk to peanut production: Asimulation study for Northern Australia. *Australian Journal of Experimental Agriculture,* **35(6):** 777–780.

Metz, B., Davidson, O., Bosch, P., Dave, R. and Meyer, L. (2007). Climate Change 2007-mitigation of climate change. Technical report, Intergovernmental Panel on Climate Change, Geneva (Switzerland). Working Group.

Midmore, D. and Poudel, D. (1996). Asian vegetable production systems for the future. *Agricultural Systems,* **50(1):** 51–64.

Minville, M., Brissette, F. and Leconte, R. (2008). Uncertainty of the impact of climate change on the hydrology of a nordic watershed. *Journal of Hydrology,* **358(1-2):** 70–83.

Mohanty, U., Acharya, N., Singh, A., Nair, A., Kulkarni, M., Dash, S., Kar, S., Robertson, A., Mitra, A., Rathore, L., et al. (2013). Real-time experimental extended range forecast system for Indian summer monsoon rainfall: A case study for monsoon 2011. *Current Science,* pages 856–870.

Mohanty, U., Nageswararao, M., Sinha, P., Nair, A., Singh, A., Rai, R., Kar, S., Ramesh, K., Singh, K., Ghosh, K., et al. (2019a). Evaluation of performance of seasonal precipitation prediction at regional scale over India. *Theoretical and Applied Climatology,* **135(3):** 1123–1142.

Mohanty, U., Sinha, P., Mohanty, M., Maurya, R. and Rao, M.N. (2019b). A review on the monthly and seasonal forecast of the Indian summer monsoon. *Mausam,* **70(3):** 425–442.

Mohanty, U., Nageswararao, M., Sinha, P., Nair, A., Singh, A., Rai, R., Kar, S., Ramesh, K., Singh, K., Ghosh, K., et al. (2019b). Evaluation of performance of seasonal precipitation prediction at regional scale over India. *Theoretical and Applied Climatology,* **135(3):** 1123–1142.

Mohanty, M., Sinha, P., Maurya, R. and Mohanty, U. (2019a). Moisture flux adjustments in regcm4 for improved simulation of Indian summer monsoon precipitation. *Climate Dynamics,* **52(11):** 7049–7069.

Monteith, J.L. (1986). How do crops manipulate water supply and demand? *Philosophical Transactions of the Royal Society of London. Series A, Mathematical and Physical Sciences,* **316(1537):** 245–259.

Mor, S. 2017. Impacts of climate variability on the wheat production in India. *Disaster Advances,* **10:** 12–18.

Muchow, R., Cooper, M. and Hammer, G. (1996). 18 characterizing environmental challenges using models. Plant adaptation and crop improvement, page 349.

Murphy, A.H., Brown, B.G. and Chen, Y.-S. (1989). Diagnostic verification of temperature forecasts. *Weather and Forecasting,* **4(4):** 485–501.

Murphy, A.H. and Winkler, R.L. (1987). A general framework for forecast verification. *Monthly Weather Review,* **115(7):** 1330–1338.

Murphy, A.H. and Brown, B.G. (1985). A comparative evaluation of objective and subjective weather forecasts in the united states. *In:* Behavioral decision making, pages 329–359. Springer.

Myers, R.H. and Myers, R.H. (1990). Classical and modern regression with applications, volume 2. Duxbury press Belmont, CA.

Nageswararao, M., Mohanty, U., Nair, A. and Ramakrishna, S. (2016). Comparative evaluation of performances of two versions of ncep climate forecast system in predicting winter precipitation over India. *Pure and Applied Geophysics,* **173(6):** 2147–2166.

Nair, A., Mohanty, U.C. and Panda, T.C. (2015). Improving the performance of precipitation outputs from global climate models to predict monthly and seasonal rainfall over the Indian subcontinent. *Comptes Rendus Geoscience,* **347(2):** 53–63.

Naiyer, A. and Md, S.U.H. (2015). Agriculture play crucial role in the life of Indian economy. *International Journal of Scientific Research and Management,* **3(3):** 2239–2245.

Nardone, A., Ronchi, B., Lacetera, N., Ranieri, M.S. and Bernabucci, U. (2010). Effects of climate changes on animal production and sustainability of livestock systems. *Livestock Science,* **130(1–3):** 57–69.

Nash, J.E. and Sutcliffe, J.V. (1970). River flow forecasting through conceptual models part i-a discussion of principles. *Journal of Hydrology,* **10(3):** 282–290.

Nguyen, D.-N., Lee, K.-J., Kim, D.-I., Anh, N. T. and Lee, B.-W. (2014). Modeling and validation of high-temperature induced spikelet sterility in rice. *Field Crops Research,* **156:** 293–302.

Nicks, A., Lane, L. and Gander, G. (1995). Weather generator. Chapter 2 in USDA–water erosion prediction project: Hillslope profile and watershed model documentation. DC Flanagan, MA Nearing, eds. NSERL report, (10).

Njoku, P., Ayuk, A. and Okoye, C. (2011). Temperature effects on vitamin C content in citrus fruits. *Pakistan Journal of Nutrition,* **10(12):** 1168–1169.

North, G.R., Cahalan, R.F. and Coakley Jr, J. A. (1981). Energy balance climate models. *Reviews of Geophysics,* **19(1):** 91–121.

Office of the Registrar General & Census Commissioner, India. (2011). Census of India 2011: Provisional Population Totals. Ministry of Home Affairs, Government of India.

Pacanowski, R. and Griffies, S. (1998). Mom3. 0 manual, noaa/geophys. FluidDyn. Lab., Princeton, NJ.

Pachauri, R.K., Allen, M.R., Barros, V.R., Broome, J., Cramer, W., Christ, R., Church, J.A., Clarke, L., Dahe, Q., Dasgupta, P., et al. (2014). Climate change 2014: Synthesis report. Contribution of Working Groups I, II and III to the fifth assessment report of the Intergovernmental Panel on Climate Change. Ipcc.

Padmanabhan, S., Chakrabarti, N. and Row, K. (1971). Forecasting and control of rice diseases. Proceedings of the National Academy of Sciences, India-*Section B: Biological Sciences,* **37(6):** 423–429.

Pai, D., Sridhar, L., Rajeevan, M., Sreejith, O., Satbhai, N., and Mukhopadhyay, B. (2014). Development of a new high spatial resolution (0.25x 0.25) long period (1901–2010) daily gridded rainfall data set over India and its comparison with existing data sets over the region. *Mausam,* **65(1):** 1–18.

Palmer, T., Alessandri, A., Andersen, U., Cantelaube, P., Davey, M., Delécluse, P., Déqué, M., Diez, E., Doblas-Reyes, F. J., Feddersen, H., et al. (2004). ´Development of a European multimodel ensemble system for seasonal-to-interannual prediction (demeter). *Bulletin of the American Meteorological Society,* **85(6):** 853–872.

Pan, J. and van den Dool, H. (1998). Extended-range probability forecasts based on dynamical model output. *Weather and Forecasting,* **13(4):** 983–996.

Panofsky, H.A. (1949). Objective weather-map analysis. *Journal of Meteorology,* **6(6):** 386–392.

Parlange, M. B. and Katz, R.W. (2000). An extended version of the Richardson model for simulating daily weather variables. *Journal of Applied Meteorology,* **39(5):** 610–622.

Parton, W.J., Stewart, J.W.B. and Cole, C.V. (1988). Dynamics of C, N, P, and S in grassland soils: *A model. Biogeochemistry,* **5:** 109–131.

Patidar, R., Mohanty, M., Sinha, N.K., Gupta, S., Somasundaram, J., Chaudhary, R., Soliya, R., Hati, K., Prabhakar, M., Reddy, K.S., et al. (2020). Potential impact of future climate change on maize (zea mays l.) under rainfed condition in central India. *Journal of Agrometeorology,* **22(1):** 18–23.

Pavan, V. and Doblas-Reyes, F. (2000). Multi-model seasonal hindcasts over the euro-atlantic: Skill scores and dynamic features. *Climate Dynamics,* **16(8):** 611–625.

Peng, P., Kumar, A., van den Dool, H. and Barnston, A.G. (2002). An analysis of multimodel ensemble predictions for seasonal climate anomalies. *Journal of Geophysical Research: Atmospheres,* **107(D23):** ACL–18.

Penning de Vries, F.W.T. and Laar, H.van (1988). Simulation of plant growth and crop production, Scientific Publication Centre, Beijing, page 356.

Peters, D., Pendleton, J., Hageman, R. and Brown, C. (1971). Effect of night air temperature on grain yield of corn, wheat, and soybeans 1. *Agronomy Journal,* **63(5):** 809–809.

Phillips, N.A. (1957). A coordinate system having some special advantages for numerical forecasting. *Journal of Meteorology,* **14:** 184–195.

Piani, C., Haerter, J. and Coppola, E. (2010). Statistical bias correction for daily precipitation in regional climate models over Europe. *Theoretical and Applied Climatology,* **99(1):** 187–192.

Piper, E., Smit, M., Boote, K. and Jones, J. (1996a). The role of daily minimum temperature in modulating the development rate to flowering in soybean. *Field Crops Research,* **47(2-3):** 211–220.

Piper, E.L., Boote, K.J., Jones, J.W. and Grimm, S.S. (1996b). Comparison of two phenology models for predicting flowering and maturity date of soybean. *Crop Science,* **36(6):** 1606–1614.

Polley, H.W., Briske, D.D., Morgan, J.A., Wolter, K., Bailey, D.W. and Brown, J.R. (2013). Climate change and north American rangelands: Trends, projections, and implications. *Range and Ecology & Management,* **66(5):** 493–511.

Prasad, C., Kulshreshtha, J. and Sinha, V. (1986). Operational research project on rice in India. Progress in Rainfed Lowland Rice, IRRI, Philippines.

Prasad, K. and Singh, S. (1996). Forecasting the spatial variability of the Indian monsoon rainfall using canonical correlation. *International Journal of Climatology: A Journal of the Royal Meteorological Society,* **16(12):** 1379–1390.

Prasad, R. and Rana, R. (2006). A study on maximum temperature during March 2004 and its impact on rabi crops in Himachal Pradesh. *Journal of Agrometeorology,* **8(1):** 91–99.

Prasanna, V. (2014). Impact of monsoon rainfall on the total foodgrain yield over India. *Journal of Earth System Science,* **123(5):** 1129–1145.

Prasanna, V. and Yasunari, T. (2008). Interannual variability of atmospheric water balance over south Peninsular India and Srilanka during northeast monsoon season. *International Journal of Climatology: A Journal of the Royal Meteorological Society,* **28(15):** 1997–2009.

Preethi, B., Kripalani, R. and Krishna Kumar, K. (2010). Indian summer monsoon rainfall variability in global coupled ocean-atmospheric models. *Climate Dynamics,* **35(7):** 1521–1539.

Preethi, B. and Revadekar, J. (2013). Kharif foodgrain yield and daily summer monsoon precipitation over India. *International Journal of Climatology,* **33(8):** 1978–1986.

Priestley, C.H.B. and Taylor, R.J. (1972). On the assessment of surface heat flux and evaporation using large-scale parameters. *Monthly Weather Review,* **100(2):** 81–92.

Pruski, F. and Nearing, M. (2002). Climate-induced changes in erosion during the 21st century for eight us locations. *Water Resources Research,* **38(12):** 34–1.

Racsko, P., Szeidl, L. and Semenov, M. (1991). A serial approach to local stochastic weather models. *Ecological Modelling,* **57(1-2):** 27–41.

Raj, R., Devaraj, M., Bose, S., Ramalingam, S. and Cas, B. (2020). Improving women's access to climate information services and enhancing their capability to manage climate risks. *APN Science Bulletin*

Rajagopalan, B., Lall, U. and Zebiak, S.E. (2002). Categorical climate forecasts through regularization and optimal combination of multiple gcm ensembles. *Monthly Weather Review,* **130(7):** 1792–1811.

Rajeevan, M., Unnikrishnan, C. and Preethi, B. (2012). Evaluation of the ensembles multi-model seasonal forecasts of Indian summer monsoon variability. *Climate Dynamics,* **38(11):** 2257–2274.

Rajeevan, M., Pai, D., Dikshit, S. and Kelkar, R. (2004). IMD's new operational models for long-range forecast of southwest monsoon rainfall over India and their verification for 2003. *Current Science,* 422–431.

Rama Rao, C., Raju, B., Rao, A., Rao, K., Ramachandran, K., Nagasree, K., Samuel, J., Ravi Shankar, K., Srinivasa Rao, M., Maheswari, M., et al. (2018). Climate change impacts, adaptation and policy preferences: A snapshot of farmers' perceptions in India. *Policy Paper,* **1(2018):** 34.

Ramadas, S., Kumar, T.K. and Singh, G.P. (2019). Wheat production in India: Trends and prospects. *In:* Recent Advances in Grain Crops Research. Intech Open.

Ramanathan, V. and Coakley, Jr, J.A. (1978). Climate modeling through radiative-convective models. *Reviews of Geophysics,* **16(4):** 465–489.

Rao, G., Ram Mohan, H., Gopakumar, C. and Krishnakumar, K. (2008). Climate change and cropping systems over Kerala in the humid tropics. *Journal of Agrometeorology,* **10:** 286–291.

Rao, S., Gopakumar, C. and Krishnakumar, K. (2013). Impacts of climate change on horticulture across India. *In:* Climate-Resilient Horticulture: Adaptation and Mitigation Strategies, pages 13–20. Springer.

Rao, S. et al. (2015). National innovations on climate resilient agriculture (nicra) research highlights (2014–15).

Ray, D.K., Ramankutty, N., Mueller, N.D., West, P.C. and Foley, J.A. (2012). Recent patterns of crop yield growth and stagnation. *Nature Communications,* **3(1):** 1–7.

Reddy, K., Maruthi, V., Kumar, M., Pankaj, P., Reddy, A. and Umesha, B. (2017). Enhancing economic water productivity under on farm reservoirs in diversified rainfed cropping systems.

Revadekar, J. and Preethi, B. (2012). Statistical analysis of the relationship between summer monsoon precipitation extremes and foodgrain yield over India. *International Journal of Climatology*, **32(3):** 419–429.

Richardson, C.W. (1981). Stochastic simulation of daily precipitation, temperature, and solar radiation. *Water Resources Research,* **17(1):** 182–190.

Richardson, C.W. and Wright, D.A. (1984). Wgen: A model for generating daily weather variables. ARS (USA).

Richardson, C. (1985). Weather simulation for crop management models.Transactions of the ASAE, **28(5):** 1602–1606.

Riha, S.J., Wilks, D.S. and Simoens, P. (1996). Impact of temperature and precipitation variability on crop model predictions. *Climatic Change,* **32(3):** 293–311.

Ritchie, J.T. (1972). Model for predicting evaporation from a row crop within complete cover. *Water Resources Research,* **8(5):** 1204–1213.

Ritchie, J. (1985). Description and performance of ceres wheat: A user-oriented wheat yield model. ARS wheat yield project, pages 159–175.

Ritchie, J.T. (1998). Soil water balance and plant water stress. *In:* Understanding options for agricultural production, pages 41–54. Springer.

Robertson, A.W., Lall, U., Zebiak, S.E. and Goddard, L. (2004). Improved combination of multiple atmospheric gcm ensembles for seasonal prediction. *Monthly Weather Review*, **132(12):** 2732–2744.

Roeckner, E., Arpe, K., Bengtsson, L., Christoph, M. and Claussen, M. (1996). L. Du Menil, M. Esch, M. Giorgetta, U. Schlese, and U. Schulzweida, The atmospheric general circulation model echam-4: Model description and simulation of present-day climate. MPI Report, 218.

Rojas-Downing, M.M., Nejadhashemi, A.P., Harrigan, T. and Woznicki, S.A. (2017). Climate change and livestock: Impacts, adaptation, and mitigation. *Climate Risk Management,* **16:** 145–163.

Roldán, J. and Woolhiser, D.A. (1982). Stochastic daily precipitation models: 1. a comparison of occurrence processes. *Water Resources Research,* **18(5):** 1451–1459.

Roudier, P., Alhassane, A., Baron, C., Louvet, S. and Sultan, B. (2016). Assessing the benefits of weather and seasonal forecasts to millet growers inniger. *Agricultural and Forest Meteorology,* **223:** 168–180.

Royce, F., Jones, J. and Hansen, J. (2001). Model–based optimization of crop management for climate forecast applications. Transactions of the ASAE, **44(5):** 1319.

Sage, T.L., Bagha, S., Lundsgaard-Nielsen, V., Branch, H.A., Sultmanis, S. and Sage, R.F. (2015). The effect of high temperature stress on male and female reproduction in plants. *Field Crops Research,* **182:** 30–42.

Saha, S., Moorthi, S., Wu, X., Wang, J., Nadiga, S., Tripp, P., Behringer,D., Hou, Y.-T., Chuang, H.-y., Iredell, M., et al. (2014). The ncep climate forecast system version 2. *Journal of Climate,* **27(6):** 2185–2208.

Saha, S., Nadiga, S., Thiaw, C., Wang, J., Wang, W., Zhang, Q., Van den Dool, H., Pan, H.-L., Moorthi, S., Behringer, D., et al. (2006). The ncep climate forecast system. *Journal of Climate,* **19(15):** 3483–3517.

Sajani, S., Nakazawa, T., Kitoh, A. and Rajendran, K. (2007). Ensemble simulation of Indian summer monsoon rainfall by an atmospheric general circulation model. *Journal of the Meteorological Society of Japan. Ser. II,* **85(3):** 213–231.

Saltzman, B. (1978). A survey of statistical-dynamical models of the terrestrial climate. *In:* Advances in Geophysics, **20:** 183–304. Elsevier.

Satapathy, S.S., Swain, D.K. and Herath, S. (2014). Field experiments and simulation to evaluate rice cultivar adaptation to elevated carbon dioxide and temperature in sub-tropical India. *European Journal of Agronomy,* **54:** 21–33.

Schaffer, B., Whiley, A.W., Searle, C., Nissen, R.J. (1997). Leaf Gas Exchange, Dry Matter Partitioning, and Mineral Element Concentrations in Mango as Influenced by Elevated Atmospheric Carbon Dioxide and Root Restriction. *Journal of the American Society for Horticultural Science* **122**: 849–855

Schlesinger, M.E. (2012). Physically-Based Modelling and Simulation of Climate and Climatic Change: Part 2, volume 243. Springer Science & Business Media.

Schneider, S.H. and Dickinson, R.E. (1974). Climate modeling. *Reviews of Geophysics,* **12(3):** 447–493.

Scholberg, J., Boote, K., Jones, J. and McNeal, B. (1997). Adaptation of the cropgro model to simulate the growth of field-grown tomato. *In:* Applications of systems approaches at the field level, pages 135–151. Springer.

Schoof, J.T. and Robeson, S.M. (2003). Seasonal and spatial variations of cross-correlation matrices used by stochastic weather generators. *Climate Research,* **24(2):** 95–102.

SCS (Soil Conservation Service) (1972). National Engineering Handbook: Hydrology Section 4: Hydrology, Supplement A. Washington DC: US Department of Agriculture, Soil Conservation Service, Engineering Division.

Sejian, V., Maurya, V.P. and Naqvi, S.M. (2010). Adaptive capability as indicated by endocrine and biochemical responses of malpura ewes subjected to combined stresses (thermal and nutritional) in a semi-arid tropical environment. *International Journal of Biometeorology,* **54(6):** 653–661.

Sejian, V. (2013). Climate change: Impact on production and reproduction, adaptation mechanisms and mitigation strategies in small ruminants: A review. *Indian Journal of Small Ruminants* (The), **19(1):** 1–21.

Sejian, V., Bhatta, R., Soren, N., Malik, P., Ravindra, J., Prasad, C. S. and Lal, R. (2015). Introduction to concepts of climate change impact on livestock and its adaptation and mitigation. *In:* Climate change Impact on livestock: adaptation and mitigation, pages 1–23. Springer.

Sejian, V., Gaughan, J., Bhatta, R. and Naqvi, S. (2016). Impact of climate change on livestock productivity. Feedipedia-Animal Feed Resources Information System-INRA CIRAD AFZ and FAO, pages 1–4.

Seligman, N. and Keulen, H. (1980). 4.10 papran: A simulation model of annual pasture production limited by rainfall and nitrogen. Simulation of nitrogen behaviour of soil-plant systems. Pudoc, Wageningen, pages 192–221.

Selvaraju, R. et al. (2012). Climate risk assessment and management in agriculture. Building resilience for adaptation to climate change in the agriculture sector, **23**: 71.

Semenov, M.A. and Porter, J. (1995). Climatic variability and the modelling of crop yields. *Agricultural and Forest Meteorology,* **73(3-4):** 265–283.

Semenov, M.A. (2007). Development of high-resolution ukcip02-based climate change scenarios in the UK. *Agricultural and Forest Meteorology,* **144(1-2):** 127–138.

Semenov, M.A., Barrow, E.M. and Lars-Wg, A. (2002). A stochastic weather generator for use in climate impact studies. User Man Herts UK.

Shabbar, A. and Barnston, A.G. (1996). Skill of seasonal climate forecasts in Canada using canonical correlation analysis. *Monthly Weather Review,* **124(10):** 2370–2385.

Shamasastry, R. (1929). Kautilyas Arthasastra, Wesleyan Mission Press. https://archive.org/details/KautilyasArthasastra/mode/2up

Sheikh, A.A., Bhagat, R., Islam, S.T., Dar, R.R., Sheikh, S.A., Wani, J.M. and Dogra, P. (2017). Effect of climate change on reproduction and milk production performance of livestock: A review. *Journal of Pharmacognosy and Phytochemistry,* **6(6):** 2062–2064.

Shivashankara, K.S., Rao, N.K.S. and Geetha, G.A. (2013). Impact of climate change on fruit and vegetable quality. *In:* Climate-resilient Horticulture: Adaptation and Mitigation Strategies, pages 237–244. Springer.

Shukla, R., Agarwal, A., Sachdeva, K., Kurths, J. and Joshi, P. (2019). Climate change perception: An analysis of climate change and risk perceptions among farmer types of Indian western Himalayas. *Climatic Change,* **152(1):** 103–119.

Simmons, A. and Bengtsson, L. (1984). Atmospheric general circulation models: Their design and use for climate studies.

Singh, J.S. (2006) Sustainable development of the Indian Himalayan region: Linking ecological and economic concerns. *Current science*, (2006): 784–788.

Singh, M.P. (2009). Rice productivity in India under variable climates. *In:* Country report presented at MARCO Symposium, pages 5–7.

Singh, A., Kulkarni, M.A., Mohanty, U., Kar, S., Robertson, A.W. and Mishra, G. (2012). Prediction of Indian summer monsoon rainfall (ismr) using canonical correlation analysis of global circulation model products. *Meteorological Applications,* **19(2):** 179–188.

Singh, B.P., Dua, V.K., Govindakrishnan, P.M. and Sharma, S. (2013). Impact of climate change on potato. *In:* Climate-resilient horticulture: Adaptation and mitigation strategies, pages 125–135. Springer.

Singh, P., Singh, N., Nedumaran, S., Bantilan, C., Byjesh, K. (2015). Evaluating adaptation options at crop level for climate change in the tropics of south Asia and west Africa. In: Climate Change Challenges and Adaptations at Farm-level: Case Studies from Asia and Africa, Eds: N. P. Singh, C. Bantilan, K. Byjesh, and S. Nedumaran. CABI Private Limited, 115–137.

Singh, K., McClean, C.J., Buker, P., Hartley, S.E. and Hill, J.K. (2017). "Mapping regional risks from climate change for rainfed rice cultivation in India. *Agricultural Systems*, **156:** 76–84.

Singh, H.P. and Singh, B. (2020). Dynamics of banana research and development in India for farm profitability enhancement. Acta Hortic. **1272**:145–152. https://doi.org/10.17660/ActaHortic.2020.1272.18

Singh, S., Singh, A., Nayak, S. (2020). Future Climate Change impacts on Crop Productivity in Coastal Regions of India: A Panel Estimation. *Climate Change*, **6(21):** 100–108.

Sinha, P., Mohanty, U., Kar, S., Dash, S., Robertson, A. and Tippett, M. (2013). Seasonal prediction of the Indian summer monsoon rainfall using canonical correlation analysis of the ncmrwf global model products. *International Journal of Climatology,* **33(7)**: 1601–1614.

Sinha, P., Mohanty, U., Kar, S., Dash, S. and Kumari, S. (2013a). Sensitivity of the gcm driven summer monsoon simulations to cumulus parameterization schemes in nested regcm3. *Theoretical and Applied Climatology,* **112(1):** 285–306.

Sinha, P., Mohanty, U., Kar, S., Dash, S., Robertson, A. and Tippett, M. (2013b). Seasonal prediction of the Indian summer monsoon rainfall using canonical correlation analysis of the ncmrwf global model products. *International Journal of Climatology,* **33(7):** 1601–1614.

Sinha, P., Mohanty, U., Kar, S. and Kumari, S. (2014). Role of the Himalayan orography in simulation of the Indian summer monsoon using regcm3. *Pure and Applied Geophysics,* **171(7):** 1385–1407.

Sinha, P., Tiwari, P.R., Kar, S., Mohanty, U., Raju, P., Dey, S. and Shekhar, M. (2015). Sensitivity studies of convective schemes and model resolutions in simulations of wintertime circulation and precipitation over the western himalayas. *Pure and Applied Geophysics,* **172(2):** 503–530.

Sinha, P., Maurya, R., Mohanty, M. and Mohanty, U. (2019). Intercomparison and evaluation of mixed-convection schemes in regcm4 for Indian summer monsoon simulation. *Atmospheric Research,* **215:** 239–252.

Smagorinsky, J. (1974). Global atmospheric modeling and the numerical simulation of climate. *Weather and Climate Modification,* pages 633–686.

Stephenson, D.B. and Dolas-Reyes, F.J. (2000). Statistical methods for interpreting monte carlo ensemble forecasts. Tellus A: *Dynamic Meteorology and Oceanography,* **52(3):** 300–322.

Stern, R. and Coe, R. (1984). A model fitting analysis of daily rainfall data. *Journal of the Royal Statistical Society: Series A (General),* **147(1):** 1–18.

Stigter, K., Chunqiang, L., Dawei, Z., Shili, W. and Yuping, M. (2008). Further identification of two agrometeorological services in hebei province. China. Information sheet resulting from a field mission in China in September/October. Available on the INSAM website (www.agrometeorology.org) under Accounts of Operational Agrometeorology

Stigter, K. (2010). Applied Agrometeorology. Springer. https://link.springer.com, https://doi.org/10.1007/978-3-540-74698-0

Stockle, C.O. and Nelson, R. (1999). ClimGen, a weather generator program. Biological Systems Engineering Dept., Washington State University, Pullman, Washington, USA.

Taylor, K.E. (2001). Summarizing multiple aspects of model performance in a single diagram. *Journal of Geophysical Research: Atmospheres,* **106(D7):** 7183–7192.

Thakur, J. (1994). Rice production constraints in Bihar (India). *In:* Workshop on Rice Research Prioritization in Asia: February, pages 21–22.

Thornton, P.K., van de Steeg, J., Notenbaert, A. and Herrero, M. (2009). The impacts of climate change on livestock and livestock systems in developing countries: A review of what we know and what we need to know. *Agricultural Systems,* **101(3):** 113–127.

Tiedtke, M. (1989). A comprehensive mass flux scheme for cumulus parameterization in large-scale models. *Monthly Weather Review,* **117(8):** 1779–1800.

Tippett, M.K., Goddard, L. and Barnston, A.G. (2005). Statistical–dynamical seasonal forecasts of central-southwest asian winter precipitation. *Journal of Climate,* **18(11):** 1831–1843.

Tippett, M.K., Barnston, A.G. and Robertson, A.W. (2007). Estimation of seasonal precipitation tercile-based categorical probabilities from ensembles. *Journal of Climate,* **20(10):** 2210–2228.

Tiwari, P.R., Kar, S., Mohanty, U., Kumari, S., Sinha, P., Nair, A. and Dey, S. (2014). Skill of precipitation prediction with gcms over north India during winter season. *International Journal of Climatology,* **34(12):** 3440–3455.

Tsuji, G.Y., Uehara, G. and Balas, S. (1994). Dssat v3. University of Hawaii, Honolulu, Hawaii. https://dssat.net/wp-content/uploads/2011/10/DSSAT-vol1.pdf.

UDSA (2022). https://www.ers.usda.gov/data-products/ag-and-food-statistics-charting-the-essentials/ag-and-food-sectors-and-the-economy/

Uehara, G. (1989). Technology transfer in the tropics. *Outlook on Agriculture,* **18(1):** 38–42.

Upadhyay, R., Singh, S., Kumar, A., Gupta, S. and Ashutosh (2007). Impact of climate change on milk production of murrah buffaloes. *Italian Journal of Animal Science,* **6(sup2):** 1329–1332.

van Roosmalen, L., Christensen, J.H., Butts, M.B., Jensen, K.H. and Refsgaard, J.C. (2010). An intercomparison of regional climate model data for hydrological impact studies in Denmark. *Journal of Hydrology,* **380(3-4):** 406–419.

Varshneya, M., Pillai, P.B., et al. (2003). Textbook of Agricultural Meteorology. https://ia800700.us.archive.org/20/items/textbookofagricu0000vars/textbookofagricu0000vars.pdf

Venkataraman, S. (2002). Tabular aids for computation of derived agrometeorological parameters on a weekly basis. *Journal of Agrometeorology,* **4(1)**: 1–8.

Venkataraman, S. (1992). Weather relations of crops. Crops and weather; Eds:Venkataraman, S. and A. Krishnan. Pp. 302–457. Indian Council of Agricultural Research, New Delhi.

Venkatasubramanian, K. (2014). "Assessment of India\u2019s integrated agrometeorological advisory service from a farmer perspective," IWMI Working Papers H046810, International Water Management Institute

Vitart, F. and Molteni, F. (2009). Dynamical extended-range prediction of early monsoon rainfall over India. *Monthly Weather Review,* **137(4):** 1480–1492.

Wang, S.Y., Bunce, J.A. and Maas, J. (2003). Elevated carbon dioxide increases contents of antioxidant compounds in field-grown strawberries. *Journal of Agricultural and Food Chemistry,* **51(15):** 4315–4320.

Wang, B., Ding, Q., Fu, X., Kang, I.-S., Jin, K., Shukla, J. and Doblas-Reyes, F. (2005). Fundamental challenge in simulation and prediction of summer monsoon rainfall. *Geophysical Research Letters,* **32(15).**

Webster, P.J., Magana, V.O., Palmer, T., Shukla, J., Tomas, R., Yanai, M. and Yasunari, T. (1998). Monsoons: Processes, predictability, and the prospects for prediction. *Journal of Geophysical Research: Oceans,* **103(C7):** 14451–14510.

Welton, G. (2011). The impact of Russia's 2010 grain export ban. Oxfam research reports, 28 June 2011.

Wheater, H., Chandler, R., Onof, C., Isham, V., Bellone, E., Yang, C., Lekkas, D., Lourmas, G. and Segond, M.-L. (2005). Spatial-temporal rainfall modelling for flood risk estimation. *Stochastic Environmental Research and Risk Assessment,* **19(6):** 403–416.

Wheeler, T. and Von Braun, J. (2013). Climate change impacts on global food security. *Science,* **341(6145):** 508–513.

Wheelock, J., Rhoads, R., VanBaale, M., Sanders, S. and Baumgard, L. (2010). Effects of heat stress on energetic metabolism in lactating Holstein cows. *Journal of Dairy Science,* **93(2):** 644–655.

Wilby, R., Tomlinson, O. and Dawson, C. (2003). Multi-site simulation of precipitation by conditional resampling. *Climate Research,* **23(3):** 183–194.

Wilks, D.S. (1989). Conditioning stochastic daily precipitation models on total monthly precipitation. *Water Resources Research,* **25(6):** 1429–1439.

Wilks, D.S. (1992). Adapting stochastic weather generation algorithms for climate change studies. *Climatic Change,* **22(1):** 67–84.

Wilks, D. (1998). Multisite generalization of a daily stochastic precipitation generation model. *Journal of Hydrology,* **210(1-4):** 178–191.

Wilks, D.S. (2011). Statistical methods in the atmospheric sciences, Volume 100. Academic Press.

Wilks, D.S. and Wilby, R.L. (1999). The weather generation game: A review of stochastic weather models. *Progress in Physical Geography,* **23(3):** 329–357.

Wilks, D.S. (1999a). Interannual variability and extreme-value characteristics of several stochastic daily precipitation models. *Agricultural and Forest Meteorology,* **93(3):** 153–169.

Wilks, D.S. (1999b). Multisite downscaling of daily precipitation with a stochastic weather generator. *Climate Research,* **11(2):** 125–136.

Williams, J.R., Jones, C. and Dyke, P.T. (1984). A modeling approach to determining the relationship between erosion and soil productivity. *Transactions of the ASAE,* **27(1):** 129–0144.

Willmott, C.J. (1981). On the validation of models. *Physical Geography,* **2(2):** 184–194.

Willmott, C.J. (1982). Some comments on the evaluation of model performance. *Bulletin of the American Meteorological Society,* **63(11):** 1309–1313.

Willmott, C.J., Ackleson, S.G., Davis, R.E., Feddema, J.J., Klink, K.M., Legates, D.R., O'donnell, J. and Rowe, C.M. (1985). Statistics for the evaluation and comparison of models. *Journal of Geophysical Research: Oceans,* **90(C5):** 8995–9005.

Wood, A. W., Leung, L. R., Sridhar, V., and Lettenmaier, D. (2004). Hydrologic implications of dynamical and statistical approaches to downscaling climate model outputs. *Climatic change,* **62(1–3):** 189–216.

World Bank (2012). "Data-Methodologies." Washington, DC: World Bank. http://data.worldbank.org/about/data-overview/methodologies accessed on November 29, 2020.

Yearbook, F.S. (2013). World food and agriculture. Food and Agriculture Organization of the United Nations, Rome, 15.

Yoshida, S. (1979). A simple evapotranspiration model of a paddy field in tropical Asia. *Soil Science and Plant Nutrition,* **25(1):** 81–91.

Yoshida, S. (1981). Fundamentals of rice crop science. The International Rice Research Institute, Manila, Philippines.

Yu, Z.P., Chu, P.S. and Schroeder, T. (1997). Predictive skills of seasonal to annual rainfall variations in the US affiliated Pacific islands: Canonical correlation analysis and multivariate principal component regression approaches. *Journal of Climate,* **10(10):** 2586–2599.

Yun, W., Stefanova, L. and Krishnamurti, T. (2003). Improvement of the multimodel super-ensemble technique for seasonal forecasts. *Journal of Climate,* **16(22):** 3834–3840.

Yun, W., Stefanova, L., Mitra, A., Vijaya Kumar, T., Dewar, W. and Krishnamurti, T. (2005). A multi-model superensemble algorithm for seasonal climate prediction using demeter forecasts. Tellus A: *Dynamic Meteorology and Oceanography,* **57(3):** 280–289.

Zhang, X.-C. (2005). Spatial downscaling of global climate model output forsite-specific assessment of crop production and soil erosion. *Agricultural and Forest Meteorology,* **135(1-4):** 215–229.

Zhang, X. and Garbrecht, J.D. (2003). Evaluation of cligen precipitation parameters and their implication on wepp runoff and erosion prediction. *Transactions of the ASAE,* **46(2):** 311.

Zhang, X.-C. and Liu, W.-Z. (2005). Simulating potential response of hydrology, soil erosion, and crop productivity to climate change in Changwu tableland region on the Loess plateau of China. *Agricultural and Forest Meteorology,* **131(3–4):** 127–142. ISSN 0168–1923, https://doi.org/10.1016/j.agrformet.2005.05.005

Zhou, M. and Butterbach-Bahl, K. (2014). Assessment of nitrate leaching loss on a yield-scaled basis from maize and wheat cropping systems. *Plant and Soil,* **374(1):** 977–991.

Zhu, H., Wheeler, M.C., Sobel, A.H. and Hudson, D. (2014). Seamless precipitation prediction skill in the tropics and extratropics from a global model. *Monthly Weather Review,* **142(4):** 1556–1569.

Index

U. Mohanty et al., *Climate Risk Management in Agriculture*,
https://doi.org/10.1007/978-3-031-51862-1

GPSR Compliance

The European Union's (EU) General Product Safety Regulation (GPSR) is a set of rules that requires consumer products to be safe and our obligations to ensure this.

If you have any concerns about our products, you can contact us on ProductSafety@springernature.com

In case Publisher is established outside the EU, the EU authorized representative is:

Springer Nature Customer Service Center GmbH
Europaplatz 3
69115 Heidelberg, Germany

Batch number: 10732076

Printed by Printforce, the Netherlands